Methods in Bioengineering

Biomicrofabrication and Biomicrofluidics

The Artech House Methods in Bioengineering Series

Series Editors-in-Chief
Martin L. Yarmush, M.D., Ph.D.
Robert S. Langer, Sc.D.

Methods in Bioengineering: Biomicrofabrication and Biomicrofluidics,
Jeffrey D. Zahn, editor

Methods in Bioengineering: Microdevices in Biology and Medicine,
Yaakov Nahmias and Sangeeta N. Bhatia, editors

Methods in Bioengineering: Nanoscale Bioengineering and Nanomedicine,
Kaushal Rege and Igor Medintz, editors

Methods in Bioengineering: Stem Cell Bioengineering,
Biju Parekkadan and Martin L. Yarmush, editors

Methods in Bioengineering: Systems Analysis of Biological Networks,
Arul Jayaraman and Juergen Hahn, editors

Methods in Bioengineering

Biomicrofabrication and Biomicrofluidics

Jeffrey D. Zahn
Department of Biomedical Engineering
Rutgers University

Editor

ARTECH
HOUSE
BOSTON | LONDON
artechhouse.com

Library of Congress Cataloging-in-Publication Data
A catalog record for this book is available from the U. S. Library of Congress.

British Library Cataloguing in Publication Data
A catalogue record for this book is available from the British Library.

ISBN-13: 978-1-59693-400-9

Cover design by Greg Lamb
Text design by Darrell Judd

10 9 8 7 6 5 4 3 2 1

Contents

Preface   *xiii*

CHAPTER 1
Microfabrication Techniques for Microfluidic Devices   1

   1.1   Introduction to microsystems and microfluidic devices   2
   1.2   Microfluidic systems: fabrication techniques   3
   1.3   Transfer processes   4
         1.3.1   Photolithography   4
         1.3.2   Molding   6
   1.4   Additive processes   6
         1.4.1   Growth of SiO_2 6
         1.4.2   Deposition techniques   7
   1.5   Subtractive techniques   10
         1.5.1   Etching   10
         1.5.2   Chemical-mechanical polishing and planarization   13
   1.6   Bonding processes   13
         1.6.1   Lamination   13
         1.6.2   Wafer bonding methods   14
   1.7   Sacrificial layer techniques   16
   1.8   Packaging processes   16
         1.8.1   Dicing   16
         1.8.2   Electrical interconnection and wire bonding   17
         1.8.3   Fluidic interconnection in microfluidic systems   18
   1.9   Materials for microfluidic and bio-MEMS applications   19
         1.9.1   Glass, pyrex, and quartz   19
         1.9.2   Silicon   19
         1.9.3   Elastomers   20
         1.9.4   Polydimethylsiloxane   20
         1.9.5   Epoxy   20
         1.9.6   SU-8 thick resists   20
         1.9.7   Thick positive resists   21
         1.9.8   Benzocyclobutene   21
         1.9.9   Polyimides   21

	1.9.10 Polycarbonate	22
	1.9.11 Polytetrafluoroethylene	22
1.10	Troubleshooting table	22
1.11	Summary	22
	References	25

CHAPTER 2

Micropumping and Microvalving 31

2.1	Introduction	32
2.2	Actuators for micropumps and microvalves	33
	2.2.1 Pneumatic actuators	35
	2.2.2 Thermopneumatic actuators	36
	2.2.3 Solid-expansion actuators	37
	2.2.4 Bimetallic actuators	38
	2.2.5 Shape-memory alloy actuators	38
	2.2.6 Piezoelectric actuators	38
	2.2.7 Electrostatic actuators	39
	2.2.8 Electromagnetic actuators	39
	2.2.9 Electrochemical actuators	40
	2.2.10 Chemical actuators	41
	2.2.11 Capillary-force actuators	42
2.3	Micropumps	42
	2.3.1 Mechanical pump	43
	2.3.2 Nonmechanical pump	48
2.4	Microvalves	51
	2.4.1 Mechanical valve	52
	2.4.2 Nonmechanical valve	54
2.5	Outlook	55
2.6	Troubleshooting	55
2.7	Summary points	56
	References	56

CHAPTER 3

Micromixing Within Microfluidic Devices 59

3.1	Introduction	60
3.2	Materials	62
	3.2.1 Microfluidic mixing devices	62
	3.2.2 Microfluidic interconnects	62
	3.2.3 Optical assembly	63
	3.2.4 Required reagents	63
3.3	Experimental design and methods	64
	3.3.1 Passive micromixers	64
	3.3.2 Active micromixers	70

3.3.3 Multiphase mixers   75

3.4 Data acquisition, anticipated results, and interpretation   77

3.4.1 Computer acquisition   77

3.4.2 Performance metrics, extent of mixing, reaction monitoring   78

3.5 Discussion and commentary   78

3.6 Troubleshooting   79

3.7 Application notes   79

3.8 Summary points   79

References   80

CHAPTER 4

On-Chip Electrophoresis and Isoelectric Focusing Methods for Quantitative Biology 83

4.1 Introduction   84

4.1.1 Microfluidic electrophoresis supports quantitative biology and medicine   84

4.1.2 Biomedical applications of on-chip electrophoresis   88

4.2 Materials   89

4.2.1 Reagents   89

4.2.2 Facilities/equipment   91

4.3 Methods   91

4.3.1 On chip polyacrylamide gel electrophoresis (PAGE)   92

4.3.2 Polyacrylamide gel electrophoresis based isoelectric focusing   96

4.3.3 Data acquisition, anticipated results, and interpretation   102

4.3.4 Results and discussion   103

4.4 Discussion of pitfalls   105

4.5 Summary notes   105

Acknowledgments   108

References   108

CHAPTER 5

Electrowetting 111

5.1 Introduction   112

5.1.1 Electrowetting theory   112

5.1.2 Droplet manipulation using electrowetting   113

5.1.3 Digital microfluidic lab-on-a-chip for clinical diagnostics   115

5.2 Digital microfluidic lab-on-a-chip design   115

5.2.1 Fluidic input port   115

5.2.2 Liquid reservoirs   116

5.2.3 Droplet pathways   116

5.3 Materials   117

5.3.1 Chemicals   117

5.3.2 Fabrication materials   118

5.4 Device Fabrication   119

5.4.1	Fabrication of single layer electrowetting chips	119
5.4.2	Fabrication of two layer electrowetting chips	119
5.4.3	Dielectric deposition	120
5.4.4	Fabrication of the top plate	121
5.4.5	Hydrophobic coating	121
5.5	Instrumentation and system assembly	121
5.5.1	Detection setup	121
5.5.2	System assembly	122
5.6	Methods	123
5.6.1	Automated glucose assays on-chip	123
5.6.2	Magnetic bead manipulation on-chip	123
5.6.3	Droplet-based immunoassay	125
5.7	Results and discussion	125
5.7.1	Testing of two layer electrowetting device	125
5.7.2	Automated glucose assay on-chip	126
5.7.3	Optimization of magnetic bead washing	127
5.8	Method challenges	129
5.9	Summary points	131
	Acknowledgments	131
	References	131

CHAPTER 6

Dielectrophoresis for Particle and Cell Manipulations — 133

6.1	Introduction: physical origins of DEP	134
6.2	Introduction: theory of dielectrophoresis	135
6.2.1	Limiting assumptions and typical experimental conditions	138
6.3	Materials: equipment for generating electric field nonuniformities and DEP forces	145
6.3.1	Electric field frequency	145
6.3.2	Electric field phase	147
6.3.3	Geometry	149
6.4	Methods: data acquisition, anticipated results, and interpretation	152
6.4.1	General considerations for dielectrophoretic devices	152
6.4.2	Electrode-based dielectrophoresis	156
6.4.3	Insulative dielectrophoresis	163
6.4.4	Summary of experimental parameters	168
6.5	Troubleshooting	169
6.6	Application notes	169
6.6.1	Particle trapping	169
6.6.2	Particle sorting and fractionation	172
6.6.3	Single-particle trapping	175
	References	177

CHAPTER 7

Optical Microfluidics for Molecular Diagnostics   183

   7.1   Introduction   184
   7.2   Integrated optical systems   185
      7.2.1   Absorbance detection   185
      7.2.2   Fluorescence detection   188
      7.2.3   Chemiluminescence detection   191
      7.2.4   Interferometric detection   192
      7.2.5   Surface plasmon resonance detection   194
   7.3   Nanoengineered optical probes   195
      7.3.1   Quantum dots   196
      7.3.2   Up-converting phosphors   197
      7.3.3   Silver-enhanced nanoparticle labeling   197
      7.3.4   Localized surface plasmon resonance   198
      7.3.5   SPR with nanohole gratings   200
      7.3.6   Surface-enhanced Raman spectroscopy   200
   7.4   Conclusions   203
   7.5   Summary   203
      Acknowledgments   204
      References   204

CHAPTER 8

Neutrophil Chemotaxis Assay from Whole Blood Samples   209

   8.1   Introduction   210
   8.2   Device design   210
   8.3   Materials   213
   8.4   Methods   213
      8.4.1   Device fabrication   213
      8.4.2   Surface treatment   215
      8.4.3   Chemotaxis assay   216
   8.5   Data acquisition   220
   8.6   Troubleshooting tips   220
      Appendix 8A   221
      References   223

CHAPTER 9

Microfluidic Immunoassays   225

   9.1   Introduction   226
      9.1.1   Microfluidic immunoassay design/operation considerations   227
      9.1.2   Example microfluidic immunoassay formats   227
   9.2   Materials   230
      9.2.1   Microfluidic device   230

9.2.2   Pumps and interconnections   231
9.3   Methods   231
9.3.1   Fabrication of flowcells   231
9.3.2   Sample and reagent delivery   233
9.4   Data acquisition and results   236
9.4.1   FMIA   236
9.4.2   CGIA   237
9.5   Discussion   239
9.5.1   FMIA   239
9.5.2   CGIA   240
9.5.3   Challenges of analyzing complex samples   241
Acknowledgments   242
References   242

CHAPTER 10
Droplet Based Microfluidics by Shear-Driven Microemulsions   245
10.1   Introduction   246
10.1.1   Advantages of droplet-based microfluidics   246
10.2   Biomedical applications of droplet microfluidics   248
10.2.1   Bioassays   248
10.2.2   Particle formation   249
10.2.3   Therapeutic delivery   252
10.3   Materials   253
10.4   Methods   255
10.4.1   Channel surface modification   255
10.4.2   Hydrophilic surface treatment   255
10.4.3   Hydrophobic surface treatment   257
10.4.4   Solution preparation   257
10.4.5   Generation of droplets   258
10.5   Data acquisition   259
10.5.1   Droplet arrays   259
10.5.2   High-speed cameras   261
10.5.3   Conventional imaging methods   262
10.6   Discussion and commentary   262
Acknowledgments   263
References   263

CHAPTER 11
MicroFACS System   267
11.1   Introduction   268
11.2   Materials   269
11.3   Methods   270
11.4   Results   274

11.5   Discussion of pitfalls   275

11.6   Statistical analysis   276

11.7   Application notes   277

11.8   Summary points   278

       Acknowledgments   279

       References   279

CHAPTER 12

Optical Flow Characterization—Microparticle Image Velocimetry (μPIV)   281

12.1   Introduction   282

12.2   Materials and methods   283

       12.2.1   Experimental setup   283

       12.2.2   Volume illumination   285

       12.2.3   Processing algorithms   287

12.3   Measurement procedures   292

       12.3.1   Step-by-step operations   292

12.4   Discussion and commentary   293

       12.4.1   Diffraction limit   293

       12.4.2   Particle size effects   294

       12.4.3   Ultimate spatial resolution   294

       12.4.4   Velocity errors   296

       12.4.5   Other flow visualization techniques based on μPIV   297

12.6   Summary points   300

12.7   Application notes   301

       12.7.1   Principles of diffusometry   301

       12.7.2 μPIV-based thermometry   303

       12.7.3   Biosensing   303

       12.7.4   Wall shear stress (WSS) measurement   303

12.8   Future developments   307

       Acknowledgments   307

       References   307

CHAPTER 13

Microtubule Motors in Microfluidics   311

13.1   Introduction   312

13.2   Materials   312

       13.2.1   Kinesin expression and purification materials   313

       13.2.2   Tubulin purification materials   314

       13.2.3   Microtubule gliding assay materials   315

       13.2.4   Microfabrication materials   316

13.3   Methods   317

       13.3.1   Kinesin expression and purification   317

       13.3.2   Tubulin purification and labeling   319

13.3.3   Standard protocol for the microtubule gliding assay   323

13.3.4   Design considerations for integrating motor proteins into
microfluidic devices   324

13.3.5   Fabricating enclosed glass channels for microtubule transport   326

13.4   Results   329

13.5   Discussion of pitfalls   331

13.5.1   Kinesin purification   332

13.5.2   Tubulin   332

13.5.3   Motility assays   333

13.5.4   Motility in microchannels   334

13.5.5   Final comments   336

Acknowledgments   336

References   336

About the Editor   339

List of Contributors   340

Index   343

Preface

In the past several decades there has been an explosion of research conducted in the field of microfluidics. Micro- and nanofabricated diagnostic devices have been termed micro-total analysis systems (μTAS), biochips, or more generically "lab on a chip," and combine sensing mechanisms (physical, optical, electrical, or chemical) with micro- and nanosystems and microfluidics. One of the main driving forces has been the idea that miniaturization of fluid-based diagnostics, chemical reaction chambers, and fluid transporting devices will lead to more efficient autonomous sensor and reaction systems. Very complex biomolecule processing reactions (cell lysis, electrophoresis, etc.) are performed autonomously permitting manipulation, characterization and analysis of cells and biological material.

While microfluidics promises to have an impact in many research fields, one of the more attractive applications of microfluidics has been towards biomedical and life-science diagnostics. Microfluidic platforms have great potential to transform the field of medicine by dramatically changing the ways in which diagnostics and clinical research are conducted in hospitals and laboratories. Such miniaturized platforms have attracted considerable research interest due to the potential of fabricating a highly integrated system that is able to perform parallel sample handling and analysis on a single chip. Microfluidic devices are attractive because they offer many advantages such as smaller reagent volume consumption, shorter reaction times, and the possibility of parallel operation. These advantages result not only in time and cost savings for diagnostic tests but can also be life saving in time critical environments such as critical medical diagnostics.

This volume describes many of the methods used in the field of microfluidics to handle, manipulate, and/or analyze cells, particles, or biological components (e.g., proteins and DNA) for microdiagnostics. Chapters 1 through 3 describe common microfabrication techniques utilized to create microfluidic devices and on-chip flow control and mixing microsystems. Chapter 4 outlines protein and DNA handling devices for electrophoretic and isoelectric separations in microchromatography columns. Chapters 5 and 6 describe electrical methods for microfluidic manipulations of droplets via electrowetting and particles via dielectrophoresis for separations and chemical reactions. The remaining chapters describe different methods commonly utilized in microfluidic devices. Chapter 7 describes methods for integrated optical characterization of microfluidic devices. Chapter 8 describes methods for controlling chemical gradients within devices. Chapter 9 describes microimmunoassay diagnostics. Chapter 10 describes multiphase microfluidics used in droplet formation for controlled chemical reactions. Chapter 11 describes particle separation and analysis in miniaturized fluorescent-activated cell-sorting (Micro-FACS) system. Chapter 12 describes flow characterization techniques in microfluidic devices. Chapter 13 describes methods for patterning and utilizing cytoskeletal filaments and cellular transport proteins within microstructures.

The contributors of each chapter are among the most notable microfluidic scientists and engineers in their respective fields, and have contributed their expertise to this volume to provide biomedical scientists and engineers with the tools they need to usher in the next generation of microdiagnostics and microtherapeutic devices to be used in clinical practice.

Jeffrey D. Zahn
October 2009

Microfabrication Techniques for Microfluidic Devices

Marcelo B. Pisani and Srinivas A. Tadigadapa

Phone: +1 (814) 865-2730, Fax +1 (814) 865-7065
The Pennsylvania State University, Department of Electrical Engineering,
University Park, PA 16802-2705
E-mail: mbpisani@psu.edu and srinivas@engr.psu.edu

Abstract

This chapter presents the basic concepts, techniques, and materials used in the fabrication of microfluidic systems. Microsystems fabrication is a broad subject in technology that has been extensively covered by a large number of specialized and seminal publications during the past few decades. The aim of this chapter is to provide the reader with a basic overview on the microfluidic device fabrication technologies. The processing steps are discussed with emphasis on the methodology rather than on the detailed science and modeling which are beyond the scope of this chapter. Through this approach, our aim is to provide the reader with a quick and practical guide on the possibilities and main techniques used to fabricate the building blocks of microfluidic systems.

Key terms	microfluidic systems materials and fabrication
	microsystems fabrication techniques
	packaging
	interconnection of microfluidic devices

1.1   Introduction to microsystems and microfluidic devices

Miniaturization is one of the key driving factors in the evolution of modern technology [1, 2]. It enables devices to have complex functionalities in a smaller volume, with less energy consumption and lower costs per implemented function. Other correlated benefits from miniaturization include faster processing time, lower manufacturing and operational costs, smaller volumes of needed samples and reactants, integration with other devices (multifunctionality), increased safety and reliability by having less external interconnects between the different parts, and high throughput. Microelectromechanical systems (MEMS) are precise, miniaturized systems with primary functionalities in both electrical and mechanical domains, with applications covering a large range of phenomena in the physical, chemical, and biological domains. The terms "micromachines" used in Japan and "microsystems" used in Europe also refer to the same field [3].

Over the last two decades, several new microsystems have been created by the fabrication and integration of small electrical and mechanical components put together as a system, using techniques that are inherited from or related to the integrated circuit technologies (IC) [4]. MEMS devices open the possibility of cointegration in a single complex system of functions like sensing, electronic signal processing, and electromechanical actuation that are traditionally performed by distinctly separated and bulky functional blocks. Depending on the specific domain of functionality and application of these devices, different terms have been coined to categorize them. As examples, RF MEMS devices are used for radio frequency telecommunication applications (e.g., switches, transmission lines, and antennas) [5], optical MEMS refers to devices interacting with light (e.g., image scanner micro mirrors, and optical switches) [6, 7] and bio-MEMS [8]/microfluidics [9] are used in systems applied to biological assays and fluid manipulation (e.g., body fluids and cell manipulation, DNA analysis [10], toxins detection, diseases diagnosis and treatment, and drug delivery [11]).

Microfluidics is the field of the microsystems technology that deals with the study and control of fluids in hydraulically/geometrically small systems, with lengths typically in the range of a few micrometers up to a few millimeters or centimeters [12]. The volume of fluid in such systems is in the range of nanoliters with flow rates ranging from a few nanoliters per second to nanoliters per hour range. These devices are of great interest in physical, chemical, and biological applications. They have stringent specifications in terms of the precision involved in their fabrication as well as in terms of the compatibility between the different materials used to fabricate them and the possible interactions between these parts and the fluids/biological systems of interest. Typical microfluidic systems consist of four main components: (1) microfluidic pumps for driving the fluids along the system (based on various actuation mechanisms), (2) microfluidic valves for controlling and directing the flow as desired, (3) microfluidic channels and chambers, which are the passive and the primary fluidic interconnecting components of these systems, and (4) active microfluidic components integrated with closed-loop temperature controllers, optical detectors, electrodes for the application of test voltages and currents, and so forth. These active devices allow the modification of the fluid, the separation of its constituents, and the detection of its components. The fabrication of each individual component of a microfluidic system can be quite complex in itself. The major goal of having a completely integrated microfluidic system consisting of all or several of the above components cointegrated in the same device has been a topic of active research during the last decade. An example of complex microfluidic system applied to DNA analysis is show in Figure 1.1.

1.2 Microfluidic systems: fabrication techniques

The fabrication of small scale microfluidic devices relies on high-precision micromachining techniques capable of creating high aspect ratio structures. Micromachining techniques can be basically divided in two major categories: bulk and surface techniques. Bulk techniques rely on the direct modification of a substrate material, usually a monocrystalline silicon wafer, glass, quartz or thick polymer matrix, used as a basic (bulk) material or support to shape the desired structures. Typical substrate thicknesses are in the range of several tens to hundreds of micrometers (Figure 1.2(a)). On the other hand, surface micromachining techniques involve the deposition of various layers of materials in the form of thin films and their definition into the desired structural shape (Figure 1.2(b)). Typical thicknesses of the various layers in surface micromachining techniques are in the range of a few nanometers to a few micrometers, much smaller than the bulk thickness, which is used to support the microfluidic device itself.

Bulk and surface micromachining techniques in turn consist of a sequence of simpler unit processes steps. These steps can be broadly classified as (1) *transfer processes,* which include the growth and deposition of films and layers of materials, (2) *additive processes,* which include etching and removal techniques, (3) *subtractive processes,* where patterned or nonpatterned structures and substrates are joined together, and (4) *bonding processes,* where the shape of the structure is copied or transferred from a template into the substrate (photolithography, hot embossing [14], nanoimprint lithography [15], etc.). Transfer processes use intermediate materials to define the desired shapes into the substrate. The systematic and sequential combination of these basic unit processes

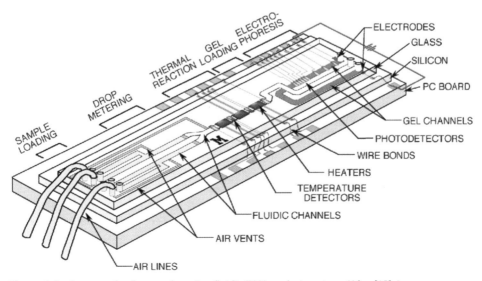

Figure 1.1 An example of a complex microfluidic DNA analysis system. (After [13] .)

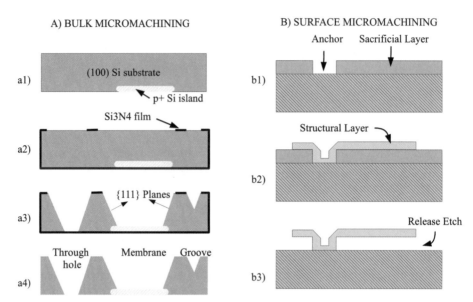

Figure 1.2 Schematic illustration examples of (a) bulk micromachining process in a silicon substrate and (b) surface micromachining process.

allows the realization of complex microfluidic systems. In the following sections we will describe the categorized fabrication processes in more detail.

1.3 Transfer processes

1.3.1 Photolithography

Photolithography is the most important and most frequently used method for the patterning of microsystems. Photolithographic processes copy designs from a template (mask) into thin ultraviolet light (UV)-sensitive polymeric films (photoresist), in a process similar to the one used to print images onto photographic films [16]. In fact, both techniques share common practices and chemistry. Photolithography can be used to create microfluidic systems out of the photoresist materials themselves or these patterned films can be used to transfer the design into the underlying substrate using etching techniques. The basic concept of photolithographic pattern transfer is illustrated in Figure 1.3. Photoresists consisting of photoactive compounds come in two tones: positive and negative. When the regions exposed to UV radiation exhibit several folds enhanced solubility in a developer solution, the resist is termed as positive photoresist (Figure 1.3(a)). In contrast, negative photoresists become cross-linked when exposed to UV radiation, and do not dissolve in the developer thereby creating an inverse pattern with respect to the masks (Figure 1.3(b)). Typical positive resist consists of an inert polymeric matrix (resin) combined with a photoactive compound (diazoquinone ester, DQN), which are typically dissolved in an appropriate solvent to create a homogeneous particle-free solution. The phenolic novolak resin matrix is inert and upon proper thermal curing can provide protection of the underlying substrate from harsh chemicals and processes used to etch the substrates.

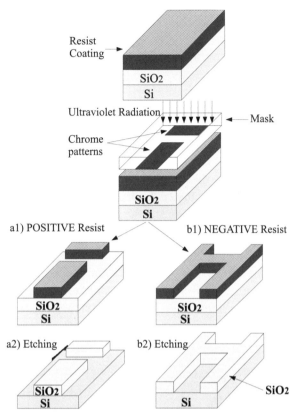

Figure 1.3 Schematic illustration of the lithographic pattern transfer process using positive (a1) and negative photoresists (b1) and subsequent results after etching (a2, b2).

Typically, the substrate is baked to dehydrate its surface and treated with a solemnization process to result in a hydrophobic surface, usually by exposing the wafers to hexamethyldisilazane (HMDS, $(CH_3)_3Si$–NH–$Si(CH_3)_3$) vapor or solution. The liquid photoresist is applied by placing a large puddle of it on the substrate wafer. The wafer is spun at high speeds up to a few thousand rpm (typically 1,000 to 4,000 rpm). Photoresists of typical thickness between fractions of a micron to a few tens of microns can be obtained depending upon the viscosity of the used solution. The film is then solidified by a thermal treatment that evaporates the solvent and then selectively exposed to UV radiation using a mask in order to induce the selective photo conversion of only the exposed material. For DQN resists, the developer is usually a weak base such as 2% tetramethyl ammonium hydroxide (TMAH, $(CH_3)_4NOH$) or sodium hydroxide (NaOH) solution, which selectively dissolves the exposed parts of the photoresist thereby transferring the pattern onto the substrate. A very short exposure to oxygen plasma, known as plasma descum, can be optionally used to slightly etch the resist and obtain clean surfaces where the resist is removed. Subsequent etch step transfers the pattern into the substrate, after which the photoresist can be completely dissolved or removed using remover solutions usually assisted by another descum using oxygen plasma (stripping step).

The ultimate resolution that can be obtained by this process is basically limited by the wavelength of the radiation (diffraction of light) and the quality of the optics used.

Typical photoresists are sensitive to UV light, mainly to the *i* (365.4 nm) or *g* (435.8 nm) spectral lines of the mercury lamps used in the lithography aligners. Lower light wavelengths define the deep UV (DUV) lithography process. Contact aligners or steppers are the most commonly used tools for photoresist exposure. These tools include a precision positioning system to align the patterns already existent on the substrate wafers with the ones on the mask, usually with an accuracy of a fraction of a micrometer. In contact mask aligners, the wafer to be printed is either in intimate contact with the mask or in close proximity to it and operates in Fresnel diffraction conditions, whereas a stepper optically projects the image of the mask onto the wafer and operates in the Fraunhofer diffraction limits. From the microfluidics standpoint, thick photoresists offer an attractive possibility using these materials themselves as part of the microfluidic system without any further need of transferring the pattern into the underlying substrate.

1.3.2 Molding

Polymeric materials with low glass transition temperature can be shaped using traditional plastic fabrication techniques like molding, hot embossing (material removal) [17, 18], injection molding [19–22], and casting [23, 24]. Table 1.1 gives an overview of the various molding processes and their capabilities.

1.4 Additive processes

1.4.1 Growth of SiO$_2$

Silicon dioxide (SiO$_2$) is the most common dielectric layer that is used on silicon substrates. Silicon dioxide has a relative dielectric constant of 3.9, and high dielectric field strength of 12 MV/cm. It is inert under atmospheric conditions, hydrophilic in nature, and is a highly biocompatible material. SiO$_2$ films can be obtained by direct oxidation of the silicon crystal at high temperatures in the range of 800°C to 1100°C. The oxidation process can be carried under dry oxidation conditions using O$_2$ gas, water (H$_2$O) vapor, or a mixture of both [25]. In general dry oxidation is a slower growth process and produces films with better stoichiometric composition whereas wet oxidation is a faster

Table 1.1 Overview of Different Molding Techniques and Possibilities

Process	Materials	Tool Costs	Cycle Time	Force & Temperature	Automation	Geometry	Min. Size / Aspect Ratio
Hot Embossing	Thermoplastics	Low/ Medium	Medium/Long (3-10 min)	High (kN)	Little	Planar	nm (nanoimprint)
	Duraplastic Thin Films			Around Tg (100–200°C)			50 small areas, 5 wafer scale
Injection	Thermoplastics	High	Short / medium (0.3–3 min)	High	Yes	Bulk, Spherical	Some 10 μm
	Duraplastics			Above Melting (150–400°C)			50 small areas, 5 wafer scale
Casting	Elastomers	Low	Long (min-h)	No forces	Little	Planar	nm
	Epoxies			Room temperature to 80°C			About 1

After [24]

growth process with greater imperfections in the silicon dioxide compositions. Oxidation process is self-limited by the diffusion of oxygen into the already formed SiO_2 film and the deposition rate decreases with the time. The obtained thickness is proportional to the square root of the processing time, following the Deal-Grove model [26]. Due to the very high temperature and excessively long time needed to obtain thick films, this technique is limited to thin films (from a few nanometers using dry oxidation processes to a few micrometers when using wet oxidation process). The high dielectric and interfacial quality of the films has made dry oxide the preferred choice as gate dielectric in metal/oxide/semiconductor field effect transistors (MOSFETs) until the latest generations of transistors. Patterned silicon dioxide layer can be used as etch masks for the preferential etching of the underlying silicon crystal. It is an ideal layer on which to lay metal electrodes and to be functionalized for the growth of biological cell cultures.

1.4.2 Deposition techniques

Deposition techniques can be used to produce thin structural films. They are based on direct placement of material on top of a substrate or by the use of physical or chemical reactions to produce stable solid products preferably with good adhesion to the underlying or desired substrate(s). The most common deposition techniques are spin-on methods, physical vapor deposition (PVD), chemical vapor deposition (CVD), and electroplating [27, 28].

Spin-on deposition

Spin-on techniques use solutions to form the desired material. Typically, the material is dissolved into a uniform liquid solution in a volatile base solvent. The liquid is directly applied onto the surface of the substrate. A thin uniform film is defined by first flooding the surface with the solution and then by rotating the substrate at high speed, resulting in a uniform thin film covering the surface of the wafer. The liquid thin film is then solidified by a thermal treatment step that densifies the material by evaporating the solvent contents and in some cases may be followed by a subsequent high temperature anneal step to densify or crystallize the film. The final thickness is a function of the viscosity of the solution, the solid/solvent content ratio, the speed and time of the spin-on step and the post-deposition thermal treatment used. Spin-on films are relatively simple to deposit, but usually exhibit low-quality electrical/mechanical properties due to the inherently low energy used in the process and to the high solvent contents of the solutions. High temperature processing steps can be used to completely eliminate the solvent contents as well as to provide the desired solid state phase of the material. However, under these conditions, the films can be highly stressed due to a dramatic volumetric reduction during the hard-bake process. Spin-on deposition is the most commonly used technique to deposit photoresists, polymer films, and spin-on glass layers.

Vacuum evaporation

In the evaporation technique both the sample and the material to be deposited are placed in a high-vacuum environment in order to avoid foreign vapor contamination. The material to be deposited is provided sufficient energy to vaporize [29]. Evaporation technique consists of heating the material to be deposited to the vapor pressure point and by placing the substrate in the trajectory of the produced vapor. Evaporation is usually performed under high vacuum conditions (~ 10^{-6} mTorr) in order to avoid contamination caused by residual gases in the environment. The mean free path of the vapor phase molecules is very large, typically of the order of several meters and the evaporating material forms a conelike solid angle emanating from a point source. Upon striking the substrate, which is at a much lower temperature, the vapor condenses on the substrate and evolves into a uniform solid thin film. The two most common heating methods are based on Joule effect heating of filaments or crucibles containing the material to be evaporated or by the use of electron beam bombardment. Low-cost thermal evaporation setup usually uses hot filament. In spite of its simplicity, this technique cannot be used for a broad range of materials due to limitations arising from reaction of the deposition material with the filament and the difficulty in achieving high temperatures required for evaporating some of the materials. Electron-beam evaporation on the other hand bombards an electronic beam, produced by a thermionic filament and accelerated by high voltages, against the target material (typically at 3 to 20 kV of voltage and a few kilowatts of power). The material to be evaporated is held in a crucible made of or coated by material such as tungsten, graphite, alumina, or boron nitride in order to support the high processing temperatures. Thermal evaporator machines have a simpler architecture but usually are not capable of evaporating materials that require high temperature. E-beam processing usually provides more efficient heating and is able to produce atoms with higher energy and therefore produces higher-quality films. It is also the method commonly used to evaporate materials with higher melting-pointlike platinum, molybdenum, and even silicon and ceramic materials (alumina, graphite, silicon, etc.).

Sputtering

The sputtering processing is based on the high-energy bombardment of a target made of the material ones desire to deposit [30, 31]. Ionized high cross-section gases like argon or nitrogen are usually used to bombard the target with high energy. This results in the generation of atomic/molecular flux of the target material and the subsequent depositon onto the surfaces of interest. The high-energy ions are provided by low-pressure gas-phase plasma discharge, usually using a DC or 13.56-MHz radio frequency source. The target material forms the cathode of the gas discharge system. Conductive materials can be deposited using either DC or RF sources by this technique since in this case the cathode does not build up any surface charge. However, with insulating target materials RF or pulsed sputtering systems remain the only alternative since the alternating electrical potential can overcome surface charging effects and capacitively couple the power through the insulating surfaces, enabling the efficient deposition of a large selection of materials including insulating films like SiO_2, AlN, and PZT. Al, Au, Cu, Ta, Ti, Pt, Pd, Ni, Si, Co, and their alloys are commonly deposited using this technique.

Chemical vapor deposition (CVD)

CVD processes are based on chemical reactions that produce thin uniform layers of stable solid products defining the desired film on top of a substrate. Usually the reactions take place inside a specially designed chemical reactor with precisely controlled pressure, temperature, and flow of reactants. A gas discharge induced by applying DC or RF electrical potential (a plasma, [32]) can also be used to produce the reactive species, defining the plasma-enhanced chemical vapor deposition technique (PECVD, [33, 34]). Chemical vapor depositions performed under low-pressure and high-temperature conditions are known as low-pressure chemical vapor deposition (LPCVD) processes and are typically used to deposit high-quality SiO_2 and Si_3N_4 films [35], with operating pressures in the range of a few mTorr to hundreds of mTorr. Many variations of CVD processing exist depending on the mechanism used to drive and control the chemical reactions taking place. APCVD reactors work at atmospheric pressure. Chemical vapor depositions are typically performed at high temperatures in the range of 400°C to 900°C. DC or plasma can be used to enhance the production of reactive species. Typical films obtained by this technique are SiO_2 (using $SiH_4 + O_2$ mixtures), Si_3N_4 (using $SiCl_2H_2 + NH_3$ or $SiH_4 + NH_3$ chemistries). Plasma-enhanced chemical vapor deposition (PECVD) films can be deposited at lower temperatures in comparison to LPCVD films because the energy for the reaction is provided by the electrical discharge itself. The quality of the films can be controlled by the nature and amount of gas phase reactants, plasma power, and deposition pressure. PECVD technique is one of the preferred methods to produce SiO_2 and Si_3N_4 films at low to moderate processing temperatures.

Electroplating

Electroplating process enables the deposition of thick metal layers by the use of electrolytic solutions containing metallic ions. These ions are reduced by electrochemical reactions providing the needed electrons to transfer the ions from the complexed state in the solution into a solid state thin and uniform film on the surface of the substrate. A conductive substrate and an external current supply provide the electrons needed for the process. A typical example is the electroplating of copper using $CuSO_4$ solution. The conductive substrate (cathode) is connected to the negative potential of a current source. The copper ions in solution are converted into a solid copper film at the cathode following the reaction Cu^{2+} (solution) $+ 2e^- \rightarrow Cu^0$ (solid) [36].

The quality of the resulting film is usually defined in terms of the uniformity of the film thickness and the resistivity of the film with respect to a given reference value. For the deposited films, these parameters are a function of (1) the current density used (typically in the range of 1–100 mA/cm^2) and (2) the chemistry of the solution (for copper: $CuSO_4$ concentration, pH, and the presence of other additives to stabilize the solution). Agitation in the form of stirring, heating (usually about 60°C), and chemical additives are usually employed to control the results in terms of deposition rate, resistivity, thickness uniformity, and step coverage. Use of stirring avoids surface reactant depletion as well as the formation of bubbles, common in electrolyte reactions. Metals commonly plated by this technique are gold, copper, nickel, indium, silver, palladium, and platinum [37].

Electroless plating

Some solutions and conductive substrates can provide the needed electrons and electrochemical potentials to enable the reduction of the ions in the solution without the application of any external current, defining the electroless plating process [38]. An example is the electroless plating of copper onto the surface of silicon wafers. Gold is also frequently deposited using this technique, sometimes requiring palladium particles or other conductive layers to act as seedling layers.

1.5 Subtractive techniques

1.5.1 Etching

In order to define complex shapes needed to fabricate MEMS devices, one needs to structure the deposited materials by transferring the patterns defined by lithography into the thin films or the substrate. Etching process involves the controlled removal of a material usually in a selective manner in order to define such structures. The part of the material that will not be removed is protected by a mask typically made of photoresist or another material resistant to the chemistry used. The etch rate is defined as the removal rate in terms of thickness per unit time, usually measured in nm/min or μm/min. An important parameter in an etching process is the selectivity, which is defined as the ratio of the etch rate of the material being structured with respect to the one being used for the masking material. High values of selectivity mean that the used etchant is capable of removing the selected material without affecting too much the mask used to protect it. Another important etching parameter is the anisotropy A, defined as $A = [1 - (R_{LATERAL}/R_{VERTICAL})]$, where $R_{LATERAL}$ is the lateral etch rate and $R_{VERTICAL}$ is the vertical etch rate. A perfectly anisotropic etch has $A = 100\%$ ($R_{LATERAL} = 0$; Figure 1.4(a)), whereas a perfectly isotropic etch has $A = 0$ ($R_{LATERAL}=R_{VERTICAL}$; Figure 1.4(b)).

Wet Etching

The simplest etching is performed by the use of liquid chemicals [39, 40]. The film to be etched is immersed in a chemical solution that reacts with it promoting its liquid phase removal. This method can be used for example to define structures on SiO_2 films by

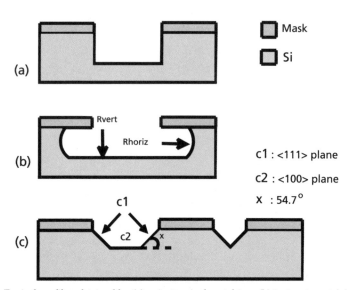

Figure 1.4 Typical profiles obtained by (a) anisotropic dry etching, (b) isotropic wet / dry etching, and (c) anisotropic wet etching of (100) crystalline silicon wafer using KOH.

exposing them to a solution of hydrofluoric acid (HF + NH$_4$F) to selectively remove SiO$_2$ without etching other films like photoresists or silicon. Other examples involve the use of phosphoric acid-based solutions to define structures in aluminum films used to fabricate electrical interconnection lines in integrated circuits.

Anisotropic wet etching of silicon is usually performed using potassium hydroxide (KOH) [41–44], tetra methyl ammonium hydroxide (TMAH) [45], and ethylene-diamine/pyrocatechol solution (EDP) [46, 47]. A solution of nitric, hydrofluoric, and acetic acids (HNO$_3$ + HF + CH$_3$COOH) can be used to isotropically etch silicon [48]. When selectivity to other materials is not of primary concern, a mixture of nitric and hydrochloric acids, usually with a HNO$_3$: 3 HCl ratio, known as acqua regia, can also be used. This mixture is used to dissolve noble metals such as gold and platinum and is selective to a small selection of metals (tantalum, iridium, osmium and titanium). With anisotropic etchants such as KOH, TMAH, and EDP, crystalline silicon shows different etch rates depending on the crystallographic orientation. Etching (100) wafer using KOH produces a pyramidal etch pit with 54.7° slope sidewalls with (111) sidewalls (Figure 1.4(c)).

Dry etching

Advanced etching processing use gas-phase reactants excited by radio frequency electrical discharges (plasmas). There are many plasma etching processes. The simplest equipments use parallel plate electrodes with capacitive power coupling (bottom part RF circuitry of Figure 1.5). This configuration, for historical reasons, is known as reactive

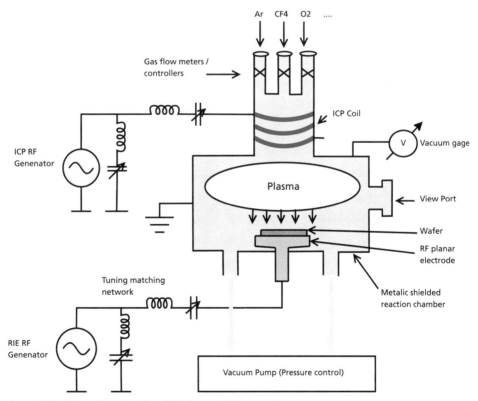

Figure 1.5 Schematic view of an ICP plasma etching reactor.

ion etching (RIE, [49]). Inductively coupled plasma configuration (ICP) uses an extra coil-shaped antenna (top part of Figure 1.5) and is able to produce plasmas with higher ion densities and such systems are capable of higher etch rates [50, 51].

The major advantage of dry etching process over wet etching is the directionality (usually associated with the presence of a vertical electric field on the surface of the sample) and the associated ability to define structures with high degree of anisotropy (Figure 1.4(a)). This enables high-fidelity copying of the patterns defined by the masks as well as the definition of small submicron structures with high aspect ratios (depth/width)—features that would not be possible to fabricate by means of using wet etching.

The specific chemistry and etching recipes to be used depend on the material to be etched, desired degree of anisotropy, etch rates, and selectivity to a given mask material. Anisotropic etching of SiO_2 is obtained by using C_4F_8 + CH_2F_2 + O_2 mixtures [52]. Anisotropic etching of silicon has been demonstrated using SF_6 + $CBrF_3$ + Ar + O_2 plasmas and graphite electrodes [53]. Cl_2 + BCl_3-based chemistries are used to anisotropically etch both silicon [54] and aluminum [55]. Deep reactive ion etching (DRIE) processes are defined as etching process that can be several microns deep and have aspect ratios (depth/width) > 10:1. An example of DRIE is the Bosch process [56] used for etching silicon. The Bosch process is based on the alternate use of fluorine rich plasma (SF_6) for etching silicon and carbon rich plasma (C_4F_8) for side-wall passivation. The partial polymerization on the sidewalls of the etched trenches helps to improve the anisotropy of the process. The process parameters (plasma power, gas flow, and pressure and cycle timing between the different gases) need to be carefully tuned to optimize scalloped sidewall profiles resulting from the process cycling (Figure 1.6).

Figure 1.6 Scanning electron microscopy (SEM) picture of vertical silicon etching profile using the Bosch process. Notice the scalloping along the sidewalls induced by the periodic passivation / etching using C_4F_8/SF_6 plasmas. (Courtesy of the Center of Micro and Nanotechnology, Ecole Polytechnique Federale de Lausanne, Switzerland.)

Deep etching of quartz is an important application of the DRIE process in the fabrication of microfluidic devices. Fluorine-containing plasmas using nickel hard mask running on state-of-the art ICP machines have been reported as a method for fabricating deep trenches in glass-related materials [57–59]. These processes have the potential to replace deep etching steps usually obtained by using concentrated HF solutions [60].

Gas phase etching of silicon

Dry, highly selective, chemical etching of silicon using fluorinated gases such as BrF_3 and XeF_2 has been used [61–63]. XeF_2 is unstable and decomposes spontaneously into pure $Xe + F_2$ when exposed to silicon. The resulting process can etch silicon isotropically with etch rates as high as 1–3 μm/min using a mixture of $XeF_2 + N_2$ and pressures in the (1–10) Torr range (without the presence of any plasma discharge) [64]. By using low-pressure gas phase etching, large structures can be released without stiction issues commonly encountered in liquid based release processes [65]. XeF_2 etching is also very selective to SiO_2, Si_3N_4, photoresist and many metals that can be used as masks.

1.5.2 Chemical-mechanical polishing and planarization

Chemical-mechanical polishing (CMP) is a technique usually applied in the fabrication of advanced high density IC interconnects, especially when using copper (which cannot be easily etched by plasma processing). It has potential applications in microfluidics when low surface roughness is required [66, 67]. This technique uses chemomechanical abrasives (usually alkaline and silica containing solution, known as slurry). The wafer is held by a carrier that applies a controlled pressure while rotating the wafer against a pad immersed in the slurry solution. This technique is used to produce highly planar and polished surfaces with very low surface roughness [68, 69]. The control parameters include (1) slurry composition (chemistry, size, amount of abrasive particles, and pH), (2) the pressure between the wafer and the pad, and (3) the rotation speeds of the wafer and the pad. The results are measured in terms of removal rate (usually in the range of a fraction of μm per minute for SiO_2 films), nonuniformity resulting due to the differences in the removal rate between different materials and by the pattern density distribution. Figure 1.7(a) shows a picture of CMP equipment and Figure 1.7(b) illustrates schematically the controlled process parameters and measured outputs.

1.6 Bonding processes

1.6.1 Lamination

Lamination is a traditional fabrication process used in the plastic industry, which can also be used for microfluidic fabrication [70]. A thin sheet of the adhesive layer can be placed in between the two substrates to be bonded. Heat is then applied to melt one of the adhesive layers and bond the materials. Laminator machines have a special design with floating plates in order to apply uniform pressure distribution on the substrates without sliding and to accommodate substrates of different thicknesses. Successful lamination process provides a complete seal of the microchannels in a reproducible manner with minimal or controlled channel deformation. Potential materials to be used

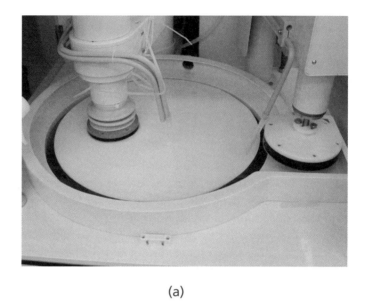

(a)

(b)

Figure 1.7 (a) Picture of a CMP machine, and (b) schematic illustration of CMP control and output process parameters.

for lamination are high-density polyethylene (HDPE), polypropylene (PP), ultrahigh molecular density polyethylene (UHMWPE), acetal, polyester [71], and polyaryletheretherketone (PEEK) for applications demanding chemical resistance. Micropumps [72] and micro channels [73] have been fabricated using this technique.

1.6.2 Wafer bonding methods

Microfluidic systems are commonly fabricated using this technique. Two wafers containing the required microfluidic structures are brought together and aligned/bonded to form a hermetically sealed interface. Wafer bonding can be performed using three different techniques: (1) wafer-to-wafer direct bonding, (2) anodic bonding, and (3) intermediate layer bonding, which includes methods such as solder bonding, adhesive bonding, and glass frit bonding [74].

Direct silicon-silicon bonding is possible by using the reaction between OH groups on the surface of oxidized silicon wafers [75]. The thin native oxide or an oxide induced by the use of solutions like piranha mixture (H_2SO_4 / H_2O_2) is usually enough to provide the bonding interfaces. High temperatures in the range of 400° to 1100°C are used to strengthen the interfacial bonds between the wafers. Direct wafer bonding methods typically need both the wafers to be made of the same material since any temperature induced expansion mismatches are not tolerated in this method. Furthermore, the method is extremely sensitive to particulates and wafer surface quality in terms of polishing roughness, warp, and cleaning. The method is extensively used in the production of silicon on glass wafers. Direct glass-to-glass bonding is also possible, but needs very clean (cleaned use of strong acid solutions and ultrasound) and moisture-free (prebaked) surfaces processed at high temperatures (typically 6 h at 600°C) [76]. Low-temperature HF-based glass-to-glass bonding processing has also been reported [77].

Wafers of silicon and sodium containing glass wafers such as Pyrex 7740 can be bonded using anodic bonding technique. The method was developed for bonding glass to Kovar (Cu-Ni Alloy) for creating glass to metal hermetic seals and was first reported by Wallis and Pomerantz in 1969 [78]. The temperature coefficient of expansion of the two materials to be bonded needs to be well matched for successful bonding [79]. The use of high voltage (typically in the range of 400–1200V) and high temperatures (typically in the range of 300°C–600°C) enables the diffusion of mobile sodium ions away from the interface creating a space charge region [80]. Most of the applied electric field exists in the thin interfacial space charge region between the two materials and pulls the two wafers together forming an extremely stable and strongly bonded interface. Figure 1.8 shows a schematic illustration of the anodic bonding setup.

Wafers can be bonded together for microfluidic applications using solder bonding [81]. Patterned electroplating of solders provides the possibility of creating complex microfluidic structures. Typical temperatures are in the range of 200°C to 300°C with minimal applied pressures can be used. It is usually performed under vacuum or under inert gas ambient (usually nitrogen) or using a mixture of nitrogen and hydrogen, typi-

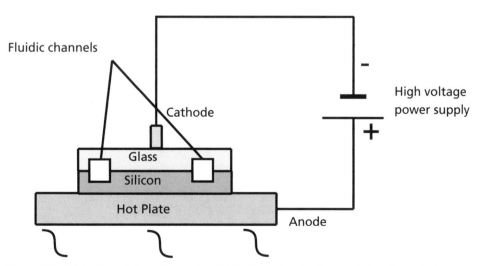

Figure 1.8 Fabrication of glass/silicon microfluidic channels using the anodic bonding technique.

cally at (95%–98%) N_2 + (5 to 2)% H_2, a mixture known as "forming gas" that helps removing oxygen and humidity from the environment and avoid the potential oxidation of the solder surfaces that usually leads to poor electrical contacts and weaker adhesion forces.

Polymer bonding

Polymer surfaces can be usually directly bonded to each other without the need of adhesives by heating them above the glass transition temperature and by subsequent cooling down process to form the desired bonds. This method is used for example to produce BCB and polyimide-based bonds. Alternative methods like microwave induced heating [82, 83] and resin gas injection have been also reported [84]. Polymers like PDMS can have poor adhesion to surfaces and usually require surface treatment to adhere to glass and other materials, usually by using a low-temperature oxygen plasma step [85–88].

1.7   Sacrificial layer techniques

Sacrificial layers are used as temporary fillers that are removed in the final steps of the processing, producing suspended structures. An example of such a process is shown in Figure 1.2(b), where a suspended structure supported by an anchor point is released when the underlying sacrificial material is removed. Dry gaseous phase release reactions are usually preferred to avoid stiction problems [65]. As examples, one can use isotropic O_2 plasma etching to release soft polymers [89]. HF vapor [91] and fluorine-based plasmas [92] can be used to release SiO_2 films [93]. Lift-off is another example of commonly used sacrificial layer technique using photoresist and wet removal solutions [27]. In wet release processes, CO_2-based supercritical drying [90] is used to avoid stiction related problems. Special photoresists processes with negative slope on the side walls are used in order to provide easy access for photoresist removal solution. An advantage of the lift-off process is that specific enchants for every material are not needed and a large range of materials or materials with poor etching selectivity can then be defined using the same base process.

Another example of sacrificial layer fabrication process specifically designed for the realization of microfluidic device using thick photoresist as sacrificial layer and plastic polymers (p-xylylene) as structural material is shown in Figure 1.9 [94].

1.8   Packaging processes

1.8.1   Dicing

Dicing is usually the first step performed before assembling and packaging of microfluidic devices. The dicing step is used to cut out and separate small individual devices out of wafers after parallel batch processing performed during fabrication. The resulting chip shape is usually planar square or rectangular with a few square millimeters to a few square centimeters in area. This process is usually performed by using a dicing saw machine having a spindle capable of rotating at very high speeds to which a thin diamond-coated blade (100–300 μm wide) is attached to precisely cut silicon, glass, and

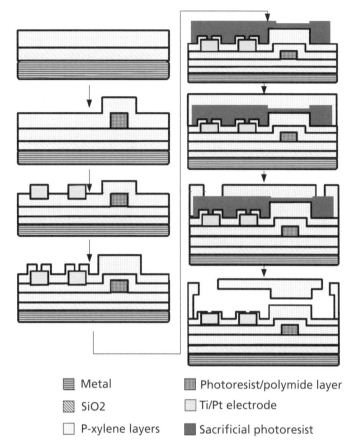

Metal	Photoresist/polymide layer
SiO2	Ti/Pt electrode
P-xylene layers	Sacrificial photoresist

Figure 1.9 Example of a fabrication process cross-section of microfluidic devices made of plastic films (p-xylylene layer) and sacrificial polymeric films. (*After* Webster and Mastrangelo [94].)

ceramic wafers. The cutting process parameters are defined by the type and size of the blade, rotation speed (typically in the range of 5 to 30 krpm), linear cutting speed (feed-rate of the wafer in mm/s), kerf and street width specification. Devices to be diced have to be protected against the debris and water that flows during the process. This is usually accomplished using a photoresist or other soft coating layer that can be readily stripped prior to the final packaging step without affecting the final device performance. Alternative and more advanced processing can cut the samples using a laser beam and water-assisted jets [95, 96] or deep reactive ion etching through almost the entire thickness of the wafer in order to form easily detachable pieces.

1.8.2 Electrical interconnection and wire bonding

Electrical interconnection of devices is usually made by means of wire-bonding techniques [97, 98]. This technique uses thin circular cross-section gold or aluminum wires as interconnects between two different pads (large metallic surfaces used to provide electrical inlets and outlets for the devices). The process is available in two variations, both of which are based on thermocompression bonding to achieve the adhesion between the interconnect wire and the bonding pads on the substrates and package. Ball bonding

(Figure 1.10(a)) uses gold alloy wires whereas wedge bonding is usually performed using aluminum alloy wires (Figure 1.10(b)). Both processes use ultrasonic energy and compression forces to promote bonding. Typical wire diameters are in the range of 25 to 100 μm with bond pad pitches as small as 50 μm. Gold ball bonding is typically performed at elevated temperatures between 80°C and 150°C on top of gold or aluminum pads [99], whereas aluminum wedge bonding can be performed at room temperature.

1.8.3 Fluidic interconnection in microfluidic systems

Packaging and interconnection is responsible for a significant part of the cost and complexity of fluidic microsystems. This is mainly due to the fact that microfluidic systems simultaneously require hermetically sealed fluidic channels and reliable and insulated electrical connections in the same device. These process steps play a pivotal role in the ultimate reliability and usefulness of these devices [101, 102]. Solutions for the packaging and interconnection of fluidic microsystems have been a topic of intense research [103–109].

At the mesoscale connection level and where higher pressures are required, O-rings are usually employed to create compression fitting seals [110]. O-rings are torus-shaped rubber that can be readily compressed between two mechanically clamped tubes to form

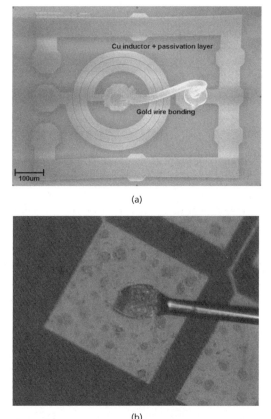

(a)

(b)

Figure 1.10 Examples of bump (a) [100] and wedge (b) wire-bonding interconnections. The square pad size is about 100 μm in both structures.

a sealed joint. They are mainly defined by their material (perfluoroelastomers like Viton, silicone, and Teflon being commonly used) along with their geometrical dimensions (namely the internal diameter and torus width). They are largely used at the mesoscale connections thanks to their low cost, simplicity, and reliability at high pressures. Efforts in making similar structure at the microscale level have been conducted [111, 112]. The access tubing is usually made of injected plastic and sealed at the joints using elastomer materials like PDMS. An example of a microfluidic system with several inlets and outlets in a glass microfluidic chip designed for molecular motor-based biochips can be found in [113]. The connections are ready for direct interface with syringe devices for biomedical applications.

1.9   Materials for microfluidic and bio-MEMS applications

Materials used in microfluidic and biological applications must fulfill a stringent list of requirements. *Biocompatibility* is probably the most important one. The entire bill of materials need to be able to operate without releasing toxic chemicals, reacting with the fluids being transported and analyzed, or activate unwanted immunological response in the biological systems under analysis [114–116]. Optical transparency is also a requirement for visualization of the flow using a microscope or video camera, enabling the identification and measurement of particles, cells, fluorescence, and the use of other colorimetric detection methods. The following are a list of materials commonly used in microfluidic fabrication technology along with their main physical and chemical properties.

1.9.1   Glass, pyrex, and quartz

Silicate glasses are by far the most commonly used material for the fabrication of microfluidic channels and very popular materials in microfluidic applications. They allow easy interface to optical monitoring devices by being highly transparent. Their ability to form strong and chemically inert bonds that can provide hermetical sealing of the channels is also a highly desirable characteristic for microfluidic applications. Together with silicon, these materials have thus far attracted much of the attention in microfluidics. Recently, they are being replaced by other low-cost alternative polymeric materials. Typical costs are between 4 to 40 U.S. cents/cm^2 for glass and quartz substrates, against 0.2 to 2 U.S. cents/cm^2 for polymers.

1.9.2   Silicon

Silicon is usually available as electronic-grade, high-quality, high-purity single crystalline material in the form of wafers (round, surface-polished slices typically of 4–12 inches in diameter and a few hundreds of micrometers to millimeters in thickness). The biggest advantage of using silicon for microfluidic applications is the availability of a mature processing technology inherited from the microelectronics IC industry as well as the possibility of defining very small structures that can be cointegrated with the electronics on the same chip [117]. Some of the disadvantages of using silicon as a structural material are linked to the polar nature of the silicon crystal resulting in undesirable adsorption of molecules in microfluidic systems. Furthermore, the higher cost of silicon as substrate material without any specific advantages from microfluidic systems stand-point makes it less attractive as a substrate material unless integration of on-chip electronic circuits is a strong requirement for the particular microsystem design. The typical cost of an average quality silicon substrate is about 0.25 U.S. cents/cm².

1.9.3 Elastomers

An elastomer [118] (also known as rubber) is a polymer with good elastic properties. They are very soft and deformable at ambient temperature (Young's modulus <~ 3 MPa). The monomers linking to form the polymer are usually made of carbon, hydrogen, oxygen, and silicon. They are primarily used as sealing materials, to mold flexible parts, and as adhesives. Most elastomers (e.g., natural rubber, polyisoprene, butyl rubber, and Viton) are thermoset, having their stable phase set by an irreversible curing process, in contrast to thermoplastic materials that can be melted and recover their initial properties after cooling down (e.g., vinyl, polyethylene, and polypropylene).

1.9.4 Polydimethylsiloxane

Silicones are polymeric materials having silicon and oxygen on their composition. They are largely inert compounds usually heat-resistant, nonstick, and rubberlike. Polydimethylsiloxane (PDMS, molecular formula $(C_2H_6OSi)_n$, density 965 kg/m³, boiling point: < 200°C, shear modulus between 100 kPa and 3 MPa, loss tangent less than 0.001) is a viscoelastic polymer (elastomer). PDMS is widely used in microfluidic technology thanks to some of its unique properties: it is low cost, nontoxic, chemically resistant, and stable against humidity and temperature variations. PDMS also presents low interfacial energy, which allows it to avoid chemical interactions with other polymers and solutions in the microfluidic channels. After polymerization and cross-linking, solid PDMS presents hydrophobic surface. A treatment using plasma oxidation is frequently used to alter the surface chemistry by adding silanol (SiOH) groups to the surface terminations and make the surface hydrophilic (wettable). PDMS is usually used as sealant [23, 119], structural microchannel material [120–122] or elastomer stamping matrix in soft lithography techniques like microcontact printing and micromolding [123].

1.9.5 Epoxy

Epoxy is a high-performance adhesive largely used in the electronics industry as well as for MEMS and microfluidic applications [124]. Some epoxies are cured by exposure to UV light and are commonly used in optical applications and dentistry. After cure, epoxies form strong bonds and smooth surfaces chemically inert to a large range of chemical substances, making them a good choice as sealing material as well as microchannels structural material. The strength of the epoxy adhesion is usually degraded at temperatures above 180°C.

1.9.6 SU-8 thick resists

SU-8 is an epoxy-based negative-tone resist [125] often used in microfluidic applications [126–131]. It is made of Epon epoxy resin dissolved in gamma-butyrolacton and a photoinitiator salt (triaryl sulfonium, representing usually about 10% in the weight of the composition). The amount of solvent sets the layer thickness that can be achieved (thicker films are obtained using less solvent content; i.e., more viscous resists). During the UV exposure the sulfonium salt decomposes into a strong acid that polymerizes the SU-8 matrix under heating provided by the baking [132]. A large range of thickness from a few microns to several hundreds of microns is possible using this resist, making it an attractive material for microfluidic applications and a suitable candidate to replace costly fabrication techniques based on the use of thick polymethyl methacrylate (PMMA) resists and X-ray synchrotron radiation lithography used in LIGA (German acronym for "Lithographie, Galvanoformung, Abformung,": (X-ray) Lithography, Electroplating, and Molding [133]) process.

1.9.7   Thick positive resists

Recent advances in the formulation of resists now offer thick positive photoresists that can be defined using standard thin film processing steps. As examples, the AZ 96 family from AZ Tech [134] and the SPR 220 from Rohm [135] can offer thicknesses in the range of 7 to 10 μm using single coatings and can be patterned using standard *i*-line exposure tools.

1.9.8   Benzocyclobutene

Benzocyclobutene (BCB) is composed of a benzene ring fused to a cyclobutane ring (chemical formula C_8H_8). It is frequently used to create photosensitive polymers. BCB-based polymer dielectrics may be spun on various substrates, with good step coverage properties that make them useful to planarize uneven topographies. Applications include wafer bonding, optical interconnects, semihermetic package sealing, passivation [136], and low-k dielectrics. BCB has been also reported as a material with good biocompatibility, being used in intracortical neural implants and other biomedical applications [137, 138].

1.9.9   Polyimides

Polyimides (PI, also known by alternative brand names: Pyralin [139] Apical, Kapton, and Kaptrex) are polymers made of imide monomers (two carbonyl groups bonded to nitrogen). It has a density of 1430 kg/m^3, Young's modulus 3200 MPa, tensile strength (75–90) MPa, glass transition temperature > 400°C, thermal conductivity of 0.53 W/(m K), and a low relative dielectric constant of 2.9. Polyimides have applications similar to BCBs and also find applications as a low dielectric constant material in high-frequency interconnects (low-k material). The higher thermal conductivity and high glass transition temperature when compared to other types of insulating polymers also makes them a good choice where heat dissipation is important or other applications requiring good thermal management or low parasitic capacitances. Microchannels and high-frequency microdevices embedded in both photosensitive and nonphotosensitive polyimides have been fabricated and reported [138, 140, 141–143].

1.9.10 Polycarbonate

Polycarbonates are plastics widely used in modern industry having good temperature and impact resistance. This plastic is particularly good to work with more conventional definition techniques (injection molding, extrusion into tubes or cylinders and thermoforming). It is also used when optical transparency is needed, having more than 80% transmission up to the 1560-nm range (short wave infrared range). It has moderated chemical resistance properties, being chemically resistant to diluted acids and alcohols. It is poorly resistant against ketones, halogens, and concentrated acids. The major disadvantage associated with polycarbonates is the low glass transition temperature ($T_g > 40°C$), but it is still largely used as low-cost material in microfluidic systems and also as a sacrificial layer [144].

1.9.11 Polytetrafluoroethylene

Polytetrafluoroethylene (PTFE, $(C_2F_4)_N$, commonly known by its Dupont commercial name Teflon [145]), is a fluorinated polymer largely used in the industry thanks to its chemical stability, being inert to virtually all chemicals. It has the lowest coefficient of friction of any known solid material, what makes it a good natural candidate for low-loss fluid flow where a high level of optical transparency is not a requirement. Being largely used as antiadherent coating, it is relatively difficult to produce reliable bonds to this material and surface plasma treatments are usually required to succeed [146].

1.10 Troubleshooting table

A lot of problems in microsystem fabrication arise from thin film contamination or unexpected incompatibility among the large range of materials used. In general these issues can be addressed by thorough cleaning of the surfaces and by careful and systematic study of the processing parameters and the materials used. Table 1.2 presents the main issues one can face in the fabrication of microfludic devices and how to address them.

1.11 Summary

This chapter presented an introductory overview of the main methods and materials used for the fabrication of microfluidic systems. Traditional IC fabrication techniques (lithography, thin film deposition, plasma etching) along with bulk and surface MEMS fabrication techniques (deep reactive etching, molding, injection, bonding, plastic microfabrication, silicon/glass/polymeric bonding, fluidic interconnection, and packaging) are the enabling technological blocks used to fabricate complex microsystems. The ultimate goal is to achieve a fabrication process flow for a microfluidic system that has been sequenced and optimized in terms of fabrication precision, compatibility, functionality in different domains, reliability, cost, performance, and biocompatibility.

Table 1.2   Troubleshooting of the Common Fabrication Steps of Microfluidic Systems

Step	Issue	Possible Causes	Solutions	Notes
Lithography	Poor resist adhesion	Thin film contamination	Acetone/isopropanol cleaning	Effective to degrease surfaces (acetone erodes rubber)
			Piranha cleaning ($H_2SO_4 + H_2O_2$)	Effective to degrease glass and silicon
			O2 plasma de-scum	Removes previous lithography residues
		Polymer/substrate interface	HMDS prime coating	Effective to SiO_2 and Si_3N_4 surfaces, Might be not effective on metals, Better when performed with heating and low pressure vapor phase
			O_2/Ar plasma treatment	Slightly rough surfaces can improve adhesion/wettability
	Poor resolution	Incorrect exposition/development parameters	Decrease exposition if features are distorted	Increasing exposition time requires decreasing development time (and vice-versa) to keep optimum resolution
			Increase exposition if small features do not appear	
	Resist erodes during etching	Poor selectivity to the used chemistry	Longer hard bake of resist	Makes the resist harder to remove, requires tuning of development/exposition
			Use thicker resist	Thick resist has less resolution
			Use of another type of resist (epoxy)	After cure, epoxy resist can be hard to remove
			Change chemistry of process	Select reactants selective to polymers, add C-H contents to plasma
			Change of plasma etching parameters	Decrease the power, change pressure (affects all other parameters: rate and profile)
Deposition	Nonuniformity	Process parameters	Lower the pressure	For plasma processing, adjust also the gas flow and plasma power
	Deposits of foreign materials	Contamination	Cleaning deposition chamber	O_2 plasma step is usually used to passivate/clean walls before depositions
	Granularity/Roughness	Contamination	Cleaning of the substrate/process chamber	Film texture is correlated with stress, adhesion, elastic modulus and resistivity (for metals)
		Process parameters	Select parameters for smaller grain size	Usually at higher power / lower pressure with sputtering systems
	Film peels off	Excessive stress	Tune process parameters for low stress	Increase of flow rate and pressure, decrease of power on sputtering systems
		Poor adhesion	Improve cleanliness and compatibility of substrate, use of adhesion layers	Metals on polymers require the use of adhesion/diffusion barriers and polymer surface treatment (O_2/Ar plasma)
Wet etching	Undesirable release of features	Etch rate is too high	Decrease time or reduce concentration	The wet chemicals might be etching underneath the mask and through the supporting films
	Undercut/poor resolution	Diffusion of liquid underneath the mask	Use of dry etching	Dry etching offers better anisotropy control

Table 1.2 (continued)

Step	Issue	Possible Causes	Solutions	Notes
Dry etching	Etch rate is too low	Incorrect chemistry	Select best etchants for the desired film	Increase plasma power, consider heating, bombardment induced defects and selectivity to the desired mask
		Incorrect plasma parameters	Increase power, change pressure	RIE process usually have maximum etch rates in the 50–150 mTorr range (much lower pressures for ICP)
	Etch rate decreases over time	Redeposition of materials	Change chemistry/process parameters	Higher power/lower pressures are less likely to exhibit this problem
	Redeposition of material	Incorrect chemistry	Check the possible solid products and their vapor pressures	Plasma etching byproducts must be volatile at the process pressure
		Erosion of the mask	Use of hard masks (metals)	Use nonerodible masks
			Change process parameters	Select chemistry and parameters to preserve the mask
Release	Stiction	van der Waals/electrostatic forces	Isopropyl alcohol finishing	Lower surface tension, easier evaporation/drying
			Critical point drier/HF vapor	Low pressure helps to control stiction
			Use of low resistivity silicon/conductive films	Helps to evade electrostatic charges in microsystems
			Use of stiffer materials	Increase in stiffness reduces displacement
	Poor adhesion after lift-off	Residual resist on the bottom line	Increase exposition/development time	Facilitates the cleaning of the exposed area
			Plasma descum	Usually needed to clean the bottom line even if exposition/development parameters are correct
	Lift-off too slow or not happening	Resist burning during deposition	Use less power for e-beam/plasma depositions	High temperatures can hard bake/burn the photoresist, making the lift-off harder
		Slow penetration of solvent on resist	Use of better remover and ultrasonic agitation	Lift-off process might need many hours to complete
Bonding	Lack of adhesion	Surface cleaningness	Degrease or use slight etching solutions/ultrasonic cleaning	Ultrasonic bath might help to get rid of contaminant films and particles, weak acid solutions can get rid of metal oxides
		Incompatible materials	Use of adhesion layers	Au and Al wires are commonly available; Cr and Ti adheres to Au and enhances the adhesion to a variety of substrates, specially glass
		Surface polarity and roughness	Plasma treatment	Check wettability, Slightly rough, polar surfaces with interdiffusion usually have better adhesion
	Partial adhesion	Substrate is nonplanar	Use higher quality wafers, control film stress	
		Nonuniform pressure during bonding	Improve bonding tool pressure uniformity	

Table 1.2 (continued)

Step	Issue	Possible Causes	Solutions	Notes
Sealing	Fluid leak	Nonhermetical sealing	Use of hermetic sealing materials	PDMS and soft polymers with correct treatments provide sealing for low to moderate pressures
		Excessive line pressure	Check ratings for tubing and fittings, eliminate system constrictions	High pressure applications might require special o'rings or metallic sealing

References

[1]   Feynman, R. P., "There's Plenty of Room at the Bottom: An Invitation to Enter a New Field of Physics," *Engineering and Science Magazine*, Vol. XXIII, No. 5, Feb. 1960.

[2]   Moore, G. E., "Cramming More Components onto Integrated Circuits," *Electronics Magazine*, Vol. 8, No. 38, Apr. 1965, pp. 114–117.

[3]   Trimmer, W. (ed.), *Micromechanics and MEMS: Classic and Seminal Papers to 1990*, IEEE Press, New York, NY, 1996.

[4]   Madou, M., *Fundamentals of Microfabrication*, Boca Raton, FL: CRC Press, 1997.

[5]   Rebeiz, G. M., *RF MEMS: Theory, Design, and Technology*, John Wiley & Sons, 2003.

[6]   Neukermans, A., Ramaswami, R., "MEMS Technology for Optical Networking Applications," *IEEE Communications Magazine*, Vol. 39, No. 1, 2001, pp. 62–69.

[7]   Lin, L.Y., Goldstein, E.L., "Opportunities and Challenges for MEMS in Lightwave Communications," *IEEE Journal of Selected Topics in Quantum Electronics*, No. 8, 2002, pp. 163–172.

[8]   Wang, W., and Soper, S. A. (eds.), *Bio-MEMS: Technologies and Applications*, CRC Press, 2007.

[9]   Tabeling, P., *Introduction to Microfluidics*, Oxford Press, 2005.

[10] Mastrangelo, C. H., Burns, M. A. Burke D. T. "Microfabricated Devices for Genetic Diagnostics," *Proceedings of the IEEE*, Vol. 86, No. 8, Aug. 1998, pp. 1769–1787.

[11] Santini, J. T., Jr., Richards, A. C., et al., "Microchips as Controlled Drug-Delivery Devices," *Angewandte Chemie*, Vol. 39, No. 14, 2000, pp. 2396–2407.

[12] Zemel, J. N., Furlan, R., "Microfluidics," in *Handbook of Chemical and Biological Sensors*, Taylor, R. F., Schultz, J. S. (eds.), Bristol and Philadelphia: Institute of Physics Publishing, June 1996, pp. 317–347.

[13] Burns M. A., Johnson, B. N., et al, "An Integrated Nanoliter DNA Analysis Device," *Science*, No. 282, 1998, pp. 484–487.

[14] Lee, G.-B., Chen, S.-H., Huang, G.-R., Sung, W.-C., and Lin, Y.-H., "Microfabricated Plastic Chips by Hot Embossing Methods and Their Applications for DNA Separation and Detection," *Sensors and Actuators B: Chemical*, Vol. 75, No. 1–2, 2001, pp. 142–148.

[15] Torres, S., Zankovych, C. M., S., et al., "Nanoimprint Lithography: an Alternative Fabrication Approach," *Materials Science & Engineering*, C, Vol. C23, 2003, pp. 23–31.

[16] Moreau, W. M., *Semiconductor Lithography: Principles, Practices, and Materials*, Plenum Publishing Co., 1987.

[17] Heckele, M., Bacher, W., and Muller, K. D., "Hot Embossing—The Molding Technique for Plastic Microstructures," *Microsystem Technologies*, Vol. 4, No. 3, 1998, pp. 122–124.

[18] Becker, H., and Heim, U. "Hot Embossing as a Method for the Fabrication of Polymer High Aspect Ratio Structures," *Sensors and Actuators A: Physical*, Vol. 83, No. 1–3, 2000, pp. 130–135.

[19] Larsson, O., et al., "Silicon Based Replication Technology of 3D-Microstructures by Conventional CD-Injection Molding Techniques," *Proceedings of the 9th International Conference on Solid-State Sensors and Actuators (Transducers'97)*, Chicago, IL, June 1997, pp. 1415–1418.

[20] Piotter, V., et al., "Injection Molding and Related Techniques for Fabrication of Microstructures," *Microsystem Technologies*, Vol. 3, No. 3, 1997, pp. 129–133.

[21] Oswald, T. A., Turng, L.-S., Gramamn, P. J. (eds.), *Injection Molding Handbook*, 2nd ed., Hanser Gardner, 2001.

[22] Kim, D. S. "Disposable Integrated Microfluidic Biochip for Blood Typing by Plastic Microinjection Moulding," *Lab on a Chip*, Vol. 6, No. 6, 2006, pp. 794–802.

[23] Chiou, C.-H., and Lee, G.-B., "Minimal Dead-Volume Connectors for Microfluidics Using PDMS Casting Techniques," *Journal of Micromechanics Microengineering*, Vol. 14, No. 11, 2004, pp. 1484–1490.

[24] Becker, H., and Gartner, C., "Polymer Microfabrication Methods for Microfluidic Analytical Applications," *Electrophoresis*, Vol. 21, No. 1, 2000, pp. 12–26.

[25] Sze, S.M. (ed.), *VLSI Technology*, McGraw-Hill, 1983.

[26] Deal, B. E., A. S. Grove, "General Relationship for the Thermal Oxidation of Silicon," *Journal of Applied Physics*, Vol. 36, No. 12, Dec. 1965, pp. 3770–3778.

[27] Wolf, S., Tauber, R. N., *Silicon Processing for the VLSI Era*, Vol. 1: Process Technology, Sunset Beach, CA, USA, Lattice Press, 1986.

[28] Muller, R. S., Kamins, T. I., *Device Electronics for Integrated Circuits*. 2nd ed, New York, John Wiley & Sons, 1996.

[29] Brodie, I., Muray, J. J., *The Physics of Microfabrication*, New York, Plenum Press, 1982.

[30] Chapman, B., *Glow Discharge Processes: Sputtering and Plasma Etching*, New York, John Wiley and Sons, 1980.

[31] Wasa, K., Hayakawa, S., *Handbook of Sputter Deposition Technology: Principles, Technology and Applications*, Park Ridge (New Jersey, USA), Noyes Publications, 1992.

[32] Braithwaite, N. St. J., "Introduction to Gas Discharges," *Plasmas Sources Science and Technology*, Vol. 9, No. 4, pp. 517–527, 2000.

[33] Reif, R., *Plasma Enhanced Chemical Vapor Deposition of Thin Films for Microelectronics*, In: Rossnagel, S. M., Cuomo, J. J., Westwood, W. D. (eds.). *Handbook of Plasma Processing Technology*, Westwood, NJ: Noyes Publications, 1989, pp. 260–307.

[34] Grill, A., *Cold Plasma in Materials Fabrication: From Fundamentals to Applications*, New York, NY: IEEE Press, 1994, Cap. 7: PECVD.

[35] Adams, A. C., "Dielectric and Polysilicon Film Deposition," in *VLSI Technology*, S. M. Sze (ed.), New York: McGraw-Hill, 1988, pp. 233–271.

[36] Steinbrüchel, C., and B. L. Chin, *Copper Interconnect Technology*, SPIE Press, Washington, 2001.

[37] Mohler, J. B., *Electroplating and Related Processes*, Chemical Publishing Co., 1969.

[38] Mallory, G. O., and Hadju, J. B. (eds.), *Electroless Plating, Fundamentals and Aplications*, American Electroplaters and Surface Finishers Society (AESF), Orlando, FL, 1990.

[39] Williams, K. R., and Muller, R.S., "Etch Rates for Micromachining Processing," *Journal of Microelectromechanical Systems*, Vol. 5, No. 4, Dec. 1996, pp. 256–269.

[40] Williams, K. R., Gupta, K., and Wasilik, M., "Etch Rates for Micromachining Processing - Part II," *Journal of Microelectrical Systems*, Vol. 12, No. 6, Dec. 2003, pp. 761–778.

[41] Stoller, A. I., "The Etching of Deep Vertical-Walled Pattern in Silicon," *RCA Review*, No. 31, 1970, pp. 271–275.

[42] Seidel, H., Csepregi, L., Heuberger, A., and Baugartel, H., "Anisotropic Etching of Crystalline Silicon in Alkaline Solutions—Part I: Orientation Dependence and Behavior of Passivation Layers," *Journal of the Electrochemical Society*, Vol. 137, No. 11, 1990, pp. 3612–3626.

[43] K. E. Bean, R.L. Yeakley and T.K. Powell, "Orientation Dependent Etching and Deposition of Silicon," *Journal of the Electrochemical Society*, Vol. 121, 1974, pp. 87C.

[44] Mayer, G. K., Offereins, H. L., Sandmaier, H., and Kuhl, K., "Fabrication of Non-underetched Convex Corners in Anisotropic Etching of (100) Silicon in Aqueous KOH With Respect to Novel Micromechanic Element," *Journal of the Electrochemical Society*, Vol. 137, No. 12, 1990, pp. 3947–3951.

[45] Tabata, O., Asahi, R., Sugyiama, S., "Anisotropic etching with quaternary ammonium hydroxide solutions," *Technical Digest of the 9th Sensor Symposium*, Tokyo, Japan, 1990, pp. 15–18.

[46] Wu, X. P., Wu Q. H., and Ko, W. H., "A Study on Deep Etching of Silicon Using Ethylene-Diamine-Pyrocathechol-Water," *Sensors and Actuators*, Vol. 9, 1986, pp. 333–343.

[47] Linde, H., and Austin, L., "Wet Silicon Etching With Aqueous Amine Gallates," *Journal of The Electrochemical Society*, Vol. 139, No. 4, 1992, pp. 1170–1174.

[48] Robbins, H., and Schwartz, B, "Chemical Etching of Silicon – II. The System HF, HNO_3, H_2O, and $HC_2C_3O_2$," *Journal of the Electrochemical Society*, Vol. 107, No. 2, 1959, pp. 108–111.

[49] Oehrlei, G. G. Reactive Plasma Processes - Reactive Ion Etching. In: Rossnagel, S. M., Cuomo, J. J., Westwood, W. D., (ed), *Handbook of Plasma Processing Technology*, Westwood (New Jersey, USA), Noyes Publications, 1989, pp. 196–232.

[50] Popov, O. A., (ed.), *High density plasma sources: Design, physics and performance*, Parkridge (New Jersey, USA), Noyes Publications, 1995.

[51] Keller, J. H., "Inductive Plasmas for Plasma Processing," *Plasma Sources Science Technology*, Vol. 5, No. 2, 1996, pp. 166–172.

[52] Tsukada, T., Nogami, H., Nakagawa, Y., Wani, E., Mashimo, K., Sato H., and Samukawa S., "SiO_2 Etching Using High Density Plasma Sources," *Thin Solid Films*, Vol. 341, No. 1–2, March 1999, pp. 84–90.

[53] Mansano, R. D., Verdonck, P. and Maciel, H. S., "Deep Trench Etching in Silicon with Fluorine Containing Plasmas," *Applied Surface Science*, Vol. 100–101, July 1996, pp. 583–586.

[54]   Sung, K. T., and Pang, S. W., "Etching of Si with Cl_2 Using an Electron Cycloron Resonance Source," *Journal of Vacuum Science and Technology A*, Vol. 11, No. 4, 1993, pp. 1206–1210.

[55]   Lutze, J. W., Perera, A. H., and Krusius, J. P., "Anisotropic Reactive Ion Etching of Aluminum using Cl_2, BCl_3, and CH_4 gases," *Journal of The Electrochemical Society*, Vol. 137, No. 1, 1990, pp. 249–252.

[56]   Larner, F., and Schilp, A., "Method of Anisotropically Etching Silicon," U.S. Patent No. 5501893, German Patent DE 4241045, 1994. (Bosh etching process).

[57]   Li, X., Abe, T., Esahi, M., "Fabrication of High-Density Electrical Feed-Throughs by Deep-Reactive-Ion Etching of Pyrex Glass", *Journal of Microelectromechal Systems*, Vol. 1, No. 6, 2002, pp. 625–630.

[58]   Pavius, M., Hibert, C., Fluckiger, P., Renaud, P., "Profile Angle Control in SiO_2 Deep Anisotropic Dry Etching for MEMS Fabrication," *Proceedings of the 17th IEEE International Conference on Micro Electro Mechanical Systems* (MEMS), 2004, pp. 669–672.

[59]   Goyal, A. Hood, V, and S. Tadigadapa, "High Speed Anisotropic Etching of Pyrex for Microsystems Applications," *Journal of Non-Crystalline Solids*, No. 352, 2006, pp. 657–663.

[60]   Iliescu, C., Chen, B., and Miao, J., "Deep Wet Etching-Through 1 mm Pyrex Glass Wafer for Microfluidic Applications," *Proceedings of IEEE MEMS 2007 Conference*, Kobe, Japan, Jan. 2007, pp. 393–396.

[61]   Winters, H. F., and Coburn, J. W., "The etching of silicon with XeF_2 vapor," *Applied Physics Letters*, Vol. 34, No. 1, Jan. 1979, pp. 70–73.

[62]   Ibbotson, D. E., Mucha, J. A., Flamm, D. L., and Cook, J. M., "Plasmaless Dry Etching of Silicon with Fluorine-Containing Compounds," *Journal of Applied Physics*, Vol. 56, No. 10, Nov. 1984, pp. 2939–2942.

[63]   Wang, X.-Q., Yang, X., Walsh, K., and Tai, Y.-C., "Gas-Phase Silicon Etching With Bromine Trifluoride," *Proceedings of the 9th International Conference on Solid State Sensors and Actuators (Transducers'97)*, Chicago, IL, USA, June 16–19, 1997, pp. 1505-1508.

[64]   Chang, F. I., Yeh, R., Lin, G., et al, "Silicon Micromachining with Xenon Difluoride," *Proceedings of SPIE Microelectronic Structures and Microelectromechanical Devices for Optical Processing and Multimedia Applications*, Vol. 2641, 1995, pp. 117–128.

[65]   Legtenberg, R., Tilmans, A. C., Elders, J., and Elwenspoek, M., "Stiction of Surface Micromachined Structures After Rinsing and Drying: Model and Investigation of Adhesion Mechanisms," *Sensors and Actuators A: Physical*, Vol. 43, No. 1–3, May 1994, pp. 230–238.

[66]   Wong, C. C., Agarwal, A., Balasubramanian, N., and Kwong, D. L., "Fabrication of Self-Sealed Circular Nano / Microfluidic Channels in Glass Substrates," *Nanotechnology*, 2007, Vol. 18, No. 13, pp. 135304–135309.

[67]   Naik, N., Courcimault, C., Hunter, H., Berg, J., Lee, J., Naeli, K., Wright, T., Allen, M., Brand, O., Glezer, A., King, W., "Microfluidics for Generation and Characterization of Liquid and Gaseous Micro- and Nanojets," *Sensors and Actuators A*, Vol. 134, No. 1, 2007, pp. 119–127.

[68]   Kaanta, C. W., Bombardier, S. G., Cote, W. J., Hill, W. R., Kerszykowski, G., Landis, H. S., Poindexter, D. J., Pollard, C. W., Ross, G. H., Ryan, J. G., Wolff, S., and Cronin, J. E., "Dual Damascene: a ULSI Wiring Technology," *Proceedings of the 8th International IEEE VLSI Multilevel Interconnection Conference*, 1991, pp. 144–152.

[69]   Patrick, W. L., Guthrie, W. L. O., Standley, C. L., and Schiable, P. M., "Application of Chemical-Mechanical Polishing to the Fabrication of VLSI Circuit Interconnections," *Journal of the Electrochemical Society*, Vol. 138, No. 6, June 1991, pp. 1978–1784.

[70]   Debjani, P., Pallandre, A., Miserere, S., Weber, J., and Viovy, J.-L., "Lamination-Based Rapid Prototyping of Microfluidic Devices Using Flexibe Thermoplastic Substrates," *Electrophoresis*, Vol. 28, No. 7, 2007, pp. 115–1122.

[71]   do Lago, C. L., da Silva, H. D. T., Neves, C. A., and Brito-Neto, J. G. A., "A Dry Process for Production of Microfluidic Devices Based on the Lamination of Laser-Printed Polyester Films," *Analytical Chemistry*, Vol. 75, No. 15, 2003, pp. 3853–3858.

[72]   Truong, T. Q., and Nguyen, N. T., "A Polymeric Piezoelectric Micropump Based on Lamination Technology," *Journal of Micromechanics and Microengineering*, Vol. 14, No. 4, 2004, pp. 632–638.

[73]   Abgral, P., "A Novel Fabrication Method of Flexible and Monolithic 3D Microfluidic Structures Using Lamination of SU-8 Films," *Journal of Micromechanics and Microengineering*, Vol. 16, No. 1, 2006, pp. 113–121.

[74]   Knechtel, R., Wiemer, M., and Frömel, J., "Wafer Level Encapsulation of Microsystems Using Glass Frit Bonding," *Microsystem Technologies*, Vol. 12, No. 5, Apr. 2006, pp. 468–472.

[75]   Maszara, W.P. et al., "Bonding of Silicon-Wafers for Silicon-on-Insulator," *Journal of Applied Physics*, Vol. 64, No. 10, 1988, pp. 4943–4950.

[76]   Stjernstrom, M., and Roeraade, J., "Method for Fabrication of Microfluidic System in Glass," *Journal of Micromechanics and Microengineering*, Vol. 8, No. 1, 1998, pp. 33–38.

[77]   Chen, L., Luo, G., Lui, K., Ma, J., Yao, B., Yan, Y., Wang, Y., "Bonding of Glass-Based Microfluidic Chips at Low- or Room-Temperature in Routine Laboratory," *Sensors and Actuators B: Chemical*, Vol. 119, No. 1, 2006, pp. 335–344.

[78]   Wallis, G., and Pomerantz, D. I., "Field Assisted Glass-Metal Sealing," *Journal of Applied Physics*, Vol. 40, No. 10, 1969, pp. 3946–3949.

[79]   Raley, N. F., Davidson, J. C., et al., "Examination of Glass-Silicon and Glass-Glass Bonding Techniques for Microfluidic Systems," *Proceedings of SPIE – The International Society of Optical Engineers*, Vol. 2639, 1995, pp. 40–45.

[80]   Wei, J., Xie, H., Nai, M. L., Wong, C. K., and Lee L. C., "Low Temperature Wafer Anodic Bonding," *Journal of Micromechanics and Microengineering*, Vol. 13, No. 2, 2003, pp. 271–222.

[81]   Goyal, A., Cheong, J., and Tadigadapa, S., "Tin-Based Solder Bonding for MEMS Fabrication and Packaging Applications," *Journal of Micromechanics and Microengineering*, Vol. 14, No. 6, 2004, pp. 819–825.

[82]   Lei, K. F, Li, W. J., Budraa, N. and Mai, J. D., "Microwave Bonding of Polymer-Based Substrates for Micro/Nano Fluidic Applications," *Proceedings of the 12th Conference on Solid State Sensors, Actuators and Microsystems (Transducers'03)*, Boston, 2003, pp. 1335–1338.

[83]   Yussuf, A. A., Sbarski, I., Hayes, J. P., Solomon, M. and Tran, N., "Microwave Welding of Polymeric-Microfluidc Devices," *Journal of Micromechanics and Microengineering*, Vol. 15, No. 9, Sep. 2005, pp. 1692–1699.

[84]   Lai, S. Yeny, H., James, L. L., Daunert, S., Madou, M. J., "A Novel Bonding Method For Polymer-Based Microfluidic Platforms," *Proceedings of SPIE: Micromachining and Microfabrication Process Technology VII*, Vol. 4557, 2001, pp. 280–287.

[85]   Ageorges, C., "Advances in Fusion Bonding Techniques for Joining Thermoplastic Matrix Composites," *Composites Science and Technology*, Vol. 32, No. 6, 2001, pp. 839–857.

[86]   Bhattacharya, S., Datta, A., Berg, J. M., and Gangopadhyay, S., "Studies on Surface Wettability of Poly(dimethyl) Siloxane (PDMS) and Glass Under Oxygen-Plasma Treatment and Correlation with Bond Strength," *Journal of Microelectromechanical Systems*, Vol. 14, No. 3, 2005, pp. 590–597.

[87]   Chow, W. W. Y., Lei, K. F., Shi, G., Li, W. J., and Huang, Q., "Microfluidic Channel Fabrication by PDMS-Interface Bonding," *Smart Materials and Structures*, Vol. 15, No. 1, 2006, pp. S112–S116.

[88]   Sun, Y., Kwok, Y. C., Nguyen N.-T., "Low-pressure, High-temperature Thermal Bonding of Polymeric Microfluidic Devices and their Applications for Electrophoretic Separation," *Journal of Micromechanics and Microengineering*, Vol. 16, No. 8, 2006, pp. 1681–1688.

[89]   Mastrangelo, C. H., and Saloka, G. S., "A Dry-Release Method Based on Polymer Columns for Microstructure Fabrication," *Proceedings of the IEEE 6th International Conference on Micro Electro Mechanical Systems (MEMS'93)*, Fort Lauderdale, FL, Feb. 7–10, 1993, pp. 77–81.

[90]   Dyck, C. W., Smith, J. H., Millera, S. L., Russick, E. M., and Adkins, C. L., "Supercritical Carbon Dioxide Solvent Extraction From Surface Micromachined Micromechanical Structures," *Proceedings of SPIE*, Vol. 2879, No. 11 (Part 1), Oct. 1996, pp. 225–235.

[91]   Overstolz, T., Clerc, P. A., Noell, W., Zickar, M., and de Rooij, N. F., "A Clean Waferscale Chip-Release Process Without Dicing Based on Vapor Phase Etching," *Proceedings of the 17th IEEE International Conference on Micro Electro Mechanical Systems (MEMS'04)*, Maastricht, the Netherlands, Jan. 25–29, 2004, pp. 717–720.

[92]   Frédérico, S., Hibert, C., Fritschi, R. Fritschi, Flückiger, P., Renaud P., and Ionescu, A. M., "Silicon Sacrificial Layer Dry Etching (SSLDE) for Free-Standing RF MEMS Architectures," *Proceedings 16th IEEE International Conference on Micro Electro Mechanical Systems (MEMS '03)*, Kyoto, Japan, Jan. 19–23, 2003, pp. 570–573.

[93]   Bühler, J., Steiner, F.-P., and Baltes, H., "Silicon Dioxide Sacrificial Layer Etching in Surface Micromachining," *Journal of Micromechanics and Microengineering*, Vol. 7, No. 1, pp. 1–13, 1997.

[94]   Webster, J. R., and Mastrangelo, C. H., "Large-Volume Integrated Capillary Electrophoresis State Fabricated Using Micromachining of Plastics on Silicon Substrates," *Proceedings of of the 9th International Conference on Solid-State Sensors and Actuators (Transducers'97)*, Chicago, IL, June 1997, pp. 503–506.

[95]   Richerzhagen, B., Perrottet, D., and Kozuki, Y., "Dicing of Wafers by Patented Water-Jet-Guided Laser: The Total Damage-Free Cut,", *Proceedings of the 65th Laser Materials Processing Conference*, 2006, pp. 197–200.

[96]   Kumagai, M., Uchiyama, N., et.al., "Advanced Dicing Technology for Semiconductor Wafer - Stealth Dicing," *IEEE Transactions on Semiconductor Manufacturing*, Vol. 20, No. 3, Aug. 2007, pp. 259–265.

[97]   Anderson, O. L., Christensen, H., and Andreatch, P., "Technique for Connecting Electrical Leads to Semiconductor," *Journal of Applied Physics*, Vol. 28, No. 8, Aug. 1957, pp. 923.

[98]   Harman, G., *Wire Bonding in Microelectronics Materials: Processes, Reliability and Yield*, 2nd ed., McGraw Hill, 1997.

[99]   Gerling, W., "Electrical and Physical Characterization of Gold-Ball Bonds on Aluminum Layers," *Proceedings of the IEEE ECC*, New Orleans, Louisiana, May 14–16, 1984, pp. 13–20.

[100] Leroy, C., Pisani, M. B., Fritschi, R., Hibert, C., and Ionescu, A. M., "High Quality Factor Copper Inductors on Wafer Level Quartz Package for RF MEMS Applications," *Proceedings of the 2006 European Solid-State Device Research Conference (ESSDERC'06)*, Montreux, Switzerland, Sept. 2006, pp. 190–193.

[101] Tadigadapa, S., Najafi, N., "Reliability of Microelectromechanical Systems (MEMS)", *Proceedings of SPIE: Reliabitily, Testing and Characterization of MEMS/MOEMS*, Vol. 4558, pp. 197–205, 2001.

[102] Han, K.-H., and Frazier, A. B., "Reliability Aspects of Packaging and Integration Technology of Microfluidic Systems," *IEEE Transactions on Device and Materials Reliability*, Vol. 5, No. 3, 2005, pp. 452–457.

[103] Galambos P., Benavides G. L., Okandan M.W., Jenkins, M.W. , and Hetherington D., "Precision Alignment Packaging for Microsystems with Multiple Fluid Connections," *Proceedings of ASME: International Mechanical Engineering Conference and Exposition*, Nov. 11–16, 2001, New York, NY.

[104] Gonzalez C., Collins S. D., Smith R. L. "Fluidic Interconnects for Modular Assembly of Chemical Microsystems," *Sensors and Actuators B: Chemical*, Vol. 49, No. 1, 1998, pp. 40–45.

[105] Gray B. L., Collins S. D., and Smith R. L., "Interlocking Mechanical and Fluidic Interconnections Fabricated by Deep Reactive Ion Etching," *Sensors and Actuators A: Physical*, Vol. 112, No. 1, 2004, pp. 18–24.

[106] Han, K.-H., McConnel, R. D., Easley, J. C., Bienvenue, J. M., Ferrance, J. P., Landers, J. P., Frazier A. B., "An Active Microfluidic System Packaging Technology," *Sensors and Actuators B: Chemical*, Vol. 122, No. 1, 2007, pp. 337–346.

[107] Jaeggi, D., Gray, B. L., Mourlas, N. J., van Drieënhuizen, B. P., Williams, K. R., Maluf, N. I., and Kovacs, G. T. A., "Novel Interconnection Technologies for Integrated Microfluidic Systems," *Proceedings of the Solid-State Sensor and Actuator Workshop*, Hilton Head, SC, June 8–11, 1998, pp. 112 – 115.

[108] Perozziello, G., Bundgaarda, F., and Geschke, O. "Fluidic Interconnections for Microfluidic Systems: A new Integrated Fluidic Interconnection Allowing Plug'n'play Functionality," *Sensors and Actuators B: Chemical*, Vol. 130, No. 2, 2008, pp. 947–953.

[109] Man, P. F., Jones, D. K., and Mastrangelo, C. H., "Microfluidic Plastic Interconnects for Multi-bioanalysis Chip Modules," *Proceedings of SPIE*, Vol. 3224, 1997, pp. 196–200.

[110] 110 Hannifin Corporation, O-ring division. *O-ring handbook* (www.parker.com/o-ring).

[111] Yao, T. J., et al., "Micromachined Rubber O-Ring Micro-Fluidic Couplers," *Proceedings of the 13th IEEE International Workshop Micro Electromechanical System* (MEMS'00), Miyazaci, Japan, Jan. 2000, pp. 745–750.

[112] Puntambekar, A., and Ahn, C. H., "Self-Aligning Microfluidic Interconnects for Glass- and Plastic-Based Microfluidic Systems," *Journal of Micromechanics and Microengineering*, Vol. 12, No. 1, 2002, pp. 35–40.

[113] Huang, Y.-M., Uppalapati, M., Hancock, W. O. and Jackson, T. N., "Microtub Transport, Concentration and Alignment in Enclosed Microfluidic Channels," *Biomedical Microdevices*, No. 9, 2007, pp. 175-184.

[114] Williams, D. F., "Revisiting the Definition of Biocompatibility," *Medical Device Technology*, Vol. 14, No. 8, Oct. 2003, pp. 10–13.

[115] Williams, D. F., "On the Mechanisms of Biocompatibility," *Biomaterials*, Vol. 29, No. 20, July 2008, pp. 2941–2953.

[116] Freitas Jr, R.A., *Nanomedicine, Volume IIA: Biocompatibility*, Landes Bioscience, Georgetown, TX, 2003 (http://www.nanomedicine.com/NMIIA.htm).

[117] Tadigadapa, S., "Integration of Micromachined Devices and Microelectronic Circuits: Techniques and Chellenges," *Proceedings of the 43rd IEEE Midwest Symposium on Circuit and Systems*, Vol. 1, 2000, pp. 224–227.

[118] Treloar, L. R. G., *The physics of rubber elasticity*, Oxford University Press, 1975.

[119] Christensen, A. M., Chang-Yen D. A., and Gale, B. K., "Characterization of Interconnects Used in PDMS Microfluidic Systems," *Journal of Micromechanics and Microengineering*, Vol. 15, No. 5, 2005, pp. 928–935.

[120] Jo, B. H., et al., "Three-Dimensional Micro-Channel Fabrication in Polydimethylsiloxane (PDMS) Elastomer," *Journal of Microelectromechanical Systems*, Vol. 9, No. 1, 2000, pp. 76–81.

[121] Wu, H., Odom T. W., et al., "Fabrication of Complex Three-Dimensional Microchannel Systems in PDMS," *Journal of the American Chemical Society*, Vol. 125, No. 2, 2003, pp. 554–559.

[122] McDonald, J. C., Duffy D. C., et al., "Fabrication of Microfluidic Systems in Poly(dimethylsiloxane)," *Electrophoresis*, Vol. 21, No. 1, 2000, pp. 27–40.

[123] Xia, Y., and Whitesides, G. M., "Soft Lithography," *Annual Review of Material Sciences*, Vol. 28, No. 1, 1998, pp. 153–194.

[124] Kim, Y.-K., Kim, E.-K., Kim, S.-W., and Ju B.-K., "Low Temperature Epoxy Bonding for Wafer Level MEMS Packaging," *Sensors and Actuators A: Physical*, Vol. 143, No. 2, 2008, pp. 323–328.

[125] Shaw, J. M., Gelorme, J. D., LaBiance, N. C., Conley, W. E., and Holmes, S. J., "Negative Photoresists for Optical Lithography," *IBM Journal of Research and Development*, Vol. 41, No. 1–2, 1997, pp. 81–94.

[126] Lorenz, H., Despont, M., Fahrni, N., LaBianca, N., Renaud, P., and Vettiger, P., "SU-8: a Low-cost Negative Resist for MEMS," *Journal of Micromechanics and Microengineering*, Vol. 7, No. 3, 1997, pp. 121–124.

[127] Lorenz, H., Despont, M., Fahrni, N., Brugger, J., Vettiger, P., and Renaud, P., "High-aspect-ratio, Ultrathick, Negative-tone Near-UV Photoresist and its Applications for MEMS," *Sensors and actuators A: Physical*, Vol. 64, No. 1, 1998, pp. 33–39, Lorenz, H., et al., "Fabrication of Photoplastic High-Aspect Ratio Microparts and Micromolds Using SU-8 UV Resist," *Microsystem Technologies*, Vol. 4, No. 3, 1998, pp. 143–146.

[128] Alderman, B. E. J., Mann, C. M., Steenson, D. P., and Chamberlain, J. M., "Microfabrication of Channels Using an Embedded Mask in Negative Resist," *Journal of Micromechanics and Microengineering*, Vol. 11, No. 6, 2001, pp. 703–705.

[129] Jackman, R. J., Floyd, T. M., Ghodssi, R., et al., "Microfluidic Systems With on-line UV Detection Fabricated in Photodefinable Epoxy," *Journal of Micromechanics and Microengineering*, Vol. 11, No. 1, 2001, pp. 263–269.

[130] Chuang, Y.-J., Tseng, F.-G., Cheng, J.-H., et al., "A Novel Fabrication Method of Embedded Micro-Channels by Using SU-8 Thick-Film Photoresists," *Sensors and Actuators A: Physical*. Vol. 103, No. 1–2, 2003, pp. 64–69.

[131] Metz, S., Jiguet, S., Bertsch, A., et al., "Polyimide and SU-8 Microfluidic Devices Manufactured by Heat-Depolymerizable Sacrificial Material Technique," *Lab on a Chip*, Vol. 2, No. 4, 2004, pp. 114–120.

[132] Anhoj, T. A., Jorgensen, A. M., Zauner, D. A., and Hübner, J., "The Effect of Soft Bake Temperature on the Polymerization of SU-8 Photoresist," *Journal of micromechanics and microengineering*, Vol. 16, No. 9, 2006, pp. 1819–1824.

[133] Becker, E. W., et al., "Fabrication of Microstructures with High Aspect Ratios and Great Structural Heights by Synchrotron Radiation Lithography, Glavanoforming, and Plastic Moulding (LIGA Process)," *Microelectronic Engineering*, Vol. 4, No. 1, 1986, pp. 35–56.

[134] Microchemicals GmbH, 2009, http://www.microchemicals.com/.

[135] Rohm and Haas, 2009, http://www.rohmhaas.com.

[136] Strandjor, A. J. G., Bogers, W. B., et al., "Photosensitive Benzocyclobutene for Stress-Buffer and Passivation Applications (One Mask Manufacturing Process)," *Proceedings of the Electronic Components and Technology Conference*, San Jose, CA,, 1997, pp. 1260-1268.

[137] Lee, K., Massia, S. and He, J., "Biocompatible Benzocyclobutene-Based Intracortical Neural Implant with Surface Modification," *Journal of Micromechanics and Microengineering*, Vol. 15, No. 11, Nov. 2005, pp. 2149-2155.

[138] Cheung, K., and Renaud, P., "Bio-MEMS for Medicine: On-chip Cell Characterization and Implantable Microelectrodes," *Solid-State Electronics*, Vol. 50, No. 4, 2006, pp. 551–557.

[139] Pyralin Polyimide Coatings for Electronics, Dupont and HD Microsystems, 1997, http://www.hdmicrosystems.com.

[140] Frazier, A. B., and Allen, M. G., "Metallic Microstructures Fabricated Using Photosensitive Polyimide Electroplating Molds," *Journal of Microelectromechanical Systems*, Vol. 2, No. 2, 1993, pp. 87–94.

[141] Metz, S., Holzer, R., and Renaud, P., "Fabrication of Flexible, Implantable Microelectrodes with Embedded Fludic Microchannels," *Proceedings of the 11th International Conference on Solid-State Sensors and Actuators (Transducers' 01)*, Munich, Germany, 2001, pp. 1210–1213.

[142] Metz, S. Holzer, R., and Renaud P., "Polyimide-Based Microfluidic Devices," *Lab on a Chip*, Vol. 1, No. 1, 2001, pp. 29–34.

[143] Pisani, M. B., Hibert, C., Bouvet, D., Beaud, P., and Ionescu, A. M.. "Copper / Polyimide Fabrication Process for Above RF IC Integration of High Quality Factor Inductors," *Microelectronics Engineering*, Vol. 73–74, No. 1, 2004, pp. 474–479.

[144] Reed, H. A., White, C. E., et al., "Fabrication of Microchannels Using Polycarbonates as Sacrificial Materials," *Journal of Micromechanics and Microengineering*, Vol. 11, No. 6, 2001, pp. 733–737.

[145] Plunkett, R.J., Wilmington, D. (to Kinetic Chemicals Inc. / Dupont), Tetrafluoroethylene Polymers, US Patent 2,230,654, filed July 1, 1939; and issued Feb. 4, 1941. French Patent (FR) 917,431, filed Nov. 14, 1945, and granted Jan. 7, 1947.

[146] Kaplan, S. L, Lopata, E. S., Smith, J., "Plasma Processes and Adhesive Bonding of Plytetrafluoroethylene," *Surface and Interface Analysis*, Vol. 20, 1993, pp. 331-336.

Micropumping and Microvalving

Nam-Trung Nguyen

School of Mechanical and Aerospace Engineering, Nanyang Technological University, 50 Nanyang Avenue, Singapore 639798, Singapore

Abstract

This chapter gives an overview about the different concepts for micropumping and microvalving. Basic actuation principles for micropumps and microvalves are first discussed. An actuator with a corresponding energy density can be chosen based on the requirements from the application. The chapter also discusses the design consideration of micropumps and microvalves and their integration with microtechnology. Based on the physics behind their operation, micropumps and microvalves are categorized under mechanical and nonmechanical concepts. Due to the lack of moving parts and the utilization of surface forces, nonmechanical concepts have advantages in microscale.

Key terms   microvalve
micropump
pneumatic
thermopneumatic
piezoelectric
electrostatic
electromagnetic
electrochemical
capillarity

2.1 Introduction

Miniaturization allows the full integration of many components into microfluidic systems for chemical and biochemical analysis, which are also called lab-on-a-chip (LOC) or micrototal analysis system (μTAS). Current microfluidic devices can be categorized as continuous-flow and droplet-based systems. In a continuous-flow microfluidic system, fluid is transported and manipulated as a continuous single-phase stream, while droplet-based systems manipulate individual droplets or a droplet train. Pumping and valving are key functions in a microfluidic system. Controlling flows of samples and reagents are achieved in such as system with the help of pumping and valving. In many commercial systems, these transport requirements are met by external pumps, valves, pressure sources and vacuum sources, or simply through passive mean such as capillary filling. Although self-contained and integrated micropumps as well as microvalves are needed for LOC applications, their successful implementation and commercialization are still delayed due to the lack of reliable and low-cost solutions. Despite the recent active research on micropumps and microvalves, finding a reliable and commercially viable solution remains a difficult task.

As mentioned above, miniaturization brings advantages to LOC applications. Miniaturization also brings new challenges to designing micropumps and microvalves. The ratio between surface area and the volume scales as $A/V \propto L^2/L^3 = 1/L$, where L is the length scale. Decreasing the length scale L increases the ratio between the surface area and the volume. That means surface effects dominate over volume effects in microfluidic devices. Miniaturization also increases the required actuation pressure. An order-of-magnitude decrease in the length scale leads to two order-of-magnitude increase in required driving pressure to maintain the same average flow velocity. Since viscous friction is a surface force, pumping based on pressure-driven flow needs a strong actuator to overcome the large friction force. The dominating surface force also brings problems with bubbles trapped in the microfluidic device. The large interfacial tension blocks the channel and causes leakage in microvalves. Furthermore, the stable bubbles act as fluidic capacitances that absorb and store the pressure generated by the actuator leading to difficulties in designing self-priming micropumps.

Reynolds number represents the ratio between inertial force and friction force $Re = uL/\nu$, where u, L, ν are the flow velocity, the characteristic length and the kinematic viscosity, respectively. Inertial force is a volume-based force, while friction force is a surface-based force. According to the square-cube law, the Reynolds number is very small in microfluidics. The common Reynolds number in microfluidic systems is less than unity, indicating a laminar flow. Dynamic pumping concepts based on inertial force do not work well in microscale.

Further challenges for micopumps and microvalves in LOC applications are the fluids to be handled. Due to the nature of analytic applications, the fluid may have a large amount of particulates, a high concentration of aggressive reagents, and possibly high pressure gradients as well as high heat flux. Most fluids have high ionic concentration and are electrically conducting. The aggressive conditions may lead to corrosion, changes in device materials due to chemical reactions and adsorption, and wear of moving parts. These problems seriously affect the reliability of micropumps and microvalves.

With the challenges presented above, a successful implementation of micropumps and microvalves should address the requirements of cost, reliability, and energy efficiency. Most LOC devices are disposable and designed for a single use. The micropumps and microvalves integrated in these devices should be cheap enough to afford disposability. While batch fabrication may help to reduce the device cost, expensive materials and clean-room processes may inhibit the cost reduction. Furthermore, microfluidic devices are usually much larger than its microelectronic counterparts. A high integration density is therefore not feasible for integrated micropumps and microvalves. Simple fabrication processes and low-cost materials such as plastics could allow cost reduction in the implementation of micropumps and microvalves in a LOC device. Thus, polymeric micromachining processes such as hot embossing, micromolding, thermal bonding, and adhesive bonding need to be considered when the pumping/valving concepts need to decided.

To address the reliability issues, the pumping and valving concepts should be simple enough to avoid possible failure of mechanical parts. While the single use may not requires the high reliability for micropumps and microvalves as compared to their macroscale stand-alone counter parts, aging of polymeric materials need to be considered in the implementation of micropumps and valves in LOC devices. Swelling due to high humidity or aging due to exposure to sunlight may reduce the shelf life of devices made of polymer.

Most LOC devices are intended to be used in portable platforms. Therefore, energy requirement and the efficiency of energy conversion in micropumps and microvalves are also crucial for their successful implementation. The choice of the right actuator is determined by the respective application. Low energy requirement and high conversion efficiency are the general guidelines for choosing actuators for micropumps and microvalves.

The existence of a number of recent excellent review papers on micropumps [1–4] and microvalves [5, 6] makes a comprehensive review on micropumps and microvalves in this chapter unnecessary. Instead of covering all possible pumping and valving concepts, the following sections will only focus on concepts that are relevant and promising for LOC applications. Actuators for micropumps and microvalves are first discussed and highlighted in their respective sections. Micropumps and microvalves are only categorized here as mechanical and nonmechanical types, because their different actuation concepts are discussed separately in Section 2.2.

2.2 Actuators for micropumps and microvalves

Besides the requirements for energy consumption and conversion efficiency, actuation pressure and response time of actuators should match the operation conditions of the fluidic system, where the micropumps and microvalves are implemented. Figure 2.1(a) shows the typical pressure range of common actuation concepts used for micropumps and microvalves. The operation concepts and realization of these actuators will be discussed in details in the later part of this section. According to the generated pressure, actuators are generally grouped in three categories: high-pressure, medium-pressure, and low-pressure actuators. Stack-type piezoelectric actuator can provide the highest pressure. Pneumatic, thermopneumatic, shape-memory-alloy, and thermomechanic actuators belong to the medium-pressure category. In these three categories, electromagnetic, disc-type piezoelectric, electrostatic, electrochemical, and chemical actuators

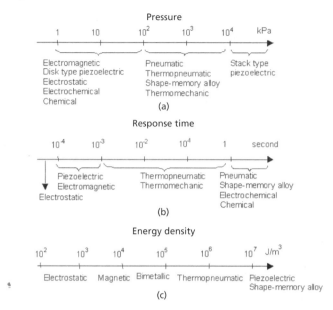

Figure 2.1 Typical ranges of operation parameters of common actuators used in micropumps and microvalves: (a) pressure, (b) response time, and (c) energy density.

provide the lowest pressure. Most LOC applications work in the low or medium pressure range. Thus, the choice of an actuation concept is relatively flexible.

The requirement on response time varies with the applications. In some LOC applications, residence time of reagents is on the order of minutes to match their reaction kinetics. Therefore, no fast switching is required. However, in high-throughput applications such as cell sorting, the required switching time for actuators may be on the order of milliseconds or faster. Figure 2.1(b) shows the typical response time of the different actuation concepts. Electrostatic, piezoelectric, and electromagnetic actuations are the fastest concepts. Actuators based on thermal concepts are in general slower due to the thermal response of the whole system. Careful design of thermal isolation and heat spreaders may increase the heating and cooling rate of thermal actuators and improve their response time. Pneumatic actuators rely on the response time of external valves that switch the pressure supply. The response of pneumatic actuators is also affected by fluidic capacitance formed by soft tubing and soft device materials. Response time of chemical and electrochemical actuators is affected by reaction kinetics. The slow diffusion usually leads to a response time on the order of minutes for chemical actuators.

Since electricity is the primary energy supply for a LOC system, common actuators converts electrical energy into mechanical energy. Therefore, the work generated by an actuator is an important parameter and selection criterion for its use in micropumps and microvalves.

The work generated by an actuator is determined by its force F_a and the displacement s_a:

$$W_a = F_a s_a \tag{2.1}$$

For a given operation condition, the work is estimated from the energy density and the volume of the actuator :

$$W_a = \dot{W}_a V_a \tag{2.2}$$

The energy density does not scale with miniaturization. As shown in Figure 2.1(c), piezoelectric actuators and shape-memory-alloy can provide the highest energy density. However, the small actuator size only provides a limited amount of total actuation energy. According to (2.1) and as a rule of thumb, a certain microactuator can either provide a large force and small displacement or a small force and a large displacement.

The energy density of thermomechanical actuators is estimated from Young's modulus E, the thermal expansion coefficient γ, and the temperature difference ΔT as:

$$\dot{W}_a = \frac{1}{2} E (\gamma \Delta T)^2 \tag{2.3}$$

The energy density of piezoelectric actuators is estimated as:

$$\dot{W}_a = \frac{1}{2} \frac{E_{el}}{(d_{33} E)^2} \tag{2.4}$$

where E_{el} is the electric field strength, E is the Young's modulus, and d_{33} is the piezoelectric coefficient. The energy density stored in electrostatic actuators can be calculated based on the energy stored in the capacitance between their two electrodes:

$$\dot{W}_a = \frac{1}{2} \varepsilon_r \varepsilon_0 E_{el}^2 \tag{2.5}$$

where ε_r and $\varepsilon_0 = 8.85418 \times 10^{-12}$ F/m are the relative dielectric constant of the material between the electrodes and the permittivity of vacuum. The energy density stored in an electromagnetic actuator is estimated as:

$$\dot{W}_a = \frac{1}{2} \frac{B^2}{\mu_r \mu_0} \tag{2.6}$$

where B is the magnetic flux density (in tesla), μ_r is the relative permeability of the magnetic material and $\mu_0 = 4\pi \times 10^{-7}$ H/m is the permeability of vacuum.

Figure 2.1 illustrates the basic actuation concepts for microvalves and micropumps. In the following subsections, the basic working concepts of the different actuators are described in details. Their implementation in microvalves and micropumps depends on the corresponding valving and pumping concepts.

2.2.1 Pneumatic actuators

Pneumatic actuation is the simplest way to provide an actuation pressure to a microvalve or to a pump chamber (Figure 2.2(a)). Since pneumatic actuation requires an external pressure source and a set of external high-speed electromagnetic valves, it is not suitable for fully integrated applications. However, the simplicity makes micropumps

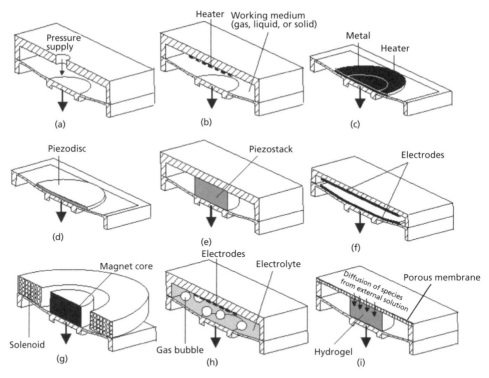

Figure 2.2 Different actuation concepts for micropumps and microvalves: (a) pneumatic, (b) thermopneumatic, (c) thermomechanic, (d) piezoelectric bimorph, (e) piezoelectric stack, (f) electrostatic, (g) electromagnetic, (h) electrochemical, and (i) chemical.

and microvalves with pneumatic actuation both inexpensive and reliable. As mentioned above, low cost and high reliability are the two main requirements for a successful implementation of these components in LOC applications. Thus, pneumatic actuators are attractive for disposable LOC devices. Pneumatic actuators can be used for a wide range of temperature. A further advantage is the response time of pneumatic actuators depends on external switching devices such as electromagnetic valves. Due to the relatively large external pressure or vacuum sources as well as long tubing, pneumatic actuators have a response time on the order from several hundred milliseconds to several seconds.

2.2.2 Thermopneumatic actuators

Thermopneumatic actuation is based on volumetric expansion of a sealed amount of gas, liquid, or solid under thermal loading (Figure 2.2(b)). This actuation concept also uses the phase change from liquid to gas or from solid to liquid to gain a larger volumetric expansion. Due to the direct heat to work conversion, the work generated in such phase-change processes exceeds the energy stored in an electrostatic or electromagnetic field.

Thermopneumatic actuators require a heat source in the actuation chamber. The heat source is usually an integrated microheater. Depending on the fabrication technology and the substrate material of micropumps and microvalves, the integrated heaters can be made of polysilicon [7], metal [8], or ITO [9].

For a maximum conversion efficiency, both pneumatic and thermopneumatic actuators should transfer most of their generated pressure to the valve seat or to the pump chamber. The amount of pressure spent to overcome the sealing membrane should be minimized. A thin membrane made of a material of low Young's modulus is therefore desirable. Polydimethylsiloxane (PDMS) is a popular material for this purpose. A thin membrane on the order of several tens of micrometers can achieve a displacement up to 100 μm with an actuating pressure on the order of only 100 kPa [10]. The membrane and actuation chambers are fabricated using soft lithography. The thickness of the membrane can be controlled by a spin-coating process. Baechi et al. reported a thermopneumatic actuator using a 130×30-μm large and 3-μm thin PDMS membrane. The actuator achieved a displacement of 3.5 μm with an input power of 240 mW [11]. Kim et al. reported a thermopneumatic actuator with a 400 × 40-μm large and 70-μm thin PDMS membrane. This actuator provided a displacement of 40 μm with a heating power of only 25 mW [9]. The typical response time of the above actuators ranges from 100 msec to 25 sec, depending on the design and the fabrication technology of the heater.

Due to its limited temperature range and limited volume of the working fluid, thermopneumatic actuators generate a lower pressure as compared to the external pressure source of a pneumatic actuator. With the same design, devices with pneumatic actuators are therefore expected to have a higher performance than those with thermopneumatic actuators.

2.2.3 Solid-expansion actuators

Solid-expansion actuators use the volumetric change of a solid body to induce stress in it. The generated force is proportional to the temperature difference ΔT between the heater and the ambient temperature:

$$F \propto \gamma_s \Delta T \tag{2.7}$$

where γ_s is the thermal expansion coefficient of the solid material. Due to the large thermal capacitance, solid-expansion actuators are slow. According to (2.3), with the higher Young's modulus of a solid material, high energy density or a high thermal stress can be expected with this type of actuator. A high operation temperature also leads to a higher energy density (2.3) and a higher force (2.7). Careful design of heaters, their location, and thermal isolation can warrant a high operation temperature.

If the high compressive stress within a structure such as a membrane is accumulated up to a critical value, the structure will buckle instantaneously. This behavior allows designing a new type of actuators called bistable actuators. In this actuator, input power is only needed in the transient period to the critical stress. Once the actuator achieves its stable positions, no actuation power is needed. The actuator holds its position using its own stress. Energy input is only needed, if the actuator moves back to its other stable position. The operation between two stable states or positions gives this actuator type its name. Initial stress is induced during the fabrication process utilizing the mismatch in thermal expansion coefficients of the different device materials. The materials are often deposited at elevated temperature. After cooling down, a prestress is generated and can be used to define the first stable state of the actuator. Using an integrated heater, the prestress is overcome, and the actuator can be switched to a second stable state [12]. The force that triggers the change of a stable state can also be induced by other actuation concepts shown in Figure 2.2.

2.2.4 Bimetallic actuators

Bimetallic actuation is based on the mismatch in thermal coefficients of expansion of two bonded solids. This concept is often called thermal bimorph actuation. The elevated temperature needed for this actuation comes from integrated heaters. Integrated temperature sensors are often integrated with the heater to monitor the operation temperature and to provide a possible feedback control. Bimetallic actuators offer an almost linear deflection dependence on heating power. Similar to other thermal actuation concepts, the drawbacks of bimetallic actuators are the high power consumption and the slow response. For silicon-based devices, a metal layer deposited by sputtering or evaporation can work with silicon to form the bimetallic system. Different material combinations such as aluminum/silicon [13] and nickel/silicon [14] was reported in the past. The concept can be extended to polymer/polymer or polymer/metal combinations.

2.2.5 Shape-memory alloy actuators

Shape-memory alloys (SMA) can return from a mechanically deformed state to their original undeformed shape upon a change of temperature. SMAs undergo phase transformations from a "soft" state (martensite) at low temperatures to a "hard" state (austenite) at higher temperatures. In an SMA actuator, the heat can be induced by passing an electric current directly through the SMA.

Titanium/nickel is the most popular metal system used in microsystems [15, 16]. Metal systems such as nickel/titanium/copper and nickel/titanium/palladium have been used as SMA actuators for microvalves [17]. These actuators consume an electrical power on the order of few hundreds milliwatts and have a response time on the order of tens of milliseconds [17].

2.2.6 Piezoelectric actuators

Applying an electric field on a piezoelectric material generates a mechanical strain, which is generally less than 0.1%. However, due to the high Young's modulus, the induced stress can be on the order of several megapascals. Therefore, piezoelectric actuators are suitable for applications that require large forces but small displacements such as micropumps. The relations between the strains φ_1, φ_2, φ_3, and the electric field strength E_3 are described with the piezoelectric coefficients d_{31} and d_{32}. The indices 1, 2, and 3 indicate the directions of polarization in space. The strains in three directions are given by:

$$\varphi_1 = \varphi_1 = d_{31}E_3$$
$$\varphi_3 = d_{33}E_3$$

(2.8)

Common piezoelectric materials for mirosystems are polyvinylidene fluoride (PVDF), lead zirconate titanate (PZT), and zinc oxide (ZnO). PZT offers high piezoelectric coefficients but is very difficult to deposit as a thin film.

Thin-film piezoelectric actuators do not deliver enough force for micropumps and microvalves. All reported devices with piezoelectric concepts used external actuators, such as piezostacks, bimorph piezocantilevers, or bimorph piezodiscs. Bimorph piezo-cantilevers and bimorph piezodiscs can deliver large displacements but low actuation forces.

Furthermore, the small strain and its corresponding small displacement can be over-come without applying a high voltage by using a piezoelectric stack [18] or hydraulic amplification [19]. Peirs et al. reported a piezoelectric stack that measures only $1.4 \times 3 \times 9$ mm but can provide a displacement of $6\,\mu m$ and a force of 140 N at an applied voltage of 100V [18]. Rogge et al. used hydraulic amplification to increase the displacement of a piezoelectric actuator by 25 times to 50 μm [19]. Li et al. achieved 40-fold hydraulic amplification using silicone oils as the working fluid [20].

2.2.7   Electrostatic actuators

Electrostatic actuation is based on the attractive force between two oppositely charged plates. The simplest approximation for electrostatic forces is the force between two plates with the overlapping plate area A, distance d, applied voltage V, relative dielectric coefficient ε_r (Table 6.4), and the permittivity of vacuum $\varepsilon_0 = 8.85418 \times 10^{-12}$ F/m :

$$F = \frac{1}{2}\varepsilon_r \varepsilon_0 A \left(\frac{V}{d}\right)^2 \qquad (2.9)$$

If an insulator layer of thickness d_i and a relative dielectric coefficient ε_i separate the two plates, the electrostatic force becomes:

$$F = \frac{1}{2}\varepsilon_r \varepsilon_0 A \left(\frac{V}{d}\right)^2 \left(\frac{\varepsilon_i d}{\varepsilon_r d_i + \varepsilon_i d}\right)^2 \qquad (2.10)$$

Electrostatic actuators have a very high dynamics and fast response time. However, electrostatic actuators have low energy density and can only generate a low force. The corresponding displacement is also small.

If the electrodes are not insulated, a direct contact between the electrodes and liquid may cause electrolysis or short circuit of the actuator. Therefore, electrostatic actuators are most suitable for handling gases. In terms of operation parameters, electrostatic actu-ators require a higher driving voltage as compared to piezoelectric actuators. Both elec-trostatic and piezoelectric actuators have low energy lost due to the absence of Joule's heating.

2.2.8   Electromagnetic actuators

Electromagnetic actuation is based on the magnetic force induced by an external mag-netic field. The magnetic field is either generated by an electromagnet or a permanent magnet. The induced force of a magnetic flux B in direction z acting on a magnet with its magnetization M_m and volume V is:

$$F = \frac{1}{2}\varepsilon_r\varepsilon_0 A\left(\frac{V}{d}\right)^2\left(\frac{\varepsilon_i d}{\varepsilon_r d_i + \varepsilon_i d}\right)^2 \tag{2.11}$$

The relation between the magnetic flux B and the magnetic field strength H is:

$$B = \mu H = \mu_0\mu_r H = \mu(1 + \chi_m)H \tag{2.12}$$

where μ, μ_0, μ_r, and χ_m are the permeability, permeability of free space ($4\pi \times 10^7$ H/m), relative permeability, and the magnetic susceptibility of the medium, respectively. Electromagnetic micropumps and microvalves often have a solenoid actuator to generate the external magnetic field. A solenoid with N turns, a length L and a driving current I can generate a field strength of:

$$H = NI / L \tag{2.13}$$

The two basic configurations of a magnetic actuator are: fixed external permanent magnet with movable integrated coil and fixed external coil with movable permanent magnet. Meckes et al. reported and actuator with integrated coil that can generate a force of 0.8 mN with a current of 25 mA [21]. Oh et al. integrated both coil and the magnet on a single device. Electroplated permalloy was used as the soft-magnetic material to form a magnetic circuit with an air gap. Passing current through the integrated coil generate a magnetic flux in the magnetic circuit, which tends to close the air gap. With an applied current of 250 mA, the actuator can hold up against a pressure on the order of 10 kPa [22].

2.2.9 Electrochemical actuators

Electrochemical actuators generate gas bubble in a liquid through electrolysis reaction. The pressure drop Δp across the generated liquid-gas in a cylindrical capillary can be estimated as:

$$\Delta p = \frac{2\sigma_{lq}\cos\theta}{r_0} \tag{2.14}$$

where σ_{lg} is the gas-liquid surface tension, θ is the contact angle, and r_0 is the channel radius. The electrolysis reaction of water follows by the generation of gaseous hydrogen and oxygen:

$$2H_2O \rightarrow 2H_2 \uparrow + O_2 \uparrow \tag{2.15}$$

The reversed reaction converts hydrogen and oxygen back to water. The reaction needs a catalyst such as platinum. Platinum is able to absorb hydrogen. The hydrogen-platinum bond is weaker than the hydrogen-hydrogen bond. Therefore, hydrogen-hydrogen bond can be broken to form water:

$$2H_2 + O_2 \xrightarrow{Pt} 2H_2O \tag{2.16}$$

The electrolysis reaction requires a very small amount of electricity and generates a gaseous volume 600 times larger than that of the original water volume. Neagu et al. used electrolysis to generate 200 kPa with only $5\,\mu A$ and 1.6V [23].

2.2.10   Chemical actuators

Chemical actuators convert chemical energy directly into mechanical energy. The swelling behavior of polymeric materials such as hydrogels can be utilized for actuation purposes. Hydrogels are polymers with high water content. Volumetric change of hydrogels is achieved with changes in temperature, solvent concentration, and ionic strength. The above changes are caused by diffusive transport processes and are relatively slow. The low Young's modulus of the polymeric material leads to a relatively low actuation pressure. Hydrogels can be designed as a photosensitive material suitable for photo lithography. Hoffmann et al. designed a hydrogel system based on PNIPAAm [poly-(N-isopropylacrylamide)] [24] that allows film coating, UV polymerization, and working conditions at room temperatures. This hydrogel system can be spin-coated with a thickness of several tens of microns and cross-linked with conventional UV exposure units [25]. The polymer has a phase transition temperature of approximately 33°C. The polymer has a large volume at temperature lower than the critical phase transition temperature. The volume reduces if the temperature increases above the transition temperature. The temperature for such an actuator can be controlled with an integrated microheater as in other thermal actuation concepts.

The volumetric increase of a hydrogel can also be induced by a solvent such as water. The temperature diffusivity is on the order of 10^{-3} cm^2/s, the diffusion coefficient of a solvent is on the order of 10^{-5} cm^2/s, the cooperative diffusion coefficient of polymer chains is on the order of 10^{-7} cm^2/s. Since the swelling behavior is determined by the slowest diffusive process, the characteristic time constant of swelling response can be estimated as [26]:

$$\tau = \frac{d^2}{D_{coop}} \tag{2.17}$$

where d and D_{coop} are the characteristic dimension of the hydrogel and the cooperative diffusion coefficient of the polymer chains, respectively. Because the diffusion coefficient is fixed, the dynamics of such a chemical actuator depends only on the size d. Using packed small particles instead of a bulk material would improve the response of the actuator.

Other hydrogels suitable for serving as chemical actuators are polymethacrylic acid tri-ethyleneglycol (pMAA-g-EG) [27], polyacrylic acid-2-hydroxymethylmethacrylate (pAA-2-HEMA) [28], and polyacrylamide-3-methaacrylamidophenylboronic acid (pAAm-3-MPBA) [29]. The volumetric change of these polymers is controlled by the concentrations of pH (potenz of hydrogen). The pH of distilled water is 7, which is neutral. A pH value below 7 means acidic and a pH value above 7 means alkaline. The hydrogel swells in alkali solution and shrinks in an acidic solution. The increase in volume results from the tendency of the polymer chains to mix with the solvent. Swelling stops when the elastic restoring force is in equilibrium with the osmotic force. The transition region of pAA-2-HEMA and pMAA-g-EG is approximately $4 < \text{pH} < 7$ [27, 28], while AAm-3-MPBA [29] changes its volume at approximately $7 < \text{pH} < 9$.

2.2.11 Capillary-force actuators

Capillary-force actuators are based on the interfacial tension between two phases. Interfacial tension can be controlled by different means such as electrocapillarity, thermocapillarity, and passive capillarity.

Electrocapillary effect is also known as electrowetting. The surface tension between two immiscible, conductive liquids, or between a solid surface and a liquid is controlled by the voltage across these phases. Ions from the liquid are adsorbed at the interface and form an electric double layer. The electric double layer is typically 1- to 10-nm thick and works a capacitance between the two conducting phases. Changing the voltage across this capacitance affects the interfacial tension σ between the two phases [30]:

$$\sigma = \sigma_0 - \frac{C}{2}(V - V_0)^2 \tag{2.18}$$

where σ_0 is the maximum value of surface tension at the applied voltage V_0, C is the capacitance per unit area of the double layer, and V is the voltage applied across the liquid interface.

Thermocapillarity is caused by the temperature dependence of interfacial tension. Higher temperature means higher kinetic energy of liquid molecules. While molecules move faster, their attractive forces decrease. The smaller attractive forces lead to a lower viscosity and a lower interfacial tension. A temperature gradient induces a gradient in interfacial tension that generates the actuation pressure.

Passive capillarity utilizes dependence of induced pressure on the radius of curvature. The actuation pressure increases with a smaller radius of curvature. A bubble inside a microchannel with different widths will move toward the larger section because of the imbalance of induced pressure.

2.3 Micropumps

The simplest design approach for micropumps is miniaturizing the existing macroscopic concepts. The next approaches are developing micropumps based on microscale effects, which are dominant and effective with miniaturization. New transport effects, such as electrokinetic effects, interfacial effects, acoustic streaming, magnetohydrodynamic effects, and electrochemical effects are nonmechanical concepts that can be used for micropumping. Thus, based on the physics behind their operation, micropumps are categorized here as mechanical pumps and nonmechanical pumps.

Mechanical pumps can be further categorized according to the way how mechanical energy is applied to the pumped fluid. Displacement pumps and dynamic pumps are the two main mechanical pumping concepts. In displacement pumps, a moving surface transfers the pressure to the pumped fluid in a periodic manner. The moving surface is often designed as a membrane that is deflected by an actuator (Section 2.2). Check valve pumps, peristaltic pumps, and valveless rectification pumps are displacement pumps. Dynamic pumps continuously add mechanical energy into the pumped fluid to increase its velocities. Due to the lower impact of inertia in microscale, dynamic pumps are not popular for implementation in microscale. Nonmechanical pumps convert another energy type into kinetic energy of the pumped fluid. For most current LOC applications, mechanical pumps are suitable due to the relatively high flow rate and flexibility in designs and fabrication.

The performance of micropumps is specified by parameters such as the maximum flow rate, the maximum backpressure, the pump power, and the efficiency. The maximum flow rate, also called the pump capacity, is the volume of liquid per unit time delivered by the pump at zero backpressure. The maximum backpressure is the backpressure at zero flow rate. The term "pump head" is also often used as a parameter for pump performance. The pump head h between the outlet and inlet of the pump represents the net work done on a unit weight of liquid:

$$h = \left(\frac{p}{\rho} + \frac{u^2}{2g} + h \right)_{\text{out}} - \left(\frac{p}{\rho} + \frac{u^2}{2g} + h \right)_{\text{in}} \tag{2.19}$$

where ρ is the fluid density. The term p/ρ is called the pressure head or flow work. The term $u^2/2g$ is called the velocity head representing the kinetic energy of the fluid. The height h is the elevation head representing the potential energy. If the static pressure at inlet and outlet are the same (e.g., atmospheric pressure), the maximum pump head can be derived from (2.19) as:

$$h_{\text{max}} = h_{\text{out}} - h_{\text{in}} \tag{2.20}$$

Based on the maximum flow rate $\dot{Q}_{\text{max}}$, the maximum back-pressure p_{max}, or the maximum pump head h_{max}, the power of the pump P_{pump} can be calculated as:

$$P_{\text{pump}} = \frac{p_{\text{max}} \dot{Q}_{\text{max}}}{2} = \frac{\rho g h_{\text{max}} \dot{Q}_{\text{max}}}{2} \tag{2.21}$$

The pump efficiency is defined as the ratio between the generated pump power and the power input at the actuator:

$$\eta_{\text{pump}} = \frac{P_{\text{pump}}}{P_{\text{actuator}}} \tag{2.22}$$

2.3.1 Mechanical pump

2.3.1.1 Check valve pumps

Check valve pumps are the most common pump type in the macroscale. Check valve pumps are also the first concepts that were attempted to be miniaturized. Figure 2.3 shows the structure of a check valve pump. A check valve pump consists of an actuator unit (see Section 2.2), a pump membrane that creates the stroke volume ΔV, a pump chamber with the dead volume V_0, and two check valves, which are opened by a critical pressure difference Δp_{crit}. The compression ratio of the pump is defined as the ratio between the stroke volume and the dead volume [31]:

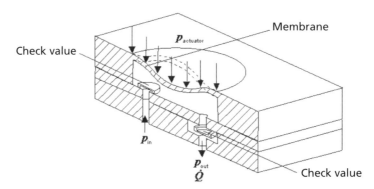

Figure 2.3 Structure of a microcheck valve pump.

$$\psi = \frac{\Delta V}{V_0} \tag{2.23}$$

A small compression ratio is desirable for a large pump stroke. Furthermore, the pump pressure p should be high enough to open the check valves:

$$|p - p_{out}| > \Delta p_{crit} \text{ and } |p - p_{in}| > \Delta p_{crit} \tag{2.24}$$

If both inlet and exit of the pump are exposed to atmospheric condition, the inlet and outlet pressures p_{in}, p_{out} are equal the atmospheric pressure p_0. The critical pressure difference Δp_{crit} of microcheck valve is on the order of several tens millibars. The criterion for the minimum compression ratio for a micropump working with gas is given as [31]:

$$\psi_{gas} > \left(\frac{p_0}{p_0 - \Delta p_{crit}} \right)^{\frac{1}{k}} - 1 \tag{2.25}$$

where p_0 is the atmospheric pressure, and k is the specific heat ratio of the gas ($k = 1.4$ for air). At a low pump frequency f and a small critical pressure difference Δp_{crit}, the relation (2.25) can be simplified as [31]:

$$\psi_{gas} > \frac{1}{k} \frac{\Delta p_{crit}}{p_0} \tag{2.26}$$

For liquids, the criterion for the compression ratio is:

$$\psi_{liquid} > \Theta \Delta p_{crit} \tag{2.27}$$

where Θ is the compressibility of the liquid (for water, $\Theta = 0.5 \times 10^{-8}$ m²/N). Since the value of Θ is very small compared to the stroke volume generated by different actuators, this criterion is easily met. In practical applications, bubble may be trapped in the pump chamber, presenting a condition similar to pumping gas. For a self-priming micropump, which is tolerant to air bubble, the criterion for gas (2.26) can be taken as the design rule.

Much of the pressure generated by the actuator is for overcoming the stiffness of the pump membrane. Pump membrane made of materials with a low Young's modulus can help to increase the pump efficiency by reducing the lost through deflection of the membrane. Furthermore, the low Young's modulus allows large deflection and large stroke volume. Pump membranes made of low-modulus materials such as polyimide [32] or silicone rubber [33] was reported.

Check valves are the next components that affect the pump performance. Figure 2.4 shows some check valve designs. The shapes of the valves are sometimes determined by their fabrication technologies. The critical pressure difference Δp_{crit} should be low, so that not much energy lost is gone into opening the check valves. Furthermore, a good check valve should have a low-pressure drop in opening state and very high-pressure drop in the closing state. Ideally, the flow resistance in opening state and closing state should be zero and infinity, respectively.

A low critical pressure difference and low-pressure drop in opening state can be achieved if the check valves are designed with a low spring constant. If the valves need to be made of hard materials such as silicon, polysilicon, and metals, cantilever design (Figure 2.4(b)) and disk design (Figure 2.4(c)) may help to reduce the spring constant. A low-spring constant leads to a low resonant frequency of the valve, which may be below the pumping frequency. In this case, the closing and opening processes of the valves are out of phase with the driving membrane. The pump is then able to deliver the fluid in backward direction.

High-pressure drop in closing state or good sealing can only be achieved with careful design of the valve seat. Design considerations of a valve seat will be discussed later in Section 2.4. As a rule of thumb, valve seat made of elastic materials such as PDMS can improve sealing properties.

A high compression ratio (2.23) is achieved with large stroke volume and small dead volume. The high stroke volume is warranted with an actuator with large displacement and a low-modulus pump membrane. A small dead volume can be obtained with a small pump chamber. Dead volumes can also be reduced by careful design of the check valves.

2.3.1.2 Peristaltic pump

Figure 2.5 shows the working principle of a peristaltic pump, which does not require passive valves for flow rectification. Three are more pump chambers work in a peristaltic

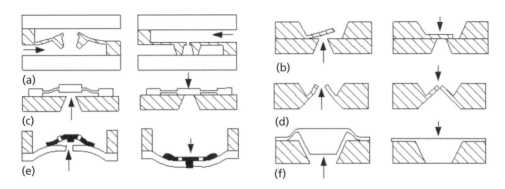

Figure 2.4 Typical check valve designs: (a) ring mesa, (b) cantilever, (c) disc, (d) V-shape, (e) membrane, and (f) floating. (*After* [34].)

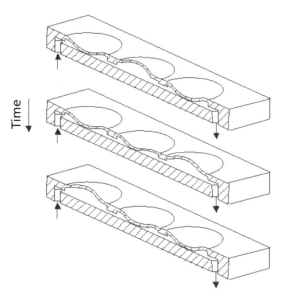

Figure 2.5 Basic structure and working principle of a peristaltic pump.

manner to squeeze the fluid in the pumping direction. Since a peristaltic pump only consists of a series of pump chambers, the fabrication technology is simpler as compared to a check valve pump. However, a peristaltic pump is prone to leakage. In the nonpumping state, one of the chambers should be closed to avoid backflow. A one-way check valve connected in series with the peristaltic pump can also be used to prevent the back flow. The performance of a peristaltic pump is improved by increasing compression ratio and increasing the number of pump chambers. Recently, pneumatic peristaltic pumps made by soft lithography are very popular because of their simplicity and compatibility to other components such as pneumatic microvalves and micromixers [10].

2.3.1.3 Valveless rectification pump

Similar to a check valve pump, a valveless rectification pump is a displacement pump with a moving pump membrane and a pump chamber. The flow rectifying check valves are replaced by fixed flow structures such as diffusers/nozzles or valvular conduits called Tesla valves. The rectification behavior is based on the different pressure losses in two flow directions and requires a relatively high Reynolds number. The pressure loss Δp at a rectification structure is given as:

$$\Delta p = \xi \frac{\rho u^2}{2} \tag{2.28}$$

where ξ is the pressure loss coefficient, ρ is the fluid density, and u is the average velocity. The ratio between the pressure loss coefficients in forward and backward directions is called fluidic diodicity η_F:

$$\eta_F = \xi_{\text{forward}} / \xi_{\text{backward}} \tag{2.29}$$

The above equation assumes the forward direction is the direction of the net flow or the pumping direction. Therefore, diodicity is always more or equal unity. If there is no rectification, the diodicity is equal 1. The flow rate of the pump can be estimated as [35]:

$$\dot{Q} = 2\Delta V f \frac{\sqrt{\eta_F} - 1}{\sqrt{\eta_F} + 1} \qquad (2.30)$$

where f is the pump frequency and ΔV is the stroke volume. Because the maximum flowrate expected from a displacement pump with ideal check valves is $\dot{Q}_{max} = 2\Delta V f$, the rectification efficiency can be defined as the ratio between the actual flow rate and the maximum flow rate [36]:

$$\eta_{rectification} = \frac{\sqrt{\eta_F} - 1}{\sqrt{\eta_F} + 1} \qquad (2.31)$$

Rectification behavior of a diffuser/nozzle depends on the opening angle. At small opening angles, the forward direction is the diffuser direction. At large opening angles, the net flow goes into the nozzle direction. The optimal opening angle for a maximum diodicity or rectification efficiency depends on the actual design of the structure.

Similar to peristaltic pumps, valveless rectification pumps are prone to backflow because of the lack of a check valve. The low fluidic resistance in both directions makes backflow even more serious. A one-way check valve connected in series with the pump can solve this problem. Maximizing the stroke volume and minimizing the dead volume also improve the performance of this pump type.

Besides a diffuser/nozzle structure, other geometries may also have rectification characteristics. The valvular conduit shown in Figure 2.6 is an example [37, 38]. The structure carries the name of its inventor Tesla [39]. A Tesla valve can provide a maximum fluidic diodicity of $\eta_F = 1.2$ and a maximum rectification efficiency $\eta_{rectification}$ [38].

Flow rectification can be achieved without mechanical parts or channel structures using temperature dependency of viscosity [40]. In this case, the inlet and exit of a pump chamber is equipped with microheaters, which are synchronized with the pumping cycle. Alternate switching the two heaters emulate the effects of check valves because the viscosity is lower at elevated temperature. Because this concept is thermal, the switching time or the pumping frequency depends on the thermal response, which is relatively slow.

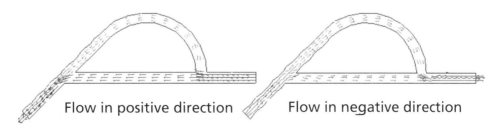

Flow in positive direction   Flow in negative direction

Figure 2.6   Valvular conduit or Tesla valve.

2.3.2 Nonmechanical pump

2.3.2.1 Electroosmotic micropump

Electroosmotic micropump is based on electroosmosis, one of the electrokinetic phenomena, which occur in a system consisting of an electrolyte, a solid, and an electric field. These phenomena are caused by surface charge and the associated electric double layer (EDL). An EDL develops between an electrolyte liquid and a charged solid surface and consists of an immobile and a mobile charge layer. The immobile layer is formed by counter-ions attached tightly to the charged surface. This immobile layer is called the Stern layer, which attracts a thicker charge layer in the liquid called the Gouy-Chapman layer. The Gouy-Chapman layer is mobile and can be moved by an electrical field. The potential at the shear surface between these two layers is called the Zeta potential ξ. The potential distribution in the liquid is determined by the Poisson-Boltzmann equation:

$$\frac{\mathrm{d}^2 \Phi}{\mathrm{d}y^2} = \frac{2zen_0}{\varepsilon} \sinh\left(\frac{ze\Phi}{k_{\mathrm{B}}T}\right) \tag{2.32}$$

where z and n_0 are the valence number and the number density of the ionic species, e is the elementary charge, k_{B} is the Boltzmann constant, and T is the absolute temperature. With the linear approximation $x = \sinh(x)$ for a small EDL, the Poisson-Boltzmann equation reduces to:

$$\frac{\mathrm{d}^2 \Phi}{\mathrm{d}y^2} = \frac{\Phi}{\lambda_{\mathrm{D}}} \tag{2.33}$$

where the Debye length λ_{D} is the characteristic thickness of the EDL

$$\lambda_{\mathrm{D}} = \sqrt{\frac{\varepsilon k_{\mathrm{B}}T}{2z^2 e^2 n_0}} \tag{2.34}$$

Under an applied electric field, the mobile layer is moved by electrostatic force. This layer drags the fluid in the entire channel and causes a pluglike velocity distribution. The momentum transfer process after applying the electrical field is on a time scale between $100\,\mu$m and 1 ms. The electroosmotic flow velocity can be estimated as:

$$u_{\mathrm{eof}} = \mu_{\mathrm{eo}} E_{\mathrm{el}} \tag{2.35}$$

The electroosmotic mobility of the fluid μ_{eo} is [41]:

$$\mu_{\mathrm{eo}} = \frac{\varepsilon \zeta}{\eta} \tag{2.36}$$

where ε is the dielectric constant of the solvent, η is the dynamic viscosity. Electroosmotic pumping is based on electrostatic force, which is a surface force and favorable in microscale. With this concept, liquid can be pumped in microchannels without a high external pressure. In LOC applications, electroosmosis is used for delivering a buffer solution, and, in combination with the electrophoretic effect, for separating molecules. From the relations derived above, the optimization parameters for an electroosmosis pump are:

- The applied electric field;
- The channel size;
- The surface charge density;
- The pumped fluid (ion concentration and pH).

An electroosmotic micropump is simply a microchannel with electrodes at two ends. The channel can be filled with porous material, or packed with microbeads or simply the pumped fluid. The flow rate of an electroosmosis micropump is usually small but its maximum back pressure is large due to the large fluidic resistance of a microchannel. For a higher flow rate, a larger cross section is needed. If the channel is filled with microbeads to maintain a high back pressure, the pores between the beads work as tortuous capillaries where the actual electroosmotic flow occurs.

2.3.2.2 Electrohydrodynamic pump

In contrast to electroosmotic pumps, electrohydrodynamic pumps utilize the electric force acting on the bulk fluid. The electric body force density (force per unit volume) is [42]:

$$ f_E = \rho_e E - \frac{1}{2} E^2 \nabla \varepsilon + \frac{1}{2} \nabla \left[E^2 \rho \left(\frac{d\varepsilon}{d\rho} \right)_T \right] \tag{2.37} $$

where ρ_e is the charge density, E is the electric field, ε is the dielectric constant of the liquid, ρ is the density of the fluid. In (2.37), first terms from left to right represent the Coloumb force, the dielectric force, and the electristrictive force, respectively. Coloumb force and dielectric force can be used for electrohydrodynamic pumping. Dielectric force requires a gradient in dielectric constants of the fluid, and thus an inhomogenous liquid. For a homogenous liquid, the Coloumb force is the only force responsible for pumping effect. Electrohydrodynamic pumping is based on three basic concepts for generation of free charge in the bulk of a dielectric liquid:

- Injection of charges from electrodes in a highly insulating liquid [43];
- Generation of charge layers based on the imbalance between the rate of dissociation of neutral molecules and the rate of recombination of ions [44];
- Induction for charges due to anisotropy of electrical conductivity [45].

An injection pump [43] consists of two electrodes, an emitter and a collector, which inject ions into a highly insulating fluid. The ions are moved by Coloumb force in the electric field between the two electrodes. Pumping occurs because the fluid is dragged by the ions due to friction. Thus, this pump type is also called ion-drag pump. Because the liquid needs to be highly insulating, this pumping concept is not suitable for LOC applications, where liquids contain relatively high concentrations of different ions.

A conduction pump [44] consists of two perforated electrodes immersed in a dielectric liquid. If the electric field exceeds a threshold on the order of 10^5 V/m, the imbalance between the rate of dissociation of neutral molecules and the rate of recombination of ions creates a charge layer next to the electrode. This charge layer is much thicker than the EDL in electroosmosis. Coloumb force acting on this layer moves the bulk liquid. This pumping concept allows the use of liquid with low conductance and may be suitable for LOC applications.

In an induction pump [45], charge is induced in an inhomogeneous liquid by the electric field of an electrode array. The inhomogeneity is created for instance by exposing the liquid to a temperature gradient. Due to the temperature dependency of conductivity, a gradient of conductivity results from the temperature gradient. A travelling electric field generated by the electrode array pumps the liquid in the propagation direction. This pumping concept is suitable for electrically conducting liquid. Instead of using an external temperature field, the temperature gradient can also be induced by Joule heating of the liquid itself. The high temperature and the relatively high field strength may be a problem for application with sensitive samples such as cells or DNAs.

2.3.2.3 Magnetohydrodynamic pump

Magnetohydrodynamics (MHD) is based on Lorentz force, which acts on the bulk liquid. The force density vector $\mathbf{f}$ is related to the current density $\mathbf{j}$ and the magnetic flux $\mathbf{B}$ as:

$$\mathbf{f} = \mathbf{j} \times \mathbf{B} \qquad (2.38)$$

The electric current and the magnetic field can be static (DC-MHD) or time- dependent (AC-MHD).

In a DC-MHD pump, the Lorentz force acts on the pumped liquid in the same manner as an applied pressure. Thus, the velocity field has the same distribution as in a pressure-driven flow. Since the liquid is electrically conducting and has a direct contact to the electrodes, problems of unintended electroosmotic flow and bubble generation by electrolysis are unavoidable. The generated bubbles may block the microchannel and interrupts the ionic current. To solve this problem, the electrodes should be placed in reservoirs outside the pumping channel. Faradaic reactions at the electrodes may lead to degradation of electrode materials. Joule heating is another problem of DC-MHD pump. A high flow rate also means high current and high temperature. Since the temperature is proportional to the square of the driving current, the temperature in the pump channel may reach an unacceptable level for biological samples. Bubble generation and electrode degradation can be avoided by using an AC current and an AC magnetic field.

2.3.2.4 Capillary pump

In microscale, surface forces dominate over body forces. Thus, effects based on surface tension are dominant over inertial effects. Surface tension can be used as an effective driving force at these scales. Surface tension driven pumping is based on capillary effects. Liquids are pumped passively by capillary effect. Active control of surface tension also allows liquid pumping. Capillary effects such as electrocapillarity and thermocapillarity (Section 2.2.11) can be used for controlling surface tension.

Passive capillarity is caused by the difference in surface tension across a liquid column. The driving pressure in passive capillarity depends on three factors:

1. Surface tensions;
2. Geometry of the interface between the different phases;
3. Geometry of the solid phase at the border line between the three phases (liquid, gas, solid).

At the contact line of a liquid/gas/solid system, the balance between the interfacial tensions between the different phases is:

$$\sigma_{si} + \sigma_{lg} \cos\theta = \sigma_{sg} \qquad (2.39)$$

where σ_{si}, σ_{lg}, and σ_{sg} are the interfacial tensions between solid and liquid, liquid and gas, and solid and gas, respectively. The contact angle θ is the angle between the tangent of the liquid/gas curvature and the solid surface. If the contact angle is less than 90°, the solid surface is hydrophilic. A microchannel with hydrophilic surfaces can draw in liquid by passive capillarity. If the contact angle is more than 90°, the solid surface is hydrophobic. A microchannel with hydrophobic surfaces prevents liquid to enter.

2.4 Microvalves

Besides micropumps, microvalves are other important components for flow control in microfluidics. Designing a valve for microscale applications needs to meet a number of challenges such as the dominant surface phenomena, the higher pressure, biocompatibility, and compatibility to fabrication technology. Mirovalves are generally categorized as passive valve and active valve. Passive valve such as check valve is actuated by the flow itself. The different check valve designs were already discussed in Section 2.3.1.1. This section only deals with active microvalves.

Active valves have two basic functions: switching and controlling. In the switching mode, the valve just needs to shut off the flow or to let it go through. In the controlling mode, the valve may work together with a sensor to maintain and control a fixed flow rate as wells as a fixed pressure. A basic mechanical active valve consists of a valve chamber to contain the fluid and its pressure, a valve seat to manipulate the fluid, and an actuator to control the position of the valve seat.

Active microvalves are categorized further based on their initial working states such as normally open, normally closed, and bistable. A normally open valve shuts off the flow in its actuated state. No external energy is required for the open state, the valve is opened either by the pressure of the flow or by a mechanical spring. A normally closed valve shuts off the flow similar to a check valve. In its actuated state, the actuator needs to overcome the pressure of the flow as well as the spring force of the valve seat to allow the fluid to flow through the valve. Bistable microvalves need actuation to switch between its two stable states.

Taking an analogy between microelectronics and microfluidics, an active valve plays the role of a transistor in a fluidic network. A transistor is used as on/off switch in digital mode and as an amplifier in analog mode. Besides the flow switching function, active valves can also work in the analog, or proportional mode. At a constant inlet pressure, the valve actuator varies the gap between the valve seat and valve opening to change the fluidic resistance and, consequently, the flow rate. The control signal at the actuator can therefore be converted to the flow rate of the fluidic network.

An active valve in digital mode can be driven with pulse width modulation (PWM) or as a digitally weighted valve array for proportional flow control. With PWM, the valve is switched open and closed states periodically. The duration of the open state determines the net flow rate through the valve. By controlling the duration, the net flow rate through the valve can be controlled.

If many valves work in parallel in an array and are weighted for their flow rates, a fluidic digital to analog converter can be realized. For instance, the flow rates of the valves in its open state are weighted in a binary manner such as $\dot{Q}$, $2\dot{Q}$, $4\dot{Q}$ ..., any multiples of $\dot{Q}$ can be achieved by binary coding of the valve states. For instance, values ranging from 0 to $255\dot{Q}$ can be realized with a set of eight binary-weighted microvalves.

The performance of a microvalve is evaluated by leakage, valve capacity, power consumption, closing force (pressure range), temperature range, response time, reliability, biocompatibility, and chemical compatibility. The ideal active valve should have zero leakage in its closed position. The leakage ratio L_{valve} is defined as the ratio between the flow rate of the closed state and of the fully open state at a constant inlet pressure:

$$L_{valve} = \frac{\dot{Q}_{closed}}{\dot{Q}_{open}} \tag{2.40}$$

The leakage according to (2.40) is often also called the on/off ratio. The valve capacity C_{valve} represents the maximum flow rate the valve can handle:

$$C_{valve} = \frac{\dot{Q}_{open}}{\sqrt{\Delta p_{open} / (L\rho g)}} \tag{2.41}$$

where $\dot{Q}_{open}$ and Δp_{open} are the flow rate and the pressure drop across the valve at the fully open position, respectively; L is the characteristic length of the valve; ρ is the fluid density; and g is the acceleration of gravity. The power consumption is the total input power of the valve in its actuated state. Power consumption is determined by the actuator in use. The closing force depends on the pressure generated by the actuator. Figure 2.1(a) shows the range and order of magnitude of the pressures generated by different actuators discussed in Section 2.2. The temperature range of the valve depends strongly on the material of the valve and its actuation concept. Pneumatic valves were often used for high-temperature applications because the pressure source is generated externally, and the operation temperature only depends on the material. The valve response time is actually the response time of the actuator in use. Figure 2.1(b) compares the response time of different actuators. The actuators and the operating conditions determine the reliability of a microvalve. In the microscale, the operation failure is often caused by particulate contamination and not by the reliability of its actuator.

2.4.1 Mechanical valve

The three basic components of a mechanic valve are the actuator, the valve spring and the valve seat. Details on actuators are discussed in Section 2.2. Figure 2.7 shows the dif-

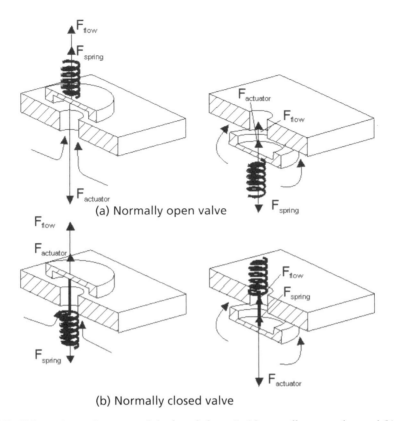

(a) Normally open valve

(b) Normally closed valve

Figure 2.7 Valve spring, valve seat, and the force balance in (a) normally open valves and (b) normally closed valves (F_{inlet}, F_{spring}, and $F_{actuator}$ are forces of fluid, spring, and actuator, respectively).

ferent arrangements of valve spring and valve seat. The spring force keeps the valve shut in normally closed valves. In the case of normally open valves, the spring keeps the valve open and works against the actuator.

Figure 2.8 depicts common forms of the valve spring and their corresponding spring constants. In normally open valves, the spring constant is to be optimized for a minimum value, which allows a larger closing force of the actuator. A small spring constant can be realized with a soft material such as rubber. The solution with soft materials offers a further advantage of excellent sealing characteristics. The leakage ratio can be improved from three to four orders of magnitude compared to those made of hard material such as silicon, glass, or silicon nitride.

If the valve is designed for bistable operation, there is no need for a valve spring because the two valve states are controlled actively. Since the non-powered state is undefined, a valve spring can still be considered for the initial, nonpowered state to assure safe operation. A bistable valve spring allows the valve seat to snap into its working position. In this case, the actuator just needs to be powered in a short period to have enough force to trigger the position change. The force generated by the spring is then high enough to seal the valve inlet.

Valve seats represent a large challenge to microvalve design and fabrication. The valve seat should satisfy two requirements: low leakage and high resistance against particulate contamination. For a minimum leakage rate, the valve should be designed with

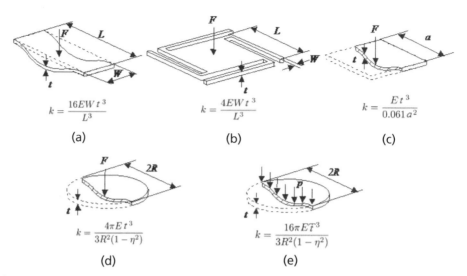

Figure 2.8 Typical valve springs and their spring constants (linear theory for small deflection, which is smaller than the structure thickness; k: spring constant, E: Young's modulus, t: structure thickness, W: structure width, L: structure length, ν: Poisson's ratio, R: radius, and a: membrane width): (a) cantilever, (b) suspended plate, (c) rectangular membrane, $\nu = 0.3$, (d) circular membrane, point load, and (e) circular membrane, distributed load.

a large sealing area, which has to be extremely flat. Softer materials such as rubber or other elastomers are recommended for the valve seat.

Resistance against particulate contamination can be realized with a hard valve seat, a strong actuator, a soft valve seat, or a particle trap. A hard valve seat and a strong actuator simply crush the contamination into finer particles. The smaller particles can be washed away during the operation of the valve or trapped in holes and trenches machined in the valve seat. A hard valve seat is fabricated by coating it with hard, wear-resistant material such as silicon nitride, silicon carbide, or diamond-like carbon. A large force can be achieved with actuators such as piezostacks. If the particles cannot be destroyed, a soft coating on the valve seat or a soft valve seat can conform to the particle shape and creates a tight seal. Designing particle traps such as holes or trenches on the valve seat or its opposite surface on the valve body can also improve the resistance of the valve against particles.

2.4.2 Nonmechanical valve

Several nonmechanical actuation concepts can provide unique solutions for microvalving. Since surface tension plays a significant role in microscale, nonmechanical valves can utilize multiphases systems such as bubbles, gas/liquid interface for valving.

Using the pressure generated by the surface tension gradient, bubbles in microchannels can be used as active valves. The implementation of bubble valve is simpler than mechanical concepts because there are no moving parts involved. Gas bubbles are generated either by evaporation or electrolysis. The heat needed for evaporation is much higher than the energy needed for electrolysis. Furthermore, bubble generation by evaporation has low energy efficiency because most of the heat is lost through conduction, convection, and radiation. As a result, for a comparable gas bubble the energy required for evaporation is four or five orders of magnitude higher that that of electrolysis. Microvalves based on electrolysis are suitable for power-saving applications, which do not require a fast response time.

Valving can also be realized with swelling of electrochemical actuators. This concept is based on the volume change of a hydrogel due to a stimulus such as temperature and pH value. Richter et al. [26] reported a microvalve with PNIPAAm as hydrogel. The hydrogel shrinks at higher temperature. The switching region is between 25°C and 35 °C. Temperature change was induced by integrated microheaters. The valve consumes several hundreds of milliwatts of power and has a response time of several tens of seconds. The microvalves reported by Liu et al. [28] use purely chemical stimuli to operate. The solution with controlling pH is fed directly into the actuator chamber filled with pAAHEMA copolymer to control the PDMS valve membrane. The response time of this valve is several minutes. The hydrogel can also be placed directly in a microchannel to block the flow if the pH value in the flow changes. A similar design was reported by Baldi et al. [29] with AAm-3-MPBA. The long response time of this valve is caused by the relatively large hydrogel amount in the actuator chamber.

2.5 Outlook

This chapter discusses the different aspects of designing micropumps and microvalves. Based on the recent review papers [1–6], a huge number of research works on micropumps and microvalves have been carried out. Despites the variety of pumping and valving solutions, only few designs have found commercial success. Cost and reliability are the main hurdles for their implementation. It is not a surprise that the simple pneumatic valve and pneumatic peristaltic pump were successfully implemented in a number LOC platforms based on PDMS. Cost would not be and issue for stand-alone microvalve and micropump, if they are used as modules for a more complex LOC system. The advantage of modularity and reconfigurablity will compensate the higher cost. Polymeric technologies for the fabrication of micropumps and microvalves as well as their actuators are considered as one the niches for future developments.

2.6 Troubleshooting

Depending on the type of the micropump in use, pumping performance can be affected by different factors such as trapped air bubbles, debris, lack of power, or bad electrical contacts. In a mechanical pump or valve with moving parts, trapped bubbles are the main causes for malfunctioning. Bubbles represent fluidic capacitances that absorb energy from the actuator and prevent the driving pressure to be transferred to the check valves. If a decrease in flow rate is observed, there is a high possibility that a bubble is trapped in the pump chamber. There are different ways for degassing the pumped liquids such as introducing bubble traps in the design of microchannel network. The best solution for avoiding bubble problems is designing a self-priming pump with a minimum dead volume.

If a micropump or microvalve stops working, a checking the power supply is the first step for troubleshooting. A loose contact in the device could be the cause for the missing electrical power. For pumps and valves based on pneumatic actuation, a pressure leak could be the cause for the malfunction.

Debris from the liquid and device materials could cause blockage in micropumps and microvalves. If the devices are made of inorganic materials such as silicon or glass, flushing the device with solvents can clear off the debris.

2.7 Summary points

- Micropumps and microvalves are key components of a microfluidic system. Depending on the requirement of the application such as pressure level, power consumption and size, a suitable actuating scheme can be selected for the pump or the valve.

- Mechanical pumps are miniaturized versions of macroscale devices. Although the pumping concept is straightforward, mechanical pumps and valves with moving parts may be prone to problems such as trapped bubbles and debris.

- Nonmechanical pumps are suitable for large-scale integration due to the simplicity in design and the lack of moving parts. However, most nonmechanical pumps cannot match the pumping power of mechanical pumps.

References

[1]   Nguyen, N. T., X.Y.Huang, and T. K Chuan, "MEMS—Micropumps: A Review," *ASME Journal of Fluids Engineering*, Vol. 124, No. 2 2002, pp 384–392.

[2]   Laser, D. J., and J. G. Santiago, "A Review of Micropumps," *Journal of Micromechanics and Microengineering*, Vol. 14, 2004, pp. R35–R64.

[3]   Woias, P., "Micropumps—Past, Progress and Future Prospects," *Sensors and Actuators, B: Chemical*, Vol. 105, No. 1, 2005, pp. 2838.

[4]   Iverson, B.D. and S.V. Garimella, "Recent Advances in Microscale Pumping Technologies: a Review and Evaluation," *Microfluidics and Nanofluidics*, Vol. 5, 2008, pp. 145–174.

[5]   Oh, K.W. and C.H. Ahn, "A Review of Microvalves," *Journal of Micromechanics and Microengineering*, Vol. 16, 2006, pp. R13–R39.

[6]   Zhang, C., D. Xing and Y. Li, "Micropumps, Microvalves, and Micromixers Within PCR Microfluidic Chip: Advances and Trends," *Biotechnology Advances*, Vol. 25, 2007, pp. 483–514.

[7]   Rich, C.A., and K.D. Wise, "A High-Flow Thermopneumatic Microvalve with Improved Efficiency and Integrated State Sensing," *Journal of Micro Electromechanical Systems*, Vol. 12, 2003, pp. 201–208.

[8]   Ruzzu, A., K. Bade, J. Fahrenberg and D. Maas, "Positioning System for Catheter Tips Based on an Active Microvalve System," *Journal of Micromechanics and Microengineering*, Vol. 8, 1998, pp. 161–164.

[9]   Kim, J.H., K.H. Na, C.J. Kang, D. Jeon and Y.S. Kim, "A Disposable Thermopneumatic-Actuated Microvalve Stacked with PDMS Layers and ITO-Coated Glass," *Microelectronics Engineering*, Vol. 73–74, 2004, pp. 864–869.

[10]   Unger, M.A., H.P. Chou, T. Thorsen, A. Scherer and S.R. Quake, "Monolithic Microfabricated Valves and Pumps by Multilayer Soft Lithography," *Science*, Vol. 288, 2000, pp. 113–116.

[11]   Baeshi, D., R. Buser and J. Dual, "A High Density Microchannel Network with Integrated Valves and Photodiodes," *Sensors and Actuators A*, Vol. 95, 2002, pp. 77–83.

[12]   Goll, C., W. Bacher, D. Maas. W. Menz and W.K. Schomburg, "Microvalves with Bistable Buckled Polymer Diaphragms," *Journal of Micromechanics and Microengineering*, Vol. 6, 1996, pp. 77–79.

[13]   Jerman, H., "Electrically Activated, Normally Closed Diaphragm Valves," *Journal of Micromechanics and Microengineering*, Vol. 4, 1994, pp. 210–216.

[14] Barth, P.W., "Silicon Microvalves for Gas Flow Control," *Proc. 8th Int. Conf. on Solid-State Sensors and Actuators (Transducer's 95)*, 1995, pp. 276–277.

[15] Wolf, R.H. and A.H. Heuer, "TiNi (shape memory) films on silicon for MRMS applications," *Journal of Micromechanics and Microengineering*, Vol. 4, 1995, pp. 206–212.

[16] Kahn, H., M.A. Huff and A.H. Heuer, "The TiNi shape-memory alloy and its application for MEMS," *Journal of Micromechanics and Microengineering*, Vol. 8, 1998, pp. 213–221.

[17] Liu, Y., M. Kohl, K. Okutsu and S. Miyazaki, "A TiNiPd Thin Film Microvalve for High Temperature Application," *Materials Science and Engineering A*, Vol. 378, 2004, pp. 205–209.

[18] Peirs, J., D. Reynaert and H. van Brussel, "Design of Miniature Parallel Manipulators for Integration in a Self-Propelling Endoscope," *Sensors and Actuators A*, Vol. 85, 2000, pp. 409–417.

[19] Rogge, T., Z. Rummler and W.K. Schomburg, "Polymer Microvalve with a Hydraulic Piezo-Drive Fabricated by the AMANDA Process," *Sensors and Actuators A*, Vol. 110, 2004, pp. 206–212.

[20] Li, H.Q., D.C. Robert, J.L. Steyn, K.T, Turner, O. Yaglioglu, N.W. Hagood, S.M. Spearing and M.A. Schmidt, "Fabrication of a High Frequency Piezoelectric Microvalve," *Sensors and Actuators A*, Vol. 111, 2004, pp. 51–56.

[21] Meckes, A., J. Behrens, O. Kayser, W. Benecke, T. Becker and G. Muller, "Microfluidic System for the Integration and Cyclic Operation of Gas Sensors," *Sensors and Actuators A*, Vol. 76, 1999, pp. 478–483.

[22] Oh, K.W., A. Han, S. Bhansali and C.H. Ahn, "A Low-Temperature Bonding Technique Using Spin-On Fluorocarbon Polymers to Assemble Microsystems," *Journal of Micromechanics and Microengineering*, Vol. 12, 2002, pp. 187–191.

[23] Neagu, C.R., J.G.E. Gardeniers, M. Elwenspoek and J.J. Kelly, "An Electrochemical Microactuator: Principle and First Results," *Journal of Micro Electromechanical Systems*, Vol. 5, 1996, pp. 2–9.

[24] Hoffmann, J., et al., "Photopatterning of Thermally Sensitve Hydrogels Useful for Microactuators," *Sensors and Actuators A*, Vol. 77, 1999, pp. 139–144.

[25] Richter, A., et al., "Electronically Controllable Microvalves Based on Smart Hydrogels: Magnitudes and Potential Applications," *Journal of Microelectromechanical Systems*, Vol. 12, 2003, pp. 748–753.

[26] Richter, A., et al., "Influence of Volume Phase Transition Phenomena on the Behavior of Hydrogel-Based Valves," *Sensors and Actuators B*, Vol. 99, 2004, pp. 451–458.

[27] Cao, X., Lai, S., Lee, L. J., "Design of a Self-Regulated Drug Delivery Device," *Biomedical Microdevices*, Vol. 3, 2001, pp. 109–118.

[28] Liu, R. H., Yu, Q., Beebe, D. J., "Fabrication and Characterization of Hydrogel-Based Microvalves," *Journal of Microelectromechanical Systems*, Vol. 11, 2002, pp. 45–53.

[29] Baldi, A., et al., "A Hydrogel-Actuated Environmentally Sensitive Microvalve for Active Flow Control," *Journal of Microelectromechanical Systems*, Vol. 12, 2003, pp. 613–621.

[30] Lee, J., and Kim, C. J., "Surface-Tension-Driven Microactuation Based on continuous Electrowetting," *Journal of Microelectromechanical Systems*, Vol. 9, No. 2, 2000, pp. 171–180.

[31] Richter, M., Linnemann, R., and Woias, P., "Robust Design of Gas and Liquid Micropumps," *Sensors and Actuators A*, Vol. 68, 1998, pp.480–486.

[32] Schomburg, W. K., et al., "Microfluidic Components in LIGA Technique," *Journal of Micromechanics and Microengineering*, Vol. 4, 1994, pp. 186–191.

[33] Meng, E., et al., "A Check-Valved Silicone Diaphragm Pump," *Proceedings of MEMS'00, 13th IEEE International Workshop Micro Electromechanical System*, Miyazaci, Japan, Jan. 23–27, 2000, pp. 23–27.

[34] Shoji, S., and Esashi, M., "Microflow Devices and Systems," *Journal of Micromechanics and Microengineering*, Vol. 4, 1994, pp. 157–171.

[35] Stemme, E., and Stemme, G., "A Valveless Diffuser/Nozzle-Based Fluid Pump," *Sensors and Actuators A*, Vol. 39, 1993, pp. 159–167.

[36] Gerlach, T., "Microdiffusers as Dynamic Passive Valves for Micropump Applications," *Sensors and Actuators A*, Vol. 69, 1998, pp. 181–191.

[37] Forster, F. K., et al., "Design, Fabrication and Testing of Fixed-Valve Micropumps," *Proc. of ASME Fluids Engineering Division*, IMECE'95, Vol. 234, 1995, pp. 39–44.

[38] Bardell, R. L., et al., "Designing High-Performance Micro-Pumps Based on No-Moving-Parts Valves," *Proc. of Microelectromechanical Systems (MEMS) ASME*, DSC-Vol. 62/ HTD-Vol. 354, 1997, pp. 47–53.

[39] Tesla, N., *Valvular Conduit*, U.S. patent 1 329 559, 1920.

[40] Matsumoto, S., Klein, A., and Maeda, R., "Development of Bi-Directional Valve-Less Micropump for Liquid," *Proceedings of MEMS'99, 12th IEEE International Workshop Micro Electromechanical System*, Orlando, FL, Jan. 17–21, 1999, pp. 141–146.

[41] Manz, A., et al., "Electroosmotic Pumping and Electrophoretic Separations for Miniaturized Chemical Analysis Systems," *Journal of Micromachanics Microengineering*, Vol. 4, 1994, pp. 257–265.

[42] Stratton, J.A., *Electromagnetic Theory*, McGraw Hill, New York, 1941.

[43]   Richter, A. and Sandmaier, H., "An Electrohydrodynamic Micropump," *Proceedings of MEMS'90, 3th IEEE International Workshop Micro Electromechanical System*, pp. 99–104.

[44]   Atten, P. and J. Seyed-Yagoobi, "Electrohydrodynamically Induced Dielectric Liquid Flow Through Pure Conduction in Point/Plane Geometry," *IEEE Trans. Dielectr. Electr. Insul.*, Vol. 10, 2003, pp. 27–36.

[45]   Fuhr, G. et al., "Microfabricated Electrohydrodynamic (EHD) Pumps for Liquids of Higher Conductivity," *Journal of Microelectromechanical Systems*, Vol. 1, 1992, pp. 141–146.

Micromixing Within Microfluidic Devices

Jeffrey D. Zahn and Alex Fok

Department of Biomedical Engineering
Rutgers, The State University of New Jersey, Piscataway, NJ, USA

Corresponding author:
Jeffrey D. Zahn, Rutgers University, Department of Biomedical Engineering, 599 Taylor Road,
Room 311, Piscataway, NJ 08854, Phone:(732) 445-4500 x6311, Fax:(732) 445-3753,
 Email:jdzahn@rci.rutgers.edu

Abstract

Microscale mixing is of upmost importance as the ability to rapidly mix samples on the microscale is vital to the successful implementation of microfluidic systems used for biochemical analysis, drug dosing, and delivery, genomic analysis such as DNA amplification or sequencing and a number of complex reactions. In most microfluidic devices, mixing occurs between inlet streams that must be dispersed to remove concentration gradients within the mixing chamber. Mixing schemes commonly employed are either based on diffusional mixing where the Brownian motion of molecules produces dispersion of the molecules by diffusing down a concentration gradient or chaotic mixing where chaotic flow patterns allow splitting, stretching, and folding of material lines ultimately dispersing different components within the flow stream.

Micromixing techniques are classified into passive and active mixing categories. Passive mixing techniques involve a fixed geometry flow channel without external flow disturbances. Passive mixing schemes include the use of successive lamination, mixing in serpentine microchannels, placing obstacles within the flow path, and augmentation of the channel lower wall in a staggered herringbone pattern to promote chaotic advection. Most passive mixing techniques require carefully designed microfluidic channels and use of complex geometries. Alternatively, active mixers use an external disturbance to the flow field to promote mixing. Active mixing schemes include using temporally controlled flow profiles with sequential flow switching between two inlets, superimposing low-frequency sinusoidally fluctuating flow on a steady state flow, electrohydrodynamic instabilities, or acoustic disturbance fields. The efficiency of micromixing is monitored by use of labeling dyes and image analysis using microscopy to determine the extent of mixing or mixing efficiency. Alternatively, specific chemical reaction kinetics may be monitored to determine the mixing efficiency.

Key terms	microfluidics, micromixers, diffusional mixing, parallel lamination, serial lamination, serpentine microchannel, chaotic advection, instability mixing, passive micromixer, active micromixer, multiphase micromixing, acoustic mixer, droplet flow

3.1 Introduction

The development of microfluidic mixers has been of upmost importance to the realization of micrototal analysis systems (μTAS), which are poised to dramatically transform the biomedical and analytical chemistry fields. The ability to mix different components in fully integrated microflow systems is necessary when carrying out common biological reactions such as enzymatic reactions for DNA manipulations (restriction endonuclease cutting of DNA, Polymerase chain reaction, DNA hybridization assays, etc.), cellular activation, cellular disruption in a lysis buffer, or drug dosing and dispensing for personalized drug delivery platforms. Such platforms require that mixing of components of a chemical reaction must be completed quickly, efficiently, and reproducibly, especially when the reactant concentrations must be tightly controlled. Since most biological processes occur in an aqueous environment, in this chapter we are primarily concerned with mixing within liquid flows. Most examples given will occur within water, although there are specific examples where multiphase mixing involving both aqueous and organic solvents is required. With the increasing importance of micromixing to microfluidics, several books on microfluidics and bio-MEMS have devoted chapters to micromixing [1, 2]. Additionally, several review papers highlight some of the recent advances in micromixing [3–6].

Microscale mixers are based on two fundamental mixing modes: diffusional mixing and chaotic mixing. Both mixing regimes are based on the convection/diffusion transport equation which for a single species is

$$\frac{\partial c}{\partial t} + \vec{u} \cdot \nabla c = D\nabla^2 c + q \tag{3.1}$$

where $\vec{u}$ is the velocity vector, c is the species concentration, D is the diffusivity, t is time, and q is a source term if the species is being produced through a chemical reaction.

The first type of mixing considered is diffusional mixing where the Brownian motion of molecules causes them to migrate directionally down a concentration gradient from an area of high molecular concentration to an area of low concentration, ultimately removing any concentration gradients to produce a constant concentration field at the outlet of a device. However, diffusion time, t, is directly proportional to W^2/D where W is the characteristic width (m) over which diffusion occurs and D is the molecular diffusivity (m^2/s) of the migrating species. Most small biomolecules (ions, glucose, urea, etc.) have diffusivities of $\sim 10^{-10}$ to 10^{-9} m^2/s while proteins have diffusivities of $\sim 10^{-12}$ to 10^{-11} m^2/s. This implies that diffusion times can easily be on the order of minutes to hours when diffusion lengths are greater than tens of micrometers. In a seminal review paper by Brody et al., [7] the approximate diffusion times for a variety of different sized molecules and particles over many length scales is given. These results are summarized in Table 3.1.

Table 3.1 Approximate Diffusion Time in Water at Room Temperature

Distance	Heat	Small Molecule	Protein	Cell
1 μm	10^{-4} s	10^{-3} s	10^{-2} s	10^{1} s
10 μm	10^{-2} s	10^{-1} s	10^{0} s	10^{3} s
100 μm	10^{0} s	10^{1} s	10^{2} s	10^{5} s
1000 μm	10^{2} s	10^{3} s	10^{4} s	10^{7} s

In devices that rely on diffusional mixing, two fluids are brought together in a converging geometry so that two fluid streams flow in parallel next to each other and the diffusional mixing of species occurs across the fluid interface. This is conducted in a laminar flow field defined by the Reynolds number (Re) where Re=$\rho UL/v$, where ρ is the solution density, U is the characteristic flow velocity within the device (often the average velocity), L is the characteristic length scale of the system (usually the hydraulic diameter), and v is the solution viscosity. The Reynolds number is a ratio of viscous flow forces to inertial flow forces. Since the flow velocity and channel dimensions are usually quite small in microdevices, mixing is usually done at low Reynolds numbers in the viscous dominated or creeping flow regime (Re<1), although some designs support Reynolds numbers of up to a few hundreds where inertial effects can be dominant. Since the mixing occurs under flow conditions, one can estimate the relative effect of the axial convection timescale compared to the transverse diffusion timescale by the Péclet number (Pe) defined as Pe=UW/D where U is the average flow velocity (m/s), and W and D are the characteristic diffusion length and diffusivity previously defined. A quick dimensional analysis reduction of (3.1) can be employed to determine how long such a device must be as the channel length must be long enough for diffusion to remove all concentration gradients that is equal to the flow velocity times the diffusion time, $L_{\text{channel}} = U * t_{\text{diffusion}} = U * W^2 / D = Pe * W$. In most continuously flowing microfluidic mixers the Péclet number ranges from 10^1 to 10^5. Thus, the channel length is directly determined by the Pe and channel width. In order to optimize diffusional mixing one has to consider the coupled effect of changing the channel width and flow velocity and the Pe and ultimately the necessary channel length. Despite its simplicity, diffusional mixing is usually not sufficient to quickly homogenize a flow system in a short period of time and usually requires a fairly long flow channel.

In addition to diffusional mixing, chaotic micromixers are employed to stretch and fold the interface between two fluid streams in order to increase interfacial area and reduce diffusional path lengths. These types of micromixers employ specialized geometries or temporal flow control in order to produce secondary flow patterns that produce advection in directions other than in the direction of the main fluid flow. These secondary flow patterns produce local flow aberrations or vortices that induce stretching and folding of the fluid interface. These types of micromixers are generally more efficient at mixing in the microscale than diffusional mixers. However, they usually require more complicated designs or active control to produce the necessary secondary flow patterns.

Micromixing techniques may be either passive or active mixing devices. Passive micromixers are fixed geometry flow channels with no external energy input beyond pumping energy in order to produce fluid flow. The mixing relies entirely on either diffusion or chaotic advection in carefully designed microfluidic channel geometries. Passive micromixers can be either fairly simple single-layer microchannels or relatively complex three-dimensional structures to produce the appropriate flow field that are more complicated to design and fabricate. Alternatively, active mixers use an external disturbance to the flow field in order to mix the fluids. These disturbances are usually either spatial and/or temporal perturbations. The use of an external perturbation to the flow field usually produces a more efficient mixer compared to passive micromixers. However, the need for integrated components to produce the external flow perturbations necessitates that the structures of active micromixers be more complicated than the passive micromixers. In addition, an external power source is needed so that active micromixers tend to be more expensive to produce and operate and are less energy efficient.

3.2 Materials

Commonly used materials are listed below by category. It should be noted that due to the large number of microfluidic setups actual materials may vary in terms of catalog number, dimensions, and so forth.

3.2.1 Microfluidic mixing devices

Micromixing occurs within a specially designed microfluidic device. The actual designs of devices will be discussed in Section 3.3. However, there are multiple device materials and standard fabrication processes used in creating micromixers. These are usually fabricated from direct wet or dry etching of glass or silicon substrates or through micromolding techniques such as soft lithography to create polydimethylsiloxane (PDMS) molded microchannels [8] or thermal compression molding [9] to create polydimethylmethacrylate (PMMA) or polycarbonate (PC) microchannels. Device materials are usually chosen for ease of fabrication, chemical resistance, and ability to interface with external components. Due to its simplicity, flexibility, and low cost, soft lithography has become a widely used rapid prototyping fabrication technique in many research laboratories. PDMS is also optically transparent in the UV/Visible light range, which makes optical characterization of micromixers possible. Micromixers must also be carefully designed for the intended application and fluid inputs. The required input parameters, (flow rate, concentration, fluid type, component concentration, etc.) must be clearly defined for optimal micromixer design.

3.2.2 Microfluidic interconnects

For most microfluidic devices, fluids are moved by either hydrostatic pressure or electrokinetics. For hydrostatic pressure-driven flow either a syringe pump with programmable flow rates or a pressure head is used. A typical syringe pump used in microfluidic applications is the Harvard Apparatus Pump 11 Pico Plus Syringe Pump (Harvard Apparatus, Holliston, MA) or similar syringe pump. Glass syringes are preferred (Hamilton Co., Reno, NV) because they produce steadier flows than plastic syringes, although plastic syringes may be used as well (Becton Dickinson, Franklin Lakes, NJ).

Small-sized tubing is used for fluidic interconnects. A variety of microfluidic nanoport tubing, fittings, and adapters may be purchased from Upchurch Scientific (http://webstore.idex-hs.com/PDF/Lit/micro_singles.pdf. Website valid on 12/12/2008. Upchurch Scientific, Oak Harbor, WA). Other tubing can be purchased from Small Parts Inc (Tygon Micro Bore PVC Tubing .010" ID x .030" OD Part # TGY-010-C) or PE-10 Catheter tubing from Becton Dickinson (PE-10 Tubing 0.011" ID x 0.025" OD). The tubing easily fits over the end of a 30-gauge needle (Becton Dickinson, Franklin Lakes, NJ) and may be easily connected to the infusion syringe via the needle's luer fitting.

Microfluidic devices require device priming to wet all surfaces and remove air bubbles from inside the flow channels. Priming can be done using a high flow rate (2–30 μl/min) to completely fill interconnection tubing either prior to or after tubing connection. This process should be carefully monitored to prevent overpressuring of fluid connections and the priming flow rate should be dramatically reduced as soon as the device is filled. If air bubbles persist within the flow channels, the devices can be filled under vacuum by submerging the entire device in a prime solution (PBS, DI water, etc.) within a beaker and placing the beaker in a bell jar and pulling vacuum to about –20 inHg or until cavitation bubbles are seen. This aids in not only filling the device but degassing the prime solution as well.

In order to connect tubing to microdevices, a small bore needle (26-gauge needle) is used to puncture a reservoir area of a PDMS channel. The tubing can be fed through the hole and held in place by the elastic properties of the PDMS. Additionally, small bore drill bits can be purchased from McMaster Carr (60 gauge 0.04" diameter carbide bits part # 2841A63; 76 gauge 0.020" diameter carbide bits part # 2841A81, McMaster Carr Cleveland, OH). Drilling can be accomplished with a microdrill press (MICROLUX Variable Speed Drill Press, Micromark Corp, Berkeley Heights, NJ). If tubing reinforcement is needed a two-part, 5-minute-set epoxy, JB Kwik, (JB Weld Company, Sulphur Springs, TX) may be used.

Other interconnects may include electrodes (Platinum electrode wire 0.5mm diameter, Sigma-Aldrich, St. Louis, MO) or microprobes (Micromanipulator Co. Carson City, NV) for electrical excitation of devices.

3.2.3   Optical assembly

Most micromixers are characterized via analyzing optical images of the devices. Thus, the devices need optically transparent windows for image acquisition and a microscopy setup for device visualization. Most researchers use a commercial epifluorescence microscope (IX71, Olympus America, Inc., Center Valley, PA; TE2000U, Nikon Instruments, Inc., Melville, NY) with standard or long working distance objective lenses (10×, 20×, 60×, 100×, etc.) depending on the needed optical quality, objective numerical aperture, and viewing window. Images are captured via a scientific grade CCD camera (e.g., Cooke Sensicam QE, Cooke Corporation, Romulus, MI) and analyzed using image processing software (e.g., ImageJ, National Institutes of Health http://rsbweb.nih.gov/ij/, accessed 12/12/2008) or specially written image processing scripts (MATLAB, The Mathworks, Natick, MA).

3.2.4   Required reagents

In general aqueous solutions are infused through the micromixing devices (DI water or PBS buffer, pH 7.4, 1X, Gibco 10010, Invitrogen Corporation, Carlsbad, CA). Most mixers are characterized by labeling an input with labeling microparticles (Fluorescent Polystyrene Microspheres, Thermo Scientific, Waltham, MA), colored dyes (food coloring, methylene blue, etc.) or florescent dyes (fluorescein, rhodamine, etc., Invitrogen Corporation, Carlsbad, CA). The microparticles are used to track flow paths while the color or fluorescence intensity of chemical dyes can be analyzed to characterize mixing parameters (see Section 3.4.2). Chemical reactions as indicators of mixing may also be tracked via optical absorbance or luminescence changes following reactions. Specific examples are discussed in Section 3.4.2.

3.3 Experimental design and methods

3.3.1 Passive micromixers

The use of passive micromixers is attractive because no external energy source beyond the pumping energy is required as an input. Passive micromixers are set up by infusing the needed input fluid streams, either by pressure driven or electrokinetic pumping, and the mixing proceeds naturally based on the mixer's flow channel design. Although many different passive micromixer designs have been demonstrated, they generally fall into three categories: parallel lamination and hydrodynamic focusing; serial lamination and split and recombine micromixers; and micromixers that produce chaotic advection.

3.3.1.1 Parallel lamination and hydrodynamic focusing

Parallel lamination micromixers are based on having fluid streams containing different solute concentrations flowing adjacent to each other. The simplest parallel lamination design consists of two inlets that converge into a central daughter channel where diffusive mixing occurs [10]. This type of mixer is often termed a Y or T inlet micromixer, noting the characteristic shape of converging inlets. Since the mixing is diffusion limited and at high Péclet number, a long mixing channel is needed to ensure complete concentration dispersion of the two streams. Thus, a serpentine- [11] or zigzag- [12] shaped microchannel is sometimes used in order to more efficiently use the microdevice surface area and decreasing the total length of the device while conserving the microchannel length. At higher flow rates, the serpentine design may also induce advection at each of the turns, improving the efficiency of the micromixer.

Another advantage of parallel lamination micromixers is the ability to produce varying outlet concentrations by modifying the relative flow rate of the two inlet sample streams. This allows concentration profiles to be varied in time, as needed. The temporal variation of concentration is accomplished by varying the relative infusion rate from two syringe pumps while keeping the total fluid flow rate through the micromixer constant. Zahn and Hsieh were able to demonstrate both step and sinusoidally varying outlet concentration profiles from a three-inlet serpentine parallel micromixer using both saline solutions mixed with DI water to control salinity [13] and glucose mixed with DI water to vary sugar concentration [14]. Each input syringe pump was computer controlled in order to vary their relative flow rate while keeping the total flow rate through the micromixer constant. As the DI infusion streams flow rate was increased relative to the saline or glucose solution, the micromixer outlet concentration decreased and vice versa.

Another method for reducing the diffusional path length in parallel micromixers is through the use of hydrodynamic focusing. Hydrodynamic focusing is accomplished by producing a three-inlet design that converges in a central daughter channel. The outer two inlets contain a sheathing flow, while a sample is infused through the central inlet (Figure 3.1). Once the three inlets converge in the daughter channel the sheathing streams, which are infused at a higher flow rate than the sample stream, focus or reduce the width of the sample stream and subsequently the diffusional path length for mixing. The width of the sample stream is tailored by the relative flow rates of the sample stream and the total flow rate. If the aspect ratio of the channel is very small (i.e., the channel is much wider than tall) then the width of the sample stream can be determined from the

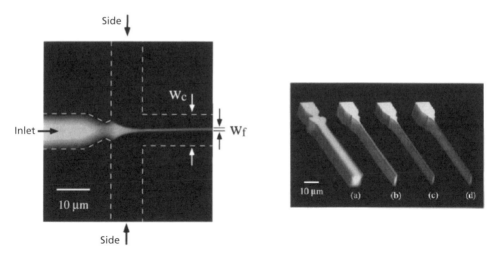

Figure 3.1 Left: hydrodynamic focusing of a fluorescein dye. Right: confocal scanning microscopy images of focused streams at different flow rate ratios ranging from (a) 0.366, (b) 0.224, (c) 0.208, and (d) 0.194. (*From:* [15]. Copyright 1998 by the American Physical Society. Used with permission.)

following relationship: $W_{sample}/W_{channel} = Q_{sample}/Q_{total}$ where W_{sample}, $W_{channel}$, Q_{sample}, Q_{total} are the focused sample width, total channel width, sample stream flow rate, and total flow rate (sample stream flow rate plus sheathing flow rates), respectively. Therefore, in order to quickly and efficiently mix using hydrodynamic focusing, the total channel width must be very small. Using a 10×10 μm in cross section microchannel, a continuous micromixer that was able to mix a fluorescein dye in 10 ms at flow rates of nl/s was demonstrated [15]. In this report focusing of the central stream containing a fluorescein dye was accomplished by changing the relative input pressures between the central sample and side sheathing inlets. The relative flow rates can be calculated by the fluidic equivalent of Ohm's law ($\Delta P = Q \cdot R$ where ΔP is the pressure drop, Q is the flow rate and R is the hydraulic resistance of the flow channel). Since the focused stream is so thin mixing into this stream occurs very rapidly. This was observed by a fluorescence quenching reaction of the fluorescein dye when the side channels contained iodine ions that quench the fluorescence emission. By monitoring the fluorescence intensity of the sample stream in time the diffusion rate of the iodine ions into the sample stream could be determined on the order of tens of microseconds depending on the inlet flow conditions.

The final approach towards creating efficient parallel lamination micromixers is to use a converging geometry of multiple inlets into a daughter mixing channel. This approach has been explored by Hessel et al. [4, 16] in single phase mixing and Wu et. al. [17] using multiphase flows. The two fluid phases are feed interdigitally so that one phase is fed into every other inlet channel and the other phase is fed into the adjacent inlet channels. This in turn creates a multilaminate flow pattern consisting of multiple lamellae with a large contact surface area over which diffusional mixing may occur while also shortening the diffusional path length between each lamella. The flow was tracked by the use of a blue food color dye or mixing between an uncolored iron ion (Fe^{3+}) and a rhodanide (SCN^-) solution, which results in the formation of a brown complex. A variety of device geometries, flow rates and hydrodynamic focusing were also explored [4, 16] to produce secondary flow patterns and modify lamellae thicknesses.

Initially, Hessel et al. tested their interdigital micromixer without any geometric focusing and noticed very little mixing occurring from 12 to 2,000 ml/hr flow rates. By adding a geometric focusing structure, the flow narrowed generating thinner lamellae improving mixing efficiency. However, the authors noted that adding the focusing geometry sometimes caused lamellae tilting instead of remaining vertical, making optical tracking of the individual lamellae more difficult. After optimizing the slit-shaped geometric-focusing device, the interdigital micromixer was operated at flow rates between 250 to 1,000 ml/hr (Re=213-850), resulting in a nearly homogenous mixing (tracked by the rhodanide reaction in the mixed sample) in a matter of milliseconds.

3.3.1.2 Serial lamination and split and recombine

Serial lamination, also termed sequential lamination or split and recombine micromixing, is used to separate two adjacent fluid streams, and recombine them at a downstream stage [18, 19]. This sequential splitting and recombining (SAR) of the fluid stream creates a multilaminate structure (creating 2^n lamellae, where n is the number of SAR stages) but in a serial manner, as opposed to the parallel approach of interdigital mixers. In order to accomplish SAR micromixing, a multistep procedure is required. First, the bilaminate or multilaminate stream is split perpendicular to the lamellae direction into two substreams that are recombined laterally at a downstream position (Figure 3.2). The splitting of the flow stream requires the fabrication of a multilayered channel geometry with a minimum of two layers in the micromixer. These multilayer mixers were produced either by laser ablation of PMMA or milled from stainless steel, aligned, layered, and held under compression in a housing. Ideally, the multilaminate structure created after a series of splitting and recombining stages should be made up of equally thick lamellae, although this is not always the case. An optimized eight-stage device, fabricated from PMMA to allow for optical inspection of the mixing process, with 2-mm wide, 4-mm high channels and a total length of 96 mm was reported by Schonfeld et al. This optimized device preserves the initial lamination and produces a highly regular lamellae pattern observed over sequential lamination stages at a low Reynolds number of 0.22 and a corresponding total flow rate of 0.21 l/hr (Figure 3.2(f)). At higher flow rates (Re > 15, total flow rate > 13.5 l/hr), the flow profiles and lamellae patterns are distorted due to secondary inertial effects that improve the mixing efficiency.

3.3.1.3 Chaotic advection

In addition to diffusion, the generation of secondary flow in directions other than the bulk channel flow can improve mixing via the generation of chaotic advection. In a chaotic flow field, fluid elements that are initially close to each other may become widely separated by stretching and folding of material lines even within a laminar flow field. At Reynolds numbers where fluid inertial effects are dominant (i.e., Re >1), then chaotic advection may be generated in planar geometries by placing obstacles in the flow field either at the wall [20] or within the channel center [21, 22]. Another way to induce advection at higher Re is to use the previously described serpentine or zigzag geometries [12]. These structured flow channels induce secondary flow structures due to inertial effects of the fluid flow field, which can cause boundary layer separation and vortices downstream of the obstacles. The vertical structures help increase the interfacial area between two flow streams while stretching and folding fluid material lines between the

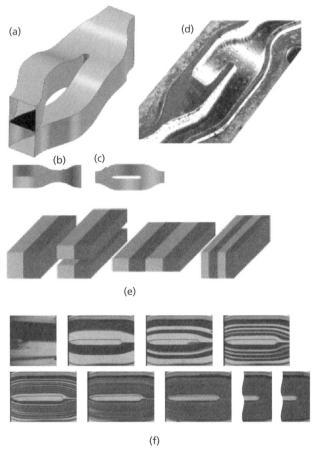

Figure 3.2 (a) Slanted view of a geometry model of a single stage in an optimized SAR mixer with the splitting layer shown in black (b) side and (c) top views of this stage. (d) Scanning-electron micrograph of the lower channel in a single stage of a SAR mixer made of stainless steel with a minimum channel cross section of 1 mm and a length of 6 mm. (e) Schematic of SAR mixing showing from left to right the initial bi-layer flow, splitting, recombining, and reshaping into a quad-layer flow. (f) Optical inspection of successive lamination in the SAR mixer from the inlet (top left) to the end of the eighth SAR stage (bottom right) using dyed (dark gray) and transparent lamellae of an 85% glycerol-water solution. The applied total volume flow rate of 0.2 l/hr corresponding to Re = 0.22. (*From:* [18]. Reproduced by permission of The Royal Society of Chemistry.)

adjacent fluid streams. The flow will become successively more chaotic with increasing Reynolds number improving the efficiency of the micromixers at higher flow rates.

In passive micromixers operating in the creeping flow regime, chaotic advection may only be generated by creating full three-dimensional flow fields. This is accomplished by creating three-dimensional flow channels using multilayered devices, as opposed to planar flow channels produced from single layer devices. An example of a chaotic micromixers based on serpentine flow channels was demonstrated by [23]. In this design, the authors created a "twisted pipe" type geometry fabricated from wet etched silicon and glass. The structure was designed as a series of C-shaped channel segments that were aligned and bonded to straight channel segments on a different silicon layer. This structure was compared to either a serpentine mixer defined in a single channel layer or a straight channel. The three-dimensional serpentine structure promotes mixing through chaotic advection at the serpentine corners by changing the flow direction as the fluid moves through each C-shaped stage flowing from the bottom layer to the top layer, around the two corners of the C-shaped structure, and from the upper level back to the lower layer. The design was tested by infusing solutions of a pH-sensitive indicator, phenolphthalein, and sodium hydroxide at Reynolds numbers ranging from 6 to 70 to demonstrate chaotic advection to passively enhance micromixing. The three-dimensional structure showed significantly higher levels of mixing than either the planar serpentine mixer or straight channel designs for all Reynolds numbers tested, with greater than 80% mixing at Re=6 and 98% mixing in the five segment mixer for Re>25. In the planar serpentine channel, the extent of mixing increases from 60% at Re=6 to a maximum of 94% mixing at Re=140, while in the straight channel the extent of mixing decreases from 60% at Re=6 to close to 0% at Re=70 due to the higher Péclet number decreasing the residence time of the fluid for passive diffusional mixing.

Two other types of chaotic micromixers were designed based on two-layer crossing channels (TLCCM) (Figure 3.3), which improves upon the previous serpentine channel design to produce fast mixing at low Reynolds numbers, Re = 0.2 [19]. These micromixers were designed as three-dimensional crossing channels perpendicular to each other and arranged in a periodic manner. These designs improve on other passive micromixer designs by combining the advection processes of fluid stretching and folding afforded by the two layer geometry with the stream splitting and recombination of SAR mixers.

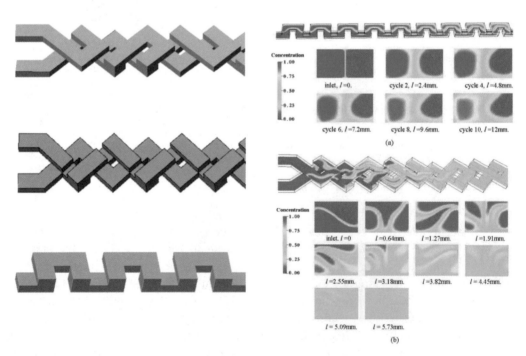

Figure 3.3 Left: Two designs of two-layer crossing channels micromixers (TLCCM). Both mixers have a channel depth of 150 μm and width of 300 μm. The bottom design shows the C-shaped serpentine micromixer described by Liu et al. [53]. Right: Computational simulation comparing the performance of the C-shaped serpentine micromixer described by Liu et al., with the TLCCM micromixer at Re= 0.2. (*From:* [19]. Reproduced by permission of The Royal Society of Chemistry.)

Other chaotic micromixers based on geometric patterning of the microchannel structures have been developed. These have focused on patterning lines or grooves at the bottom of a microchannel into which the incoming fluid can flow. The effect of these grooves is to twist the incoming flow streamlines, creating vortices and hence chaotic advection within the flow stream. Johnson et al., produced a trapezoid-shaped T inlet microchannel that was ~70 μm wide at the top, ~30 μm wide at the bottom, and ~30 μm deep via laser ablation of polycarbonate [24]. At the bottom of the channel they ablated slanted wells that were 14 μm wide, ~85 μm deep, and spaced 35 μm center to center. In order to demonstrate mixing they established an electroosmotic flow of a rhodamine dye into the top inlet and buffer solution into the bottom inlet at a velocity of 0.06 cm/s. At these flow velocities they saw a significant increase in mixing (extent of mixing >75%) compared to the same channels without the grooves (Figure 3.4). Computational simulations showed that the rhodamine-dyed phase was guided along the microgrooves in a helical pattern interspersing it with the undyed phase to improve mixing.

Another chaotic micromixer was developed by Stroock et al. Here a staggered herringbone mixer (SHM) was fabricated by a two-layer process to define the main microchannel and a series of herringbone shaped grooves that were embedded into the channel floor, periodically alternating the direction of the herringbone pattern [25, 26]. By varying the shape of the grooves as a function of axial position, the fluid experiences a repeated sequence of rotational and extensional local flows (Figure 3.5). The herringbone structure is made up of a short arm that generates a small rotational flow, and a longer arm that generates a larger rotational flow. The first half of a mixing cycle patterned the series of grooves with the short arm on one side of the channel, while in the second half of the cycle, the short arm in the grooves are switched to the other side of the channel. This switching of the structure of the grooves changes the extent of rotation experienced locally enhancing the chaotic stirring of the two phases. Using this

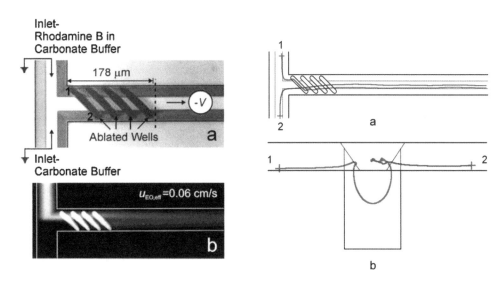

Figure 3.4   Left: (a) white light microscopy image of a T inlet microchannel with a series of laser ablated wells. Fluorescence images of electroosmotic flow past the ablated wells at flow rates of (b) 0.06 cm/s. Right: (a) top-view and (b) side-view images of fluid streamlines within the four-well micromixer determined from computational simulations showing how the microwells twist the incoming flow to improve dispersion. (Reprinted with permission from [24]. Copyright 2002 American Chemical Society.)

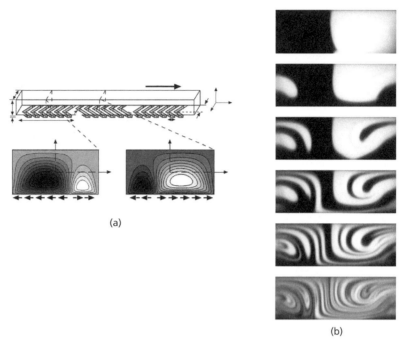

Figure 3.5 Schematic of one-and-a-half cycles within a staggered herringbone micromixer (a) SHM mixer induces two rotational flows. (b) Confocal micrographs of vertical cross sections in the channel at various axial positions; inlet, and after cyles 1–5. (*From:* [26]. Reproduced by permission of The Royal Society.

design, a high extent of mixing (>95%) was observed for Reynolds numbers up to 100 and two fluids were able to fully mix within a 3-cm long SHM even when Péclet numbers were of the order of 10^5.

3.3.2 Active micromixers

In contrast to passive micromixers, active micromixers improve mixing by actively producing spatial or temporal perturbations within the flow structure. A variety of perturbations have been used, including temporally varying pressure fields to produce chaotic flows or Taylor dispersion, electrical energy to promote flow instabilities, and the use of spatially defined acoustic wave fields to produce pressure perturbations within the mixing chamber. These active micromixers can generally produce faster and more complete mixing compared to passive micromixers, yet require extra energy to produce flow perturbations, which may lead to more complicated channel designs, transducer interfacing, or complicated control mechanisms in order to obtain the most efficient mixing.

3.3.2.1 Temporal pulsing/Taylor dispersion

In the creeping flow regime, three degrees of freedom within the flow field are required to produce a chaotic flow field. In passive micromixers, this is accomplished via a three-dimensional flow channel structure. Planar flow channels may also be used to induce chaotic advection but since the flow will be planar with only two degrees of freedom, the third degree of freedom to produce chaotic advection is afforded by temporal control of the flow field. An early demonstration of the use of temporal pulsing to produce chaotic advection was demonstrated by Evans et al. [27]. This device was designed as a pulsed double dipole mixer (Figure 3.6). In this mixer design, two pump/valve designs using patterned polysilicon heaters were patterned adjacent to a batch mixing chamber on either side of the chamber. The resistive heating of the polysilicon was designed to produce bubbles which could be used either as a pressure source (bubble pump) or a valve to block flow. By switching the generation of the pressure source between the two sides of the bubble pump and using bubble valves to direct flow, a double source/sink pressure field could be produced that would cycle the direction of fluid flow. This cycling leads to a chaotic flow field.

Another use of temporal pulsing technique was demonstrated using on-chip bubble pumps by Deshmukh et al. and using external syringe pumps by Niu and Lee [28–30]. In these cases, a converging two-inlet geometry was used, but the flow from each inlet was pulsatile and 180 degrees out of phase with the other inlet. Thus, a periodic perturbation was created staggering the introduction of each fluid so that they occupy the entire channel width, resulting in the two fluids being layered axially along the length of the channel. This layering of the two fluid phases creates an axial concentration gradient so that axial diffusion tends to mix the two streams. Since the flow is pressure-driven, the axial diffusion is enhanced by distortion of the fluid-fluid interface as it flows in the characteristic parabolic flow profile down the length of the channel (Figure 3.7). This mixing, known as Taylor dispersion, can be modeled using an "effective" diffusion coefficient or Taylor-Aris dispersion coefficient [31]

$$D_{\text{eff}} = D\left(1 + \frac{Pe^2}{48}\right) \tag{3.2}$$

where D is the diffusivity of the component to be mixed in free solution and Pe is the imposed Péclet number.

3.3.2.2 Electrohydrodynamic mixing

Electrohydrodynamic mixing (sometimes termed Electokinetic Instability, EKI, chosen due to the relevance of the work to electrokinetic microdevices) is due to flow instabili-

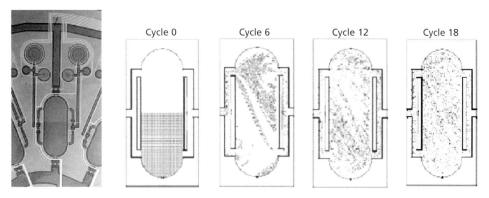

Figure 3.6 Left: Micrograph of a planar mixing chamber actuated by bubble pumps and valves. Right: Simulation showing material dispersion from 0 pulse cycles through 18 cycles. (From: [27]. Copyright 1997 IEEE. Used with permission.)

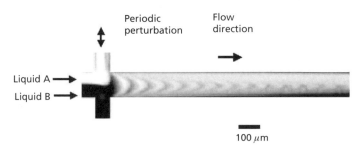

Figure 3.7 Mixing enhancement by Taylor dispersion using sinusoidal pulsing of two liquid inlets. (From [30]. Copyright (2003) Institute of Physics Publishing Ltd. Used with permission.)

ties that develop at a fluid-fluid interface when electrical stresses are applied. These electrical stresses develop at the interface primarily due to a conductivity or permittivity gradient between two fluid phases in an applied electric field. Electrohydrodynamic instabilities are a complex phenomena, modeled using classical linear stability analysis considering momentum transport equations and both kinematic and interfacial stress boundary conditions [32].

Some of the early work on electrokinetic (EK) instability mixing was conducted by Santiago and colleagues [33]. Mixing was produced using AC electrical excitation at a frequency of 10 Hz. With increasing applied voltage from 1 to 8 kV, the flow became unstable and material line stretching and folding was observed (Figure 3.8, top). It is thought that the observed instability may be due to a slight conductivity difference due to the fluorescent dye used for flow visualization which has been modeled to show that flow instability may be promoted when the conductivity ratio between two fluid streams is as low as 1.01 [34].

Further mixing experiments were conducted using a two layer stratified aqueous electrokinetic flow with a conductivity gradient where one layer of fluid was at least 10 times more conductive than the other [35, 36] Here the application of a critical electric field, produces the electrokinetic instability, seen to grow with characteristic instability waves, as shown in Figure 3.8, bottom. In this instance, it was noted that the disturbance originates where the two fluids meet, increases in magnitude with increasing electric field and is convected downstream within the microchannel.

3.3.2.3   Acoustic micromixers

Acoustic micromixers are based on establishing standing acoustic waves within a mixing chamber to induce acoustic streaming, or flow vortices due to the local pressure and velocity gradients established by the sound waves generated from an acoustic actuator. The mixing chambers are etched in silicon or glass, with a piezoelectric actuator (lead zirconate titanate (PZT) or zinc oxide) patterned over an actuation membrane. Liu et al. demonstrated an active acoustic micromixing technique using acoustic streaming around a small bubble within their mixing chamber. This micromixing technique was used to enhance DNA transport to a microarray during hybridization studies [37]. The experiment was conducted using a 16-mm × 16-mm, 200-μm deep, mixing chamber with a 15 mm diameter PZT actuator glued on top. The PZT was excited at 5 kHz and 10 V_{pp} to enhance DNA mixing within the chamber and transport the DNA to the

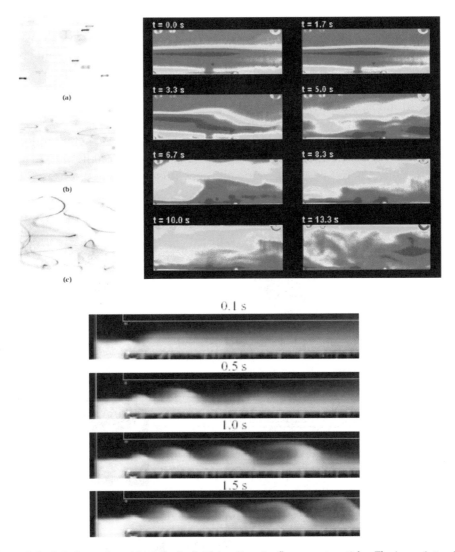

Figure 3.8   Left: Imaged particle paths for 0.49-μm diameter fluorescent particles. The image intensity scale is inverted to better visualize the particle paths (a) Stable oscillations at 1 kV. (b) Particles oscillate in plane with a transverse component at a larger applied voltage of 4 kV. (c) Particle trajectories in the unstable state at 8 kV. Here the flow field is three-dimensional, as particles traverse the depth of field of the microscope moving in and out of focus. Right: Images obtained from the electrokinetic instability micromixer during the mixing process with an initially stable interface and transition into instability mixing. (Reprinted with permission from [33]. Copyright 2001 American Chemical Society.) Bottom: Time series images of instability waves due to an electrokinetic instability in a conductivity gradient at an applied field of 1.25 kV/cm. The two solutions are Borate buffers of 1 and 10 mM that are introduced into the device at a stable electric field of 0.25 kV/cm. The high-conductivity stream is seeded with a neutral fluorescent dye.(Images taken from [35]. Copyright 2005 Cambridge University Press. Used with permission.)

microarray surface. Following the PZT excitation, the DNA hybridization fluorescence signal saturated faster (less than 1 hour with active mixing compared to 2 hours without), was at least twofold larger and more uniform across the hybridization array compared to purely diffusive incubation.

Another example of acoustic micromixing was demonstrated by Jang et al. [38]. In this report, a PZT disc was coupled to a silicon membrane above a glass batch mixing chamber (Figure 3.9(a)). The authors demonstrated acoustic streaming by exciting the piezodisc at the higher resonant modes of the structure (first mode, 9 kHZ, second mode, 15 kHz, third mode, 39 and 45 kHz). The primary mode produced minimal acoustic streaming, but the higher modes produced distinctive circulation patterns (Figure 3.9(b)). Mixing was further enhanced by resonant mode hopping where the piezoelectric would be excited at one resonant frequency for a period of time followed by switching to a second resonant frequency, which produces a different circulation pattern (Figure 3.9(c)). By utilizing resonant mode hopping the authors were able to demon-

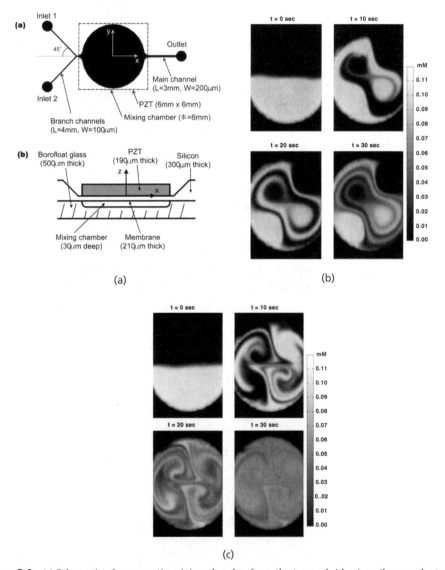

Figure 3.9 (a) Schematic of an acoustic mixing chamber from the top and side view. (Image adapted from [38].) Active acoustic micromixing at (b) 39 kHz and (c) mode hopping between 45 kHz and 39 kHz for 5 seconds. (Reprinted from [38]. Copyright 2007, with permission from Elsevier.)

strate complete mixing in a 6-mm diameter mixing chamber in ~30 seconds compared to minimal mixing even after 7 minutes without acoustic energy.

3.3.3 Multiphase mixers

The use of multiple fluid phases to enhance mixing on the microscale has also been explored. The micromixers use two distinct immiscible fluid phases that are coinfused through a mixing channel. Two-phase flow is the simultaneous flow of two immiscible fluid phases and is the simplest case of multiphase flow. Within the microfluidic device, the mass transfer rate is determined, to a large extent, by the fluid flow pattern. The flow profile and flow patterns that develop in multiphase systems is a strong function of the relative flow rates, differences in physical properties of the phases (density, viscosity, interfacial tension, etc.) and can vary from stable stratified flow profiles to connected droplets to isolated droplets within the flow channel [39, 40]. The tendency of immiscible liquids to form dynamic droplet emulsion patterns as a function of capillary number and flow velocity ratios between a continuous and noncontinuous phase has been exploited from a T-section microfluidic channel [39] to obtain a tight control of the emulsion droplet size distribution. One advantage of creating isolated droplets in a flow channel (at low flow rates) is that as droplets flow within a microchannel, internal recirculation of fluid within the droplet fluid phase occurs. If multiple species are incorporated as the droplet is formed, the internal circulation promotes stretching of the interface between these species enhancing mixing within the droplet. Mixing has been demonstrated using water droplets in air by merging droplets with different species and actively controlling their movements via electrowetting [41]. Mixing has also demonstrated where three water streams, either dyed with red, green or no dye were continuously injected into microchannel with a continuous phase consisting of a water immiscible perfluorodecaline (PFD) solution [42]. The water phase spontaneously broke up into droplets of ~250 pl volume separated by PFD. Each flowing droplet can be considered as an individual microreactor. As the droplets flow within a microchannel the internal recirculation quickly disperses the dye inside the drop. This method was further refined for protein crystallization studies [43] with protein crystals being produced in 7.5-nl droplets, formed in a fluorinated oil continuous phase, containing 23 mg/ml protein (thaumatin, bovine liver catalase and glucose isomerase) in 0.05M pH 6.5 adenosine deaminase (ADA) coinfused with 0.3M sodium potassium tartrate (0.03M pH 7.5 in Hepes) as opposed to volumes of at least 100 nl using conventional approaches.

Mixing between phases has also been explored as an approach towards developing a miniaturized DNA extraction and purification module using a two or three inlet converging microchannel geometry [44, 45] using either a stratified or droplet flow. Here, two phases are coinfused through a microfluidic network where one of the components is phenol:chloroform:isoamyl alcohol (25:24:1) and the other is a 0.5% SDS aqueous solution. Using a droplet flow profile between the phenol and water phases was shown to improve extraction efficiency because of a convective enhancement to the mass transport due to the internal fluid circulation within the droplet as it moves through the microchannel. To demonstrate this effect, a sulphorhodamine dye was added to the aqueous phase. The rhodamine dye naturally partitioned from the aqueous phase into the organic phase so it was utilized as a fluorescent marker to visualize component extraction from the aqueous phase into the organic phase. Figure 3.10(a) shows a fluo-

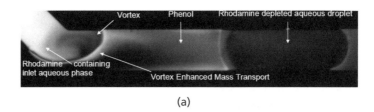

(a)

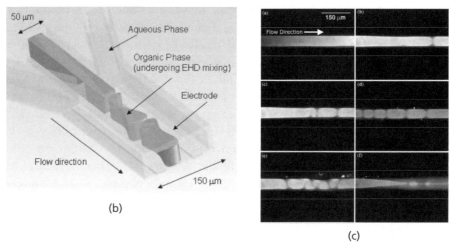

(b)

(c)

Figure 3.10   (a) Experiment results of two phase extraction using rhodamine dye. Internal circulation of the fluorescent dye helps the mass transfer from the aqueous to the phenol phase. At the same time, the fluorescent dye in the previous aqueous droplet moving through the channel has almost been depleted. (Image taken from[44]. Reproduced with permission from ASME.) (b) Schematic of a three-inlet geometry with electrodes and instability mixing. (c) Disruption of a stable stratified flow (top left) via electrohydrodynamic instability with increasing electric field strength from $E_o = 8.2 \times 10^5$ V/m, 250 kHz to $E_o = 9.0 \times 10^5$ V/m at 10 MHz. (Reprinted with kind permission from [46]).

rescent microscopy image of an aqueous droplet forming within the microchannel. By initially labeling the aqueous phase with rhodamine, the fluid-fluid interface as well as the extraction between two phases can be tracked. A significant amount of fluorescent dye transfers into the organic phase even before the droplet formation process is complete. The "flare" of the fluorescent trace extending from the interface toward the organic phase indicates that the fluorescent dye is quickly being transported across the organic/aqueous interface. Downstream of the forming droplet, the previous droplet which has broken off has been almost completely depleted of fluorescent dye [44]. Further experiments conducted with DNA and protein solutions have demonstrated the ability to extract protein into the organic phase while keeping the DNA partitioned in the aqueous phase using both stratified and droplet flows [45].

More recently, an electrohydrodynamic approach for destabilizing the stratified phenol-aqueous flow to increase the interfacial area between the two phases to improve extraction efficiency has been investigated [46]. Here, a three layer stratified system consisting of a low conductivity organic layer sandwiched between two high conductivity aqueous layers was used with patterned electrodes spaced 75 µm apart in the aqueous phase above and below the organic phase (Figure 3.10(b)). As noted earlier, the conductivity gradient between the aqueous and organic phase allows electric charge and hence electrical stresses to develop in an applied electric field. At flow rates of 5, 2.5, and 5 µl/min for the water-phenol-water system, and an applied electric field ranging from 8.2×10^5 V/m at 250 kHz and 9.0×10^5 V/m at 10 MHz, a significant destabilization of the flow is seen. Since, the two fluid layers are immiscible, mixing in this context was not defined as molecular dispersion to remove concentration gradients but rather as an increase in interfacial area between the two phases from droplet formation to reversible dispersion of the two phases (Figure 3.10(c)).

3.4 Data acquisition, anticipated results, and interpretation

Following the successful design, fabrication, and testing of a micromixer, the device must be characterized for how efficiently it mixes the samples. As shown in Section 3.3, most of the characterizations are based on the ability to optically visualize and track a mixing interface using a microscopy setup. Since standard microscopy only allows the generation of a two-dimensional image at the objective focal plane, with volume illumination through the depth of the microdevice, several assumptions are commonly made. One major assumption is that concentration gradients do not exist through the depth of the device so that concentration variations only exist within the imaging plane. This assumption limits the types of investigations that can be used to quantify mixing and can make image analysis difficult if this assumption cannot be justified, such as in the parallel laminating micromixer where the authors noted tilting of the lamellae [16]. If full three-dimensional chaotic flow fields are produced, an advanced 3-D imaging technique such as confocal microscopy may be used to evaluate gradients though the depth of the microchannel as in [25]. Finally, this limitation may require researchers to only evaluate the outlet of their micromixer, where complete mixing is assumed to have occurred.

3.4.1 Computer acquisition

As noted, a high-quality microscopy setup with a high-resolution digital camera should be used to capture images. Image acquisition may vary spatially along a flow channel or temporally in the case of active micromixers. Images are acquired according to camera manufacturer's instructions and software packages. Usually, a region of interest within the mixing the channel is focused under microscopy and images are captured in either phase contrast mode (for brightfield tracking of colored dyes) or under epifluorescence microscopy when using fluorescence dyes or particle tracers. Images should ideally have constant illumination across the imaging field as intensity variations within the images can complicated image analysis. Captured images are analyzed within a region of interest within the image (usually a rectangular ROI within the mixing channel) using image analysis packages such as ImageJ or by converting image pixel intensities into bins within a matrix array in MATLAB where the array position represents the pixel position and the array value represents the intensity. Once the image intensity is arrayed it can be quantified using standard performance metrics (Section 3.4.2).

3.4.2 Performance metrics, extent of mixing, reaction monitoring

Once images are converted to a pixel intensity array, the images can be quantified for either mixing variance or extent of mixing. One assumption made is that image intensity is linearly proportional to species concentration for the optical tracking dye. The metrics most commonly used to quantify mixing are the mixing variance coefficient (MVC) and the extent of mixing. The MVC is defined as

$$\sigma = \sqrt{\frac{\sum_{i=1}^{N}(C_i - C_{\text{mix}})^2}{N}} \approx \sqrt{\frac{\sum_{i=1}^{N}(I_i - I_{\text{avg}})^2}{N}} \tag{3.3}$$

where σ is the MVC, C represents the tracked species concentration I represents the image intensity, subscript i and average represent the image bin number and average of all bins respectively, and N is the total number of bins in the image. It should also be noted that this definition is independent of bin size, although single pixel bins are most commonly used. As can be seen from this definition, a fully mixed solution should have a MVC of 0, since the image concentration (or intensity) should be equal to the average concentration (or intensity) at every bin location within the evaluated ROI. A completely unmixed solution on the other hand has a MVC of 0.5.

A variation of the MVC is to define the extent of mixing (EOM) as a percentage as (MVC)$\times$100%. In this case, again, an unmixed solution would have an EOM of 50% and a completely mixed solution should have an EOM of 100%.

These two metrics essentially define the image variance from a completely mixed solution. One advantage of using such metrics as opposed to quantifying spatially varying pixel intensities is that it reduces errors that arise from nonuniform channel illumination.

As noted in specific examples in Section 3.3, chemical reactions may also be used to quantify mixing between two species. These can include optical characterization of reactions such as fluorescein quenching [15], the generation of a colored product such as in the rhodanide reaction [4, 16], and pH indicator dyes [23]. The mixer may also facilitate transport to a surface as in the case of DNA hybridization to a microarray chip [37]. Finally, the micromixing design can be evaluated on its ability to properly mix reactants to obtain the desired result such as in the case of protein crystallization studies [43]. The use of chemical reactions for evaluating mixer performance is usually determined based on the ultimate micromixer application.

3.5 Discussion and commentary

Many different micromixer designs have been demonstrated. The ultimate choice of which approach is taken is made by a researcher based on their needs. A diffusion-based Y or T mixer is the simplest design requiring the simplest fabrication, only two fluid inputs and a single outlet and one only needs to consider the required diffusional residence time of the fluid for complete mixing. However, these designs require long residence times and may not provide complete mixing within the device. If faster mixing times are required one may consider hydrodynamic focusing approaches, which limits the mixing volume or active mixing approaches that increases the complexity of the devices. Finally, if precise dosing of reactants is needed, often batch mixing chambers or droplet based approaches are considered.

3.6 Troubleshooting

Performance limitations of micromixers are generally characterized by several factors: incomplete mixing, speed of mixing, and accurate dosing. If incomplete mixing is seen, the researcher must reevaluate their design to make sure the mixing channel is properly designed. If diffusional mixing is used, lowering the species flow rate through the mixing channel will increase residence time to improve mixing. Alternatively, significantly increasing flow rates to higher Re >100 will allow secondary advection effects to improve mixing. However, this is not recommended for devices designed to operate in the creeping flow regime (Re<1) since the dramatic increase in flow rate required to increase the Re significantly may cause overpressuring of the fluid interconnects or device. If a high speed of mixing is required, devices should be designed with the shortest diffusional path or promoting the largest amount of chaotic advection possible to introduce as much internal circulation and folding of the flow structure as possible in a short period of time. If accurate dosing is required, batch mixing chambers that can be precisely filled and monitored may be advantageous. These include the batch chambers shown for temporal pulsing, acoustic mixing, or the discrete droplets where each droplet is a well controlled individual mixing/reaction chamber.

3.7 Application notes

MATLAB script for converting .tif image files into a number array:

```
filePath_1='C:\FILE PATH FOR IMAGE FILES\';


file_name=[filePath_1 'IMAGE FILE NAME.tif'];
A=imread(file_name);
```

3.8 Summary points

- Micromixing is an important step required in many microdevice applications. These devices are design to provide rapid and reproducible mixing of small sample volumes continuously or in discrete batch devices.
- Passive micromixers can achieve mixing in simple planar design via diffusional mixing or more complicated three-dimensional designs. These designs require no energy input beyond the establishment of a fluid flow, either electrically or via hydrostatic pressure supplied from a syringe pump.
- Active micromixers are more complex designs requiring active control of devices or active energy input to create spatial or temporal disturbances to the flow field to enhance mixing.
- The choice of mixer type depends on the end user needs for either ease of fabrication, rapid mixing, high throughput, or accurate dosing.

- Mixing may be quantified via microscopy and image analysis of tracer particle paths, marker dyes, or through monitoring of chemical reactions.

References

[1] Nguyen, N.-T.; Wereley, S. T., *Fundamentals and Applications of Microfluidics* Second ed., Boston, MA: Artech House, 2006.

[2] Wang, W., and S. A. Soper, *BioMEMS Technologies and Applications*. Boca Raton, FL: CRC Press, 2007.

[3] Nguyen, N. T.; Wu, Z. G., "Micromixers—a review," *Journal of Micromechanics and Microengineering*, Vol. 15, No. 2005, pp. R1–R16.

[4] Hessel, V., H. Lowe, and F. Schonfeld, "Micromixers—A Review on Passive and Active Mixing Principles," *Chemical Engineering Science*, Vol. 60, No. 8–9, 2005, pp. 2479–2501.

[5] Hardt, S., K. S. Drese, and V. Hessel,et al., "Passive Micromixers for Applications in the Microreactor and Mu TAS Fields," *Microfluidics and Nanofluidics*, Vol. 1, No. 2, 2005, pp. 108–118.

[6] Campbell, C. J., and B. A. Grzybowski, "Microfluidic Mixers: From Microfabricated to Self-Assembling Devices," *Philosophical Transactions of the Royal Society A: Mathematical, Physical and Engineering Sciences*, Vol. 362, No. 1818, 2004, pp. 1069–1086.

[7] Brody, J. P., P. Yager, and R. E. Goldstein,et al., "Biotechnology at Low Reynolds Numbers," *Biophys. J.*, Vol. 71, No. 6, 1996, pp. 3430–3441.

[8] Duffy, D. C., J. C. McDonald, and O. J. A. Schueller, et al., "Rapid Prototyping of Microfluidic Systems in Poly(dimethylsiloxane)," *Analytical Chemistry*, Vol. 70, No. 23, 1998, pp. 4974–4984.

[9] Heckele, M., W. Bacher, and K. D. Muller, "Hot Embossing—The Molding Technique for Plastic Microstructures," *Microsystem Technologies*, Vol. 4, No. 3, 1998, pp. 122–124.

[10] Wu, Z., and N.-T. Nguyen, "Convective–Diffusive Transport in Parallel Lamination Micromixers," *Microfluidics and Nanofluidics*, Vol. 1, No. 3, 2005, pp. 208–217.

[11] Jeon, N. L., S. K. W. Dertinger, and D. T. Chiu, et al., "Generation of Solution and Surface Gradients Using Microfluidic Systems," *Langmuir*, Vol. 16, No. 22, 2000, pp. 8311–8316.

[12] Mengeaud, V., J. Josserand, and H. H. Girault, "Mixing Processes in a Zigzag Microchannel: Finite Element Simulations and Optical Study," *Analytical Chemistry*, Vol. 74, No. 16, 2002, pp. 4279–4286.

[13] Hsieh, Y. C., and J. D. Zahn, "On-Chip Microdialysis System with Flow-Through Sensing Components," *Biosens Bioelectron*, Vol. 22, No. 11, 2007, pp. 2422–2428.

[14] Hsieh, Y. C., and J. D. Zahn, "On-Chip Microdialysis System with Flow-Through Glucose Sensing Capabilities," *Journal of Diabetes Science and Technology,* Vol. 1, No. 3 2007, pp. 375–383.

[15] Knight, J. B., A. Vishwanath, and J. P. Brody,et al., "Hydrodynamic Focusing on a Silicon Chip: Mixing Nanoliters in Microseconds," *Physical Review Letters*, Vol. 80, No. 17, 1998, pp. 3863–3866.

[16] Hessel, V., S. Hardt, and H. Lowe, et al., "Laminar Mixing in Different Interdigital Micromixers: I. Experimental Characterization," *Aiche Journal*, Vol. 49, No. 3, 2003, pp. 566–577.

[17] Wu, Z. G., and N. T. Nguyen, "Rapid Mixing Using Two-Phase Hydraulic Focusing in Microchannels," *Biomedical Microdevices*, Vol. 7, No. 1, 2005, pp. 13–20.

[18] Schonfeld, F., V. Hessel, and C. Hofmann, "An Optimised Split-and-Recombine Micro-Mixer with Uniform 'Chaotic' Mixing," *Lab on a Chip*, Vol. 4, No. 1, 2004, pp. 65–69.

[19] Xia, H. M., S. Y. Wan, and Shu, M., et al., "Chaotic Micromixers Using Two-Layer Crossing Channels to Exhibit Fast Mixing at Low Reynolds Numbers," *Lab on a Chip*, Vol. 5, No. 7, 2005, pp. 748–755.

[20] Wong, S. H., P. Bryant, and M. Ward, et al., "Investigation of Mixing in a Cross-Shaped Micromixer eith Static Mixing Elements for Reaction Kinetics Studies," *Sensors and Actuators B: Chemical*, Vol. 95, No. 1–3, 2003, pp. 414–424.

[21] Wang, H. Z., P. Iovenitti, and E. Harvey,et al. "Optimizing Layout of Obstacles for Enhanced Mixing M Microchannels," *Smart Materials & Structures*, Vol. 11, No. 5, 2002, pp. 662–667.

[22] Wang, H. Z., P. Iovenitti, and E. Harvey, et al., "Numerical investigation of mixing in microchannels with patterned grooves," *Journal of Micromechanics and Microengineering*, Vol. 13, No. 6, 2003, pp. 801–808.

[23] Liu, R. H., Stremler, M. A., and Sharp, K. V., et al., "Passive Mixing in a Three-Dimensional Serpentine Microchannel," *Journal of Microelectromechanical Systems*, Vol. 9, No. 2, 2000, pp. 190–197.

[24] Johnson, T. J., D. Ross, and L. E. Locascio, "Rapid Microfluidic Mixing," *Analytical Chemistry*, Vol. 74, No. 1, 2002, pp. 45–51.

[25] Stroock, A. D., S. K. W. Dertinger, and A. Ajdari,et al., "Chaotic Mixer for Microchannels," *Science*, Vol. 295, No. 5555, 2002, pp. 647–651.

[26]   Stroock, A. D., and G. J. McGraw, "Investigation of the Staggered Herringbone Mixer with a Simple Analytical Model," *Philosophical Transactions of the Royal Society of London Series A-Mathematical Physical and Engineering Sciences*, Vol. 362, No. 1818, 2004, pp. 971–986.

[27]   Evans, J., D. Liepmann, and A. Pisano, "Planar Laminar Mixer." In *Proceedings of the IEEE 10th Annual Workshop of MEMS*, Nagoya, Japan, 1997, pp 96–101.

[28]   Deshmukh, A., D. Liepmann, and A. P.Pisano, "Continuous Micromixer with Pulsatile Micropumps." In *IEEE Workshop on Solid-State Sensors and Actuators*, Hilton Head, SC, 2000, pp 73–76.

[29]   Deshmukh, A., D. Liepmann, and A. P. Pisano, "Characterization of a Micro-Mixing, Pumping, and Valving System." In *Proceedings of the 11th International Conference on Solid State Sensors and Actuators (Transducers '01)*, Munich, Germany, 2001; pp 950–953.

[30]   Niu, X., and Y. K. Lee, "Efficient Spatial-Temporal Chaotic Mixing in Microchannels," *Journal of Micromechanics and Microengineering*, Vol. 13, No. 3, 2003, pp. 454–462.

[31]   Probstein, R. F., *Physicochemical Hydrodynamics: An Introduction*, 2nd Edition, Hoboken, NJ: Wiley-Interscience, 2003.

[32]   Melcher, J. R., *Continuum Electromechanics*. Cambridge, MA: MIT Press, 1981.

[33]   Oddy, M. H., J. G. Santiago, and J. C. Mikkelsen, "Electrokinetic Instability Micromixing," *Analytical Chemistry*, Vol. 73, No. 24, 2001, pp. 5822–5832.

[34]   Oddy, M. H., and J. G. Santiago, "Multiple-Species Model for Electrokinetic Instability," *Physics of Fluids*, Vol. 17, No. 6, 2005, pp. 064108.

[35]   Chen, C. H., H. Lin, and S. K. Lele, et al., "Convective and Absolute Electrokinetic Instability with Conductivity Gradients," *Journal of Fluid Mechanics*, Vol. 524, No. 2005, pp. 263–303.

[36]   Lin, H., B. D. Storey, and M. H. Oddy, et al., "Instability of Electrokinetic Microchannel Flows with Conductivity Gradients," *Physics of Fluids*, Vol. 16, No. 6, 2004, pp. 1922–1935.

[37]   Liu, R. H., R. Lenigk, and P. Grodzinski, "Acoustic Micromixer for Enhancement of DNA Biochip Systems," *Journal of Microlithography Microfabrication and Microsystems*, Vol. 2, No. 3, 2003, pp. 178–184.

[38]   Jang, L.-S., S.-H. Chao, M. R. Holl, and D. R. Meldrum, "Resonant Mode-Hopping Micromixing," *Sensors and Actuators A: Physical*, Vol. 138, No. 1, 2007, pp. 179–186.

[39]   Dreyfus, R., P. Tabeling, and H. Willaime, "Ordered and Disordered Patterns in Two-Phase Flows in Microchannels," *Physical Review Letters*, Vol. 90, No. 14, 2003, pp. 144505.

[40]   Reddy, V. and J. D. Zahn, "Interfacial Stabilization of Organic-Aqueous Two-Phase Microflows for a Miniaturized DNA Extraction Module," *Journal of Colloid and Interface Science*, Vol. 286, No. 1, 2005, pp. 158–165.

[41]   Paik, P., V. K. Pamula, and R. B. Fair, "Rapid Droplet Mixers for Digital Microfluidic Systems," *Lab on a Chip*, Vol. 3, No. 4, 2003, pp. 253–259.

[42]   Song, H., J. Tice, and R. Ismagilov, "A Microfluidic System for Controlling Reaction Networks in Time," *Angewandte Chemie International Edition*, Vol. 42, No. 7, 2003, pp. 768–772.

[43]   Zheng, B., L. S. Roach, and R. F. Ismagilov, "Screening of Protein Crystallization Conditions on a Microfluidic Chip Using Nanoliter-Size Droplets," *Journal of the American Chemical Society*, Vol. 125, No. 3,7 2003, pp. 11170–11171.

[44]   Mao, X., S. Yang, and J. D. Zahn, "Enhanced Organic-Aqueous Liquid Extraction Using Droplet Formation in a Microfluidic Channel." In *Proceedings of ASME International Mechanical Engineering Conference and Exposition Fluids Engineering 2006 Symposium* ASME, Ed. ASME: Chicago, Il, 2006; pp IMECE2006–16084.

[45]   Morales, M., and J. D. Zahn, "Development of a Diffusion Limited Microfluidic Module for DNA Purification Via Phenol Extraction." In *Proceedings of ASME International Mechanical Engineering Conference and Exposition Fluids Engineering 2008 Symposium* ASME: Boston, MA, 2008; pp IMECE2008–68086.

[46]   Zahn, J. D., and V. Reddy, "Two Phase Micromixing and Analysis Using Electrohydrodynamic Instabilities," *Microfluidics and Nanofluidics*, Vol. 2, No. 5, 2006, pp. 399–415.

On-Chip Electrophoresis and Isoelectric Focusing Methods for Quantitative Biology

Lavanya Wusirika, Zohora Iqbal, and Amy E. Herr*

Department of Bioengineering, University of California, Berkeley
*e-mail:aeh@berkeley.edu

Abstract

Equilibrium and nonequilibrium electrophoresis both benefit from miniaturization. Inclusion of cross-linked polymer gels as molecular sieving matrices increases the customization of assays. Cross-linked gels allow localization of gel features within the microdevice, as well as suppress electroosmotic and pressure-driven flow. Here we describe straightforward, inexpensive methods for using polyacrylamide gels as a separation matrix for both electrophoresis and isoelectric focusing. For electrophoresis, we introduce a new method for establishing gradients in chemical constituents along the separation axis. We are developing urea gradient methods to study protein unfolding and refolding using rapid separations. We also describe a method for establishing gradients in pH along the separation channel, as a means to stabilize and accomplish isoelectric focusing. Our focus in the chapter is to provide a practical guide to fabricating both protein analysis formats. Both methods are important as means to characterize the physicochemical properties of proteins and other biological macromolecules.

Key terms   microfluidics, point-of-care diagnostics, lab-on-a-chip, on-chip polyacrylamide gel electrophoresis (page), electrophoresis, isoelectric focusing, protein folding and unfolding, chemical gradient

4.1 Introduction

Electrophoresis encompasses a powerful suite of separation techniques suitable for analysis of myriad biochemical constituents. Both equilibrium and nonequilibrium electrophoretic methods have found use in macroscale (slab gel) and microscale (capillary) implementations. Over the last two decades, microfluidic electrophoretic methods have gained importance as a means to rapidly and effectively analyze biological samples. From the basic science perspective, microfluidic electrophoretic assays enable rapid quantitation in short separation distances. The tools allow researchers to generate multiple measurements (including replicates) with, often, very small sample volumes. At the far end of the spectrum, the popularity and nascent success of microfluidic technology for clinical and point-of-care use stems from key design and performance attributes, summarized here:

- Utilizes directed, high-efficiency electrophoretic protein transport—fast assays and preparation are critical for multiple concurrent or serial analyses.
- Integrates multiple sophisticated steps. Micro- and nanofabrication methods allow chip designs that obviate labor intensive and time-consuming steps by incorporating preparatory steps—attributes key for multiplexed protein measurements in a translatable instrument.
- Provides compatibility for a wide-range of diagnostic fluids. To date, most commonly reported for serum, microfluidic assays combined with appropriate preparatory functions are adaptable to a wide-range of diagnostic fluids (urine, cerebral spinal fluid, synovial fluid, saliva).
- Affords simultaneous multi-analyte detection of tens of proteins (multiplexing) —important for "profiling" and other measurements that require concurrent monitoring of multiple species.
- Consumes low volumes (2 μL)—important for sparing consumption of precious, costly, or limited-volume samples.
- Allows translational and remote studies owing to a compact, robust physical form-factor—essential for translating the diagnostic into clinical settings, for use in the field, and for analysis at the point-of-care.

Recent progress has demonstrated that microfluidic electrophoretic methods are not only useful in medical research and pre-clinical settings, but also potentially useful in clinical settings and at the point-of-care. Details regarding use of on-chip electrophoretic methods for bioengineering and biomedically relevant measurements will be presented in this chapter.

4.1.1 Microfluidic electrophoresis supports quantitative biology and medicine

Historically, the biological sciences have faced challenges in three key areas: (1) sensitive protein quantitation,(2) adequate throughput (number of samples from unique samples, as well as measurement of multiple analytes in a single sample), and (3) robust measurement reproducibility. New technologies capable of rapid measurement of biological fluids are sorely needed to achieve the required throughput and quantitation necessary to meet the data fidelity demands of bioinformatics and computational modeling (i.e., cell signaling studies) [1]. To further speed analysis and improve reproducibility, significant innovation in well-controlled sample preparation is also required [2]. Broadly speaking, the speed and performance of microfluidic electrophoretic methods, coupled with sample preparation strategies, could go far to position biology and medicine on par with the established quantitative physical sciences (Figure 4.1).

4.1.1.1 Electrophoretic separations benefit from microfluidics

Electrophoresis (including slab gel, capillary, and microfluidic) allows for the determination of electrophoretic mobilities for species. The separation is conducted in gel or "separation columns" (i.e., capillary or microchannel) under an applied axial electric field. Species with different electrophoretic mobilities migrate through the channel at different velocities. Thus, mobility differences result in migration speed (i.e., velocity) differences that lead to the ability to resolve (or differentiate between) two migrating analyte species. Burgreën and Nakache present a brief historical review of electrokinetic theory [3].

For electrophoretic analysis, the migration speed (velocity) of a charged analyte is given by:

$$U = \mu E$$

where U is the electrophoretic velocity, μ is the electrophoretic mobility of an analyte, and E is the applied electric field. Use of photopatterned molecular sieving gels enhances differences in μ between analytes. The short separation lengths required make development of devices having literally hundreds of separation channels for parallel analyses feasible. Short separation times also make implementation of serial analyses practicable. Both equilibrium and nonequilibrium electrophoresis afford significant throughput for repeat analysis of a single sample and concurrent analysis of multiple samples.

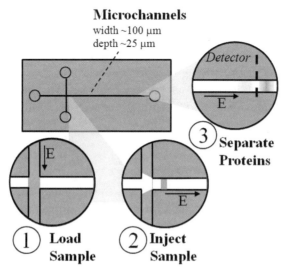

Figure 4.1 Microfluidic implementations support low dead-volume integration of sample injection channels and analysis channels for on-chip electrophoresis. The three fundamental steps in analysis are presented as (1) load sample (green), (2) electrokinetically inject sample, and (3) separate analyte species using electrophoresis.

During analysis, electrophoretic protein characterization benefits twofold from microfluidics: (1) injection of sample into an electrophoretic separation channel is a well-controlled, low-dispersion process translating into readily resolvable protein bands (high separation resolution), and (2) exceptionally high microchannel surface area-to-volume ratios dissipate Joule heating effectively making high electric fields feasible. The separation resolution between two analyte bands is defined as a ratio of the peak-to-peak separation distance, ΔL, to an average measure of the peak width, σ. The separation resolution, SR, of two analyte bands can be defined as:

$$SR = \frac{\Delta L}{\sigma}$$

Common practice allows replacement of σ, with 4σ, the full base width of the Gaussian band. When $SR = 1$, two species are said to be baseline resolved.

To quantify the quality and efficiency of a separation, factors that increase the width of the concentration distribution (σ) of injected species are important to consider. Such parameters are often said to contribute to "band-broadening." Aside from the contribution of diffusion to the variance of the band (σ_{diff}^2), the very act of injecting the sample plug into a capillary or microchannel defines a finite initial variance associated with the sample plug (σ_{inj}^2), as does interaction of the species with the channel wall (σ_{wall}^2) and temperature nonuniformities (σ_T^2). Assuming that a Gaussian distribution can represent the axial sample concentration distribution, the various dispersion contributions can be summed to estimate the overall variance of the sample band, assuming statistically independent processes:

$$\sigma_{total}^2 = \sigma_{inj}^2 + \sigma_{diff}^2 + \sigma_{wall}^2 + \sigma_T^2$$

The low dead-volume channel intersections enable electrophoretic separations to be completed over short distances, as the small intersections also allow sharp initial sample injection plug definition (small σ_{inj}^2). The low injection-related dispersion translates into shorter analysis time and high throughput.

Minimizing the channel cross-sectional area limits Joule heating arising from migration of charged ions (current) along the channel length under an applied electric potential. Use of channels with high surface area-to-volume ratios efficiently dissipate heat, as compared to slab gel and large bore tubing based systems that require cooling. Use of microfluidic and nanofluidic channels, as well as capillary tubes benefits from high surface area-to-volume ratios.

4.1.1.2 Multiplexed analyses

As compared to capillary-based systems, the adaptability afforded by microfabrication techniques enables design and construction of more complex channel networks, with more complex channel features. Interconnected channel networks, beyond a simple cross-t configuration within a single system, and arrays of parallel systems are possible. Owing the capability to fabricate multiple channels in roughly the same footprint used for a single channel analysis, researchers are using multiplexed electrophoretic assays to analyze a single sample under multiple separation conditions (Figure 4.2) or multiple

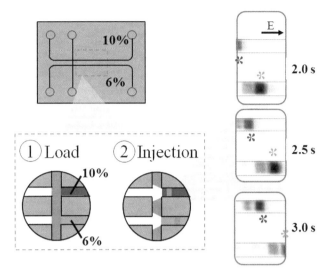

Figure 4.2 Concurrent protein separations (multiplexed assays) are enabled by microfluidic formats. Left: Schematic of microfluidic chips with two parallel separation channels. Here the top channel has a 10% T polyacrylamide sieving gel, while the bottom channel has a larger pore size 6% T gel. Right: Image sequence shows concurrent separation of protein in the small pore size (top channel) and larger pore size (bottom channel) gels.

samples concurrently. Concurrent multiplexed analyses improve throughput, as well as the fidelity of data obtained during a particular analysis.

4.1.1.3 Multidimensional analyses

Since the middle of last century, multidimensional separations have been pursued as a means to increase the separation capacity of bioanalytical systems [4]. The number of peaks (i.e., resolved, distinct analyte bands) in a given separation column or peak capacity, n, can be expressed as:

$$n = L / (4 \cdot SR \cdot \sigma)$$

Here σ is again the average standard deviation of adjacent zones and L is the length of the separation column. In essence, the peak capacity can be used to estimate how many focused bands, for a given SR, can fit into a separation channel. The peak capacity describes the information density accessible using a particular separation system. For proteomic discovery studies, peak capacities in the mid-to-high thousands are necessary. To achieve such high peak capacities requires that more than one separation "dimension" is employed to parse a given complex sample.

In multidimensional separations, the total capacity, n_{total}, is (for orthogonal separation mechanisms), the product of the peak capacities of each successive dimension [5]. Usually, a multidimensional combination strategy includes two different modes of separation, although three [6] or more [7] have been proposed, often with mass spectrometry as one of the dimensions. A significant potential for microfabrication is in multidimensional systems design affording automated operation. Conventional multidimensional analysis can be labor-intensive. In both slab- and capillary-based formats, two or more independent separation mechanisms are often coupled, thus creating what is known as a multidimensional (or *n*-dimensional) separation system. Coupling of techniques allows species that would not necessarily be resolved by either of two tandem techniques to become fully resolved by the combined system. The sequential coupling of discrete, independent separation mechanisms is of particular relevance to assays that contain numerous sample species, such as those important to proteome analysis.

A shortcoming of single dimensional microfluidic electrophoretic analyses has been in the realization of a limited peak capacity, especially for protein analysis. The observation is not surprising, as conventional single-dimensional assays are rarely able to achieve high peak capacity. Microfluidic multidimensional systems are an active field of research. Isotachophoresis (ITP), isoelectric focusing (IEF), capillary electrophoresis (CE), SDS-PAGE, and a myriad of other such chromatographic or electrophoretic approaches are just some of the technologies that have been combined in various permutations to improve and optimize the available power of resolution [8–14]. Specific applications demonstrated include analysis of oligomers [15, 16], metabolites [17, 18], and proteins [19, 20]. Excellent reviews can be found by [21–24].

4.1.2 Biomedical applications of on-chip electrophoresis

4.1.2.1 Proteomics

Proteomics necessitates that protein samples be collected, separated, detected, and ultimately compared to a control in order to identify discrepancies between the samples and, thus, potentially identify disease mechanisms. Multidimensional separation techniques are crucial to proteomics, as the high peak capacity approaches separate complex sample mixtures effectively and efficiently. An automated 2-D system is essential for performance and reliability [25]. Ultimately, the integration of these multiple sample analysis steps in a single device will aid the accurate identification and quantification of specific proteins present in complex samples. Such analysis is crucial to the recognition of an expressed phenotype [8].

The field of microfluidic multidimensional systems is nascent, at best. Few reports of substantial peak capacity have been made using microfluidic systems. While adversely impacting biomarker discovery (which requires measurements of thousands of proteins in a sample), limited peak capacity does not adversely impact biomarker verification, qualification, and subsequent validation—which requires that a small subset of proteins (~10) be measured with high reproducibility in numerous patient samples. Consequently, microfluidic electrophoretic assays can provide an efficient means to validate putative biomarker panels discovered a priori using established proteomic discovery methods.

4.1.2.2 Clinically relevant assays

Transformation from curative medicine to predictive, personalized, and possibly preemptive medicine depends on the availability and accessiblity of point-of-care diagnostic tools. Supporting a "bench-to-bedside" paradigm, translation of diagnostic tools from centralized laboratories to near-patient settings is critically needed. Rapid diagnostic assays, including those based on microflluidic electrophoresis, could be useful for routine use in near-patient settings. Examples of target applications include (1) prediction of disease onset, (2) stratification of disease, (3) indications of disease progression, (4) guidance of treatment decisions (i.e., identify drug resistance or potential adverse reactions), and (5) monitoring of treatment efficacy notable advances in clinical diagnostic instrumentation fueled by progress in [26–28]. Electrophoretic protein analysis—in particular affinity-based electrophoretic and electrochromatographic assays—play a key role in clinical and point-of-care usage of microfluidic electrophoresis.

4.2 Materials

Protocols for fabricating and conducting electrophoretic protein analysis in microchannels containing cross-linked polyacrylamide gels are the focus of this section. Materials needed are grouped and described below.

4.2.1 Reagents

4.2.1.1 Polyacrylamide gel electrophoresis (PAGE)

Reagents requiring room temperature storage:

Filtered 1M sodium hydroxide (NaOH)

- Product #: S318-500, Fisher Scientific, 500 grams tablet form;
- Safety concerns: causes skin and eye burns (skin irritant);
- Directions: make solution as needed, stable for 3 weeks.

16M urea

- Product #: 9510, OmniPur, 500 grams powder form;
- Safety concerns: avoid contact with skin, eyes, and clothing (skin irritant);
- Directions: mix as needed.

DI water
PBS buffer
- pH 7.4, 1X, Product #: 10010, Gibco, from Invitrogen.
10X tris-glycine buffer
- Product #: LC2672, Novex, from Invitrogen.
Methanol

- Product #: A412P-4, Fisher Scientific, 4 liters;
- Safety concerns: flammable.

Acetic acid

- Product No: A9967, Sigma-Aldrich, 500 grams;
- Safety concerns: combustible corrosive.

Reagents requiring refrigeration (2°C –8°C):
Silane

- 3-(trimethoxysilyl)propylmethacyrlate, Product #: 440159, Sigma-Aldrich, 100 ml;
- Safety concerns: irritant.

VA-086US photoinitiator

• 2,2'-Azobis (2-methyl-N-(2-hydroxyethyl) propionamide), Wako chemicals, 25g;
• Safety concerns: flammable, do not inhale.

29:1 acrylamide/bis-acrylamide 30% stock solution

• Product #: A3574, 100 ml, 0.2 uM filtered;
• Safety concerns: neurotoxin, carcinogen, can cause reproductive damage, do not inhale or touch without gloves.

Protein sample
  Bovine serum albumin

• Product #: 15561-020, Invitrogen, 150 mg (50 mg/ml), fluorescently labeled;
• Directions: store as aliquots of 20 ul of 1 uM BSA at −20°C.

4.2.1.2   Isoelectric focusing

Reagents requiring refrigeration (2°C–8°C):
  Silane

• 3-(trimethoxysilyl)propyl methacrylate, 98%, Product #: 440159, Sigma-Aldrich;
• Sensitivity: light, humidity, and heat;
• Safety concerns: an irritant in contact; wear gloves, use in fume hood.

VA-086 photoinitiator

• 2,2'-Azobis[2-methyl-N-(2-hydroxyethyl)propionamide], Wako Chemicals;
• Sensitivity: light.

30% solution of acrylamide/bis-acrylamide

• Product information: mix ratio 29:1, T30%, C3.3% Product #: A3574, Sigma-Aldrich;
• Safety concerns: toxic, carcinogen, may cause heritable genetic damage and impair fertility, toxic in contact with skin (readily absorbed through skin) or if swallowed;
• Sensitivity: light.

Biolyte 3/10 ampholyte 20% (100X)

• Product #: 163-2094, Bio-Rad;
• Safety concerns: may cause irritation upon contact;
• Concentration: 20% (the given concentration) for fabrication, and 2% (diluted down 10:1) for storage and conducting isoelectric focusing.

Protein samples
Trypsin inhibitor from soybean, Alexa Fluor 488 conjugate, Product #: T23011, Invitrogen

• Concentration: 0.625mg/mL 2% Biolyte solution, centrifuge protein solution for 40 seconds before use, take only the supernatant;
• Storage: 20-μL aliquots, at −20°C;

- Sensitivity: light, heat.

  Lectin, FITC labeled from Lens culinaris (lentil), Product #: L9262, Sigma-Aldrich

- Concentration: 1 mg/mL 2% Biolyte solution;
- Storage: 20-μL aliquots, at –20°C;
- Sensitivity: light, heat.

4.2.2   Facilities/equipment

Equipment and infrastructure

- Water sonicator with degassing vacuum attachment
- Spectrolinker XL-1500 UV cross-linker oven (Spectronics Corp.)
- Black-Ray Long Wave UV lamp (100 AP)
- Microchip manifold system (custom built) or Nanoport fittings (Upchurch Scientific)
- High-voltage sequencer—HVS448 3000V high-voltage power supply, Labsmith
- Sequence software (version 1.140), Labsmith
- ImageJ software (downloaded from: http://rsbweb.nih.gov/ij/)
- Microscope with camera (e.g., epifluorescence microscope, Olympus IX-70 or IX-71; cooled scientific grade CCD camera, Roper Scientific; SVM340 Synchronized Video Microscope, Labsmith)
- WinView software (for Olympus) or UScope software v1.013 (for Labsmith)

Consumables

- Glass or quartz microfluidic chips (Figure 4.3)
- 1.7 mL Eppendorf tubes or similar
- Platinum wire for electrodes (Sigma-Aldrich, 0.5 mm diameter, 99.9% metals basis, 267228-1G, 25 cm

Safety equipment

- Face shield for UV lamp
- Disposable gloves, eyewear, lab coat, and other personal protective equipment

4.3   Methods

Here we present a step-by-step guide to fabricating polyacrylamide gels in glass microfluidic chips with two intersecting channels (simple-t geometry in Figure 4.4), running electrophoretic protein separations, and analyzing the data. Two protocols are given, one for PAGE analysis of proteins and one for PAGE-based isoelectric focusing analysis of proteins. An outline of both protocols is given in Figure 4.5.

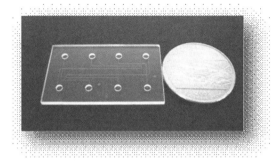

Figure 4.3 Photograph of a standard glass microfluidic device. A U.S. quarter is shown at right for scale.

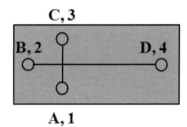

Figure 4.4 Schematic of glass chip used in the protocol presented here.

4.3.1 On chip polyacrylamide gel electrophoresis (PAGE)

4.3.1.1 Fabrication

Chip preparation

1. If no cleaning of the microchannels is necessary:
   (a) Visually inspect the chip; if debris present, remove cover and flush channels with NaOH, allowing the filled channels to sit filled with NaOH for 10 minutes. Follow NaOH rinse with three DI water rinses.
   (b) If no debris present, do not perform the rinse indicated in (a)
2. Silane coating of glass microchannels:
   (a) Prepare a silane solution by mixing silane and acetic acid first in a clean and dry Eppendorf tube; add filtered DI water according to amounts shown in Table 4.1. Prepare fresh solution for each fabrication run.
   (b) Mix thoroughly by hand pipetting, taking care not to entrain air in the solution (Do not centrifuge).
   (c) Puncture top of plastic tube with a needle. During degassing, interface a tube to vacuum with the small hole in the tube top. The interface allows efficient degassing of the small volume fluid during the next sonication step.
   (d) Immediately before loading the silane solution into chip, degas and sonicate until no bubbles present.
   (e) Pipette silane solution into microchannels through well at the terminus of the highest fluidic resistance channel. Allow full wicking.
   (f) Cover all holes with excess silane, to ensure fluid does not evaporate out of wells and channels during next steps.

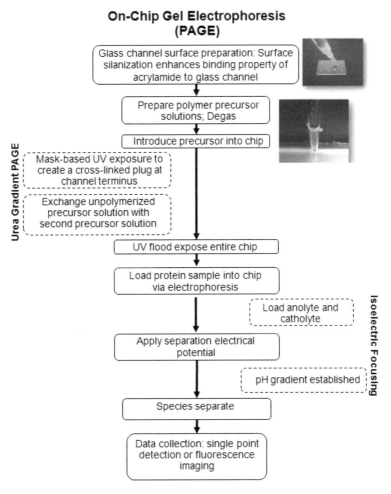

Figure 4.5 Overview of fabrication process used for on-chip electrophoresis and isoelectric focusing protocols described in this chapter. Photos show loading of chip with solution (top) and degassing of polymer precursor solution prior to introduction into the chip.

Table 4.1 Recipe for Silane Coating Solution

	Silane	Acetic Acid	DI water
Number of parts	2	3	5
Volume (μL)	100	150	250

(g) Cover chip with the pipette tips cover and incubate for no more than 30 minutes.

(h) After incubation, use vacuum to remove silane solution from all wells and the channels.

(i) Rinse channels twice with filtered methanol and filtered DI water; each time, cover entire chip surface with fluid to clean chip surfaces as well as channels (important for subsequent imaging).

(j) Vacuum to thoroughly dry chip.

(k) Visually inspect to ensure no debris or excess fluid is left in channels. If debris is present, rerun cleaning procedure.

Monomer solution preparation
(a) Weigh 1 mg of VA-086 photoinitiator into 1.7 mL plastic tube.
(b) Mix the acrylamide solution according to Table 4.2 for a desired percentage acrylamide gel. Prepare fresh for each fabrication run.
(c) Cut the 250-μL pipette tips so that they fit into the wells of the channels (do this now so that there is no delay in loading acrylamide into channels during later steps). Set tips aside.
(d) Degas and sonicate acrylamide mixture until no bubbles left, usually about 60 minutes. (This step is critical.) Keep the vial attached to the tubing and vacuum if not using right away.

Microchannel loading
(a) Pipette 4 uL of degassed acrylamide mixture into the well leading to the longest channel. Allow full wicking. Pipette 4 uL of fluid into all wells.
(b) Place the cut pipette tips in the wells. Twist a little for a snug, but not too tight fit. Fill tips with equal amounts of acrylamide mixture (to prevent evaporation).
(c) Equalize liquid levels by adding or removing fluid (to minimize effects of bulk flow).
(d) Visually inspect to check for bubbles and debris.
(e) If bubbles are present, remove the tips; vacuum out the channels and reload degassed acrylamide mixture.

Stage 1 photopolymerization (Proceed to "Storage" step if gel gradient is not desired)
(a) Wearing a face shield, preheat UV lamp for 5 minutes. Turn on air and vacuum so that it does not overheat.
(b) Clean the mask with methanol and dry with N2 gas.
(c) Place chip with tips on top of mask, so that well 2 is directly over the rectangle on the chrome side of the mask.
(d) Tape chip into place. Check under microscope to mask layout matches channel layout. (This is to ensure polymerization only occurs on the desired well).
(e) Place mask on top of lamp and expose to UV for 12 minutes.
(f) If making a uniform gel in your chip, jump to Storage step.
(g) Remove mask. Let chip cool. Turn off lamp. After 2 minutes, turn off the cooling air supply to the lamp.

Gradient generation
(a) Vacuum each well except the well containing polymerized gel.

Table 4.2 Recipe for Polyacrylamide Separation Gels

30% Acrylamide/Bis Stock (29:1)	H$_2$O	10X Tris/Glycine Buffer	% T
100 uL per mg of VA-086	350 uL per mg of VA-086	50 uL per mg of VA-086	6
59.5 uL per mg of VA-086	405.5 uL per mg of VA-086	45 uL per mg of VA-086	3.5
133 uL per mg of VA-086	339.3 uL per mg of VA-086	37.7 uL per mg of VA-086	8

(b) Place a drop of second acrylamide solution on wells 1 and 3 (Figure 4.4).

(c) Vacuum well 4 for 30 seconds.

(d) Place drop of second acrylamide solution on wells 3 and 4.

(e) Vacuum well 1 for 30 seconds.

Stage 2 photopolymerization

(a) Place chip into the Spectrolinker UV oven right away or time interval to establish gradient in gel properties between the two regions.

(b) Polymerize for 600 seconds.

(c) Visual inspection to evaluate gel fabrication success (Inspect for: Bubbles? Debris? Full Polymerization? Channel defects?)

Storage

(a) If successful, store polymerized chip in 1X Tris/Glycine buffer at 4°C. If not, store in DI water at 4°C for later ashing (gel removal via heating) and reuse.

4.3.1.2 Electrophoresis

System setup

(a) Mount chip in custom manifold or use Nanoport fittings (Upchurch Scientific).

(b) Following Figure 4.4, pipette 10 ul of 1X Tris/Glycine buffer into holes #2, 3, and 4. Pipette 10 ul of 1 uM of fluorescent BSA into well 1. Make sure there are no air bubbles in system.

(c) Mount chip in manifold down to microscope stage (Olympus or Labsmith microscopes). Use 10X magnification.

(d) Cut four platinum wires of about 1.5 cm (platinum wire electrodes can be reused indefinitely). Connect the wires to the electrodes A to D for the high voltage system.

(e) Place electrode A in well 1; B in well 2, etc. (Clean up any excess buffer that spills from the wells as fluid leaks can cause a short in the system.)

(f) Turn on imaging system (microscope, camera, any lighting) and high-voltage power supply.

Imaging measurements using Labsmith system

While the protocol is written for imaging with the Labsmith microscopy system, the protocol can be modified to any imaging system. We do recommend an epifluorescence, inverted microscope for ease of concurrent fluidic and imaging access.

(a) Open UScope software to image chip during electrophoresis.

(b) Move the microscope using the controls on the Labsmith or with the arrow keys on the keyboard.

(c) Open Sequencer software to control voltage supply to the electrodes.

(d) Go to Tools Simple Sequence Wizard. If protein is negatively charged (like BSA), input the voltage program Steps A through C found in Table 4.3 and press OK. It should ask whether to send this to HVS448 3000V. Click OK.

(e) In the main control page, check the boxes to monitor the desired channels. The output range should be −1500V to 1500V.

(f) Press the Toggle Offline/Online button on the top menu.

(g) Go to Actions Run Step A to load the proteins; Step B to perform the separations; Step C to clean out the channels in preparation for another load/separation sequence.

(h) Note the general range of current during loading.

(i) When finished, store chip in manifold at 4°C. Cover wells to limit evaporation. If chip will not be reused, then remove from manifold and store in a centrifuge tube filled with buffer.

Measurements using fluorescence microscopy (Olympus IX-70)

(a) The Sequencer software is the same here as on the LabSmith workstation. So follow the same procedures.

(b) Turn on the CCD camera for imaging following the instructions for your camera system.

(c) Using WinView software, use a 100-ms exposure time, a 0.2s delay time, and 0 ADC offset. The time between frames is the sum of the exposure and delay times.

(d) Change the window size to show just the t-junction and about 1 mm of the separation junction.

(e) Program Steps are the same as shown in Section 4.3.1.2, "Imaging measurements using Labsmith system."

(f) Watch the current carefully to avoid current spikes above 20 uA. Current spikes have been observed to cause gel breakdown. If current spike occurs, disable voltages on all electrodes.

(g) When finished, store chip in manifold at 4°C. Cover wells to limit evaporation. If chip will not be reused, then remove from manifold and store in a centrifuge tube filled with buffer.

4.3.2 Polyacrylamide gel electrophoresis based isoelectric focusing

4.3.2.1 Fabrication

Sections A through D are protocols for channel preparation and cleaning that are typically repeated several times during IEF gel fabrication. Following the preparatory protocols are step-by-step procedures for performing IEF and data analysis.

(A) Loading the channels with liquids

1. Other than silane solution and acrylamide solution, filter all other solutions prior to introducing them into the microchannels.

   • Debris of all sizes gathers at the wells, which may enter microchannels when fluid is introduced into the well. Apply vacuum for a short time at each well (using vacuum with pipette tip whose tip has been cut off) so that any loose debris may be

Table 4.3 Currents and Voltages Used for Loading And Separation Experiments

Steps	Electrode A	Electrode B	Electrode C	Electrode D
Step A = Load	0V	0V	450V	0V
Step B = Separate	0 uA	−800V	0 uA	0V
Step C = Clean	0V	0V	450V	−200V

Electrodes A–D correspond to wells A–D as shown in Figure 4.4.

removed. Always vacuum at all the wells on the device before introducing any solution to the channels.

- Fill 3-mL syringe with solution to be loaded into device. Use in-line luer-lock filter to yield filtered solutions.
- Aliquot solution droplets from the syringe and filter system directly into the well, or filter out the solution first into an Eppendorf tube, then pipette solution into the well. If using droplets directly from the syringe, ensure that no air bubble is trapped in the well, hindering the solution from entering the channel. If a bubble is trapped, use a pipette tip to move the solution around in the well a bit until the bubble is released.

2. Place a drop of solution into the highest resistant well first (the longest channel) and allow it to wick through all the way to the wells of the rest of the channel. Once the solution reaches a well, place a drop of solution at that well.

(B) Removing solution from the microchannels

1. Remove excess solutions from each well using tubing fitted with a 250-μL pipette tip whose tip is cut off to about ¾ cm. Other end of tubing is connected to a vacuum source.
2. Place vacuum at well D or M for at least 2 minutes continuously
3. Place the vacuum at the other wells for about a minute each.
4. Use a microscope to examine the channels, to ensure there are not small droplets of solution left behind.

(C) Rinsing microchannels

1. Load the channel with solution.
2. With the large droplet of solution still present on all wells, apply vacuum with pipette tip to one well (preferably the one with most debris in its channel if there are debris in the channels. This would induce the debris to move out rapidly). Move the pipette tip around (in circular motion) and allow the solution to be pulled through the system.
3. After repeating this process at each well, especially at the wells near which debris may be located, remove solution out of the device.

(D) Degassing solution

1. Place the solution to be degassed into a 1.5 mL eppendorf tube.
2. With a needle (22G1), make a hole at the center of the tube lid.
3. Place the eppendorf tube under a vacuum setup with pipe, with the hole at the center of the vacuum pipe.

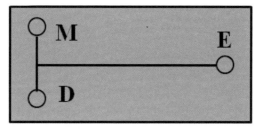

Figure 4.6   Schematic of glass chip used in the protocol presented here for IEF.

4. Fill the sonicator with water just below the solution level in the eppendorf tube.

5. Turn on the sonicator, and tap the bottom of the eppendorf tube at various frequencies, keeping the tube in upright position with it attached to the vacuum.

6. Note: It helps to hold the eppendorf tube, while supporting the vacuum pipe as well, so that the tube does not become detached, which may result in contamination from the water in the sonicator.

7. Degas the solution for the prescribed period of time. However, if you see bubbles still rising from the solution, sonicate and agitate the solution for a longer period of time. Make sure there is a good vacuum seal on your eppendorf tube and you are tapping the tube in an area near where you can see the starting position of the vibration waves.

(E) Fabrication of 6% polyacrylamide gel in microdevice

1. Use the microscope to visually inspect the device channels. Even in a new chip, debris is often present.

2. Clean the device to ensure uniformity of the channels, as well as to get rid of any debris.

- Load the channels with 1M NaOH solution, and set the device aside for 10 minutes. During this time, keep the device covered with a glass petri dish to minimize evaporation of the solution.

- Remove the NaOH solution from the channels, and rinse twice with DI water.

- Look at the channels of the device under the microscope and make sure channels are debris-free.

- Most debris is easily removed with this first run, however, if there are still some left, you can repeat the cleaning procedure, or just rinse with DI water. Load water in the well farthest from the debris, and apply vacuum at the well most closest to the debris.

- If the debris is insignificantly small, less than 1/50 of a channel width, and debris has not moved from its original position, it generally does not interfere with the fabrication process and it is acceptable to move onto the next step. Otherwise, start with a new device entirely.

Silanize the channels

- Aliquot 50 μL of silane into a 1.5-mL eppendorf tube.

- Add to silane 75 μL of acetic acid, and gently aspirate with a pipette to mix thoroughly (approximately 30 seconds) until the solution is clear.

- Add 125 μL of DI water, and gently aspirate as in the previous step.

- Making the given amount of silane solution is more than sufficient, however, if you would like to make a different total volume, make sure to maintain the ratios listed in Table 4.4.

- Sonicate the silane mixture for approximately 5 minutes. This is generally sufficient to degas silane solution.

- Load the channels of the device with the silane mixture. Cover all wells with plenty of excess silane solution and set the device aside for exactly 30 minutes. Make sure to keep the device covered with a glass petri dish during this period.

- Remove the silane solution from the device.
- Rinse device with methanol twice, followed by two DI water rinses.
- Visually inspect the channels under the microscope to ensure no debris has accumulated and all the solutions have been removed (no droplets, however small, are left behind).

1. If there are solution drops in the device, rinse one more time with methanol, and two more times with DI water.
2. Keep the device under vacuum for a longer period of time, approximately 6 minutes at a well.

Prepare acrylamide/bis-acrylamide solution

- Weigh out 1.1 mg of photoinitiator VA-086 in a 1.5-mL eppendorf tube.
- Add and mix in 100 μL of 30% acrylamide/bis-acrylamide solution per 1.1 mg of photoinitiator.
- Add and mix in 350 μL of DI water per 1.1 mg of photoinitiator.
- Add and mix in 50 μL of 20% Biolyte buffer per 1.1 mg of photoinitiator.
- Often it is hard to measure out exactly 1.1 mg of photoinitiator. However, make sure the ratio of VA-086 to the additives are maintained as listed in Table 4.5.
- Degas solution for about 15 minutes. Keep solution under vacuum until it is ready to be placed in device.

Preparing device for polymerization step

- Cut three 250-μL pipette tips to fit to each of the three wells.
1. Cut about ¾ cm from the bottom of the pipette tip.
2. Cut off the excess top so that the entire tip is about 2 cm long.
3. Make sure to try each at their designated wells. It should fit tight enough that the entire device can be picked up by only holding on to the pipette tip fitted into the well.
4. After the trial, put the fitted pipette tips aside.

- Remove the acrylamide mixture from the vacuum.
1. To avoid bubbles after the gel fabrication, it is very important to have degassed the acrylamide solution effectively. Conduct all the steps as rapidly and efficiently as possible so that the degassed acrylamide is not exposed for a prolonged period of time.
- Load acrylamide mixture into the wells. This time, instead of loading with large drops, only fill the wells just so that the fluid is approximately level with the walls of the well.

Table 4.4 Recipe for Silane Coating Solution

	Silane	Acetic Acid	DI water
Ratio used	5	3	1
Volumes	50 μL	75 μL	125 μL

Table 4.5 Calculations for Mixing Acrylamide Aolution

	VA-086	30% Acrylamide/bis stock (29:1)	DI H2O	20% Biolyte	n% T
Ratio	1.1 mg	100 μL	350 μL	50 μL	6
Measured (Calculations)	x mg	$\frac{100\,\mu L}{1.1\,mg}*x$	$\frac{350\,\mu L}{1.1\,mg}*x$	$\frac{50\,\mu L}{1.1\,mg}*x$	6

- Fit in the cut pipette tips so that fluid from the well rises up the pipette tip and there is no bubble trapped in the system. There must be total fluidic connection. Be careful not to jam in the pipette tips too tightly as that might prevent the fluid flow, since this fluid is somewhat viscous.
- Once again, using 20-μL pipette, load 20 μL of acrylamide solutions into each fitted pipette tip, making sure to pipette out at the very bottom where the tip is gently touching the glass well base so that there is full fluidic connection.
- Look under the microscope for bubbles or new debris within the channels. If there is, remove (vacuum out) the solution and reload the acrylamide solution as described above. It is best to degas the polyacrylamide solution again. If the channels look clear and bubble free, move onto the next step.

Cross-linking acrylamide using ultraviolet light

- Place the device in the UV oven, and set the time for 600 seconds.
- Once the polymerization is complete, take off the pipette tips. At each well, place a large drop of 2% Biolyte solution to keep the gel from drying out.
- Check the channels of the device again under microscope to ensure no bubbles have formed.

1. If bubbles are present, the device may still be useful if bubbles only occupy one of the three channels available, or if the bubbles are insignificant enough to not cause any blockage to the flow of proteins and solutions in the gel.

Storing device

- A new gel can be placed in a 50-mL tube with about 30 mL (or enough to submerge the device fully) of 2% Biolyte solution. Store the device in 4°C to 8°C.
- Otherwise, the chip can be directly set up in the manifold, and the wells filled up to 60 μL of 2% Biolyte each. Set up of the manifold is described in step 1 of Section 4.3.2.2. Place the manifold with device in it in a Ziploc bag, with a wet paper towel to minimize evaporation of the solution from the wells. Tape four pipette tips upright at the four corners of the manifold to avoid the solution in the well from coming in contact with the Ziploc bag.

4.3.2.2 Isoelectric focusing

1. Set up device in manifold (fluid fixturing). Insert chip into your custom manifold or attach Nanoport wells to device using low-temperature adhesive.
2. Load each of the three wells with 2% Biolyte buffer.
3. Visually inspect channels for debris.

4. Clean the surfaces of the device with a cotton swab dipped in methanol to avoid any debris on the surface to obstruct light and imaging.

5. Load proteins in the manifold, here for analysis using channel D to E.

   (a) Proteins to be loaded into the device: lectin (pI = 7.8, 8.0, 8.3—lectin has three subspecies) and trypsin inhibitor (pI = 4.5). Since the pH of Biolyte solution is approximately neutral, all subspecies of lectin will be positively charged, while trypsin inhibitor will be negatively charged. Note the charge of the proteins under the given conditions, and estimate magnitude and direction of migration.

   (b) Remove the Biolyte solution from well E

   (c) In well E, load 7 μL of diluted lectin solution (pipette in with a 20-μL pipette, touch the glass surface of the device with the pipette tip before releasing the solution).

   (d) Remove the Biolyte solution from well D.

   (e) In well D, load 7 μL of diluted trypsin inhibitor in the same manner as lectin was loaded.

6. Loading proteins into channels

   (a) Clean two pieces of 2.5 cm of platinum wires with alcohol reagent, and rinse well with water.

   (b) Choose one pair of electrodes from the Labsmith High Voltage Sequencer, and note the letter associated with the pair.

   (c) Place in the platinum wires in each of the two electrodes from the sequencer.

   (d) Place in the positive and the ground platinum electrodes into well E, and well D, respectively, each electrode touching the bottom surface of the well.

   (e) Open Sequencer program on the computer controlling the high voltage sequencer.

   (f) Apply 100V/cm electric field (401.5V) for about 10 minutes. Monitor the current and observe the channels using imaging software. The current is generally around 4 μA. If the current hovers around 1 μA, then there is a lack of fluidic connection. Disable high voltage, take out solutions, and reload (a bubble might be in the way). Once the channels show a higher intensity than the background while imaging, disable the high voltage.

   (g) Apply voltage until most of the channel across which voltage is applied is slightly fluorescent under exposure time of 500 ms. The difference is apparent when comparing to the background intensity, though it may not be visible if observed by eye through the microscope eyepieces.

   (h) Remove the excess proteins from the wells. Using 2% Biolyte, rinse out the excess proteins from the wells using 20 μL of 2% Biolyte at a time. Repeat twice.

7. Current stabilizing period

   (a) In well D: load 40 μL of acid buffer, 40-mM sodium phosphate.

   (b) In well E: load 40 μL of base buffer, 20-mM sodium hydroxide.

   (c) Place in positive electrode in well D, and ground in well E.

   (d) Before applying voltage, begin the current trace in the Sequencer program. From Labsmith menu, choose file—>trace measurements—>save to start current recording.

   (e) Apply 175.5V/cm electric field. It is really important at this step to monitor the current carefully, as there is current spike. Current increasing beyond 20 μA, often degrades the polyacrylamide gel. Therefore, before current reaches $20\,\mu A$, lower the voltage applied.

(f) When the current stabilizes to a small value of approximately $4.5\,\mu A$, prepare to look for focused bands.

(g) Camera Settings: Camera setup—hardware—shutter type and open compensation = 1 ms, and exposure time = 500 ms.

(h) Take images in WinView, and save in Tiff 8-bit format for analysis in ImageJ.

4.3.3 Data acquisition, anticipated results, and interpretation

(A) PAGE protein mobility acquisition

1. Open ImageJ software and the folder in which the Separation/Injection movie was saved.
2. Drag the desired movie from its folder into the space below the selection options in the ImageJ window. A new window with the separation movie will open (Figure 4.7).
3. With the rightmost rectangle selection enabled in ImageJ, draw a thin, long rectangle that spans the entire frame of the movie. This rectangle should be positioned in the center of the protein plug (Figure 4.7).
4. In the ImageJ toolbar, choose Analyze Plot Profile, which will show a graph of grayscale of pixel versus distance from the top left corner of image (i.e., origin). If this graph has a curve which saturates at 250 gray value (Figure 4.7), then the image captured is too bright. Reducing the exposure time to about 50 ms or lower will supply a Gaussian distribution that does not saturate the pixels.
5. In the ImageJ toolbar, choose Plugins Compile and Run plugins (in the ImageJ folder) Stack_Peak_Tracker_
6. A dialog box opens with a list of numbers. The first number in the output string is the distance (from the origin) of the brightest point of the protein plug. The next number is where the plug is in the next frame and so on.
7. Copy the list of numbers into an Excel spreadsheet. Graph the points as a line of number of pixels versus frame number. Remove points that do not correspond to

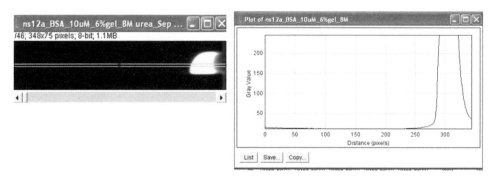

Figure 4.7 Two screen-shots of ImageJ-based analysis. Frame at left show image of injected plug of fluorescently labeled protein. Electrophoretic separation channel spans the length of the image. A small rectangle shows selection of the pixels near the center of the microchannel for analysis of protein mobility. Frame at right shows intensity along yellow rectangle plotted in ImageJ from frame at left. This image is too bright since the intensity curve saturates.

the meaningful data (e.g., peak has exited imaging frame). Plot the trend line. Pixels should be converted to distance units after distance calibration of the system (Figure 4.8).

4.3.4 Results and discussion

4.3.4.1 On-chip PAGE for urea gradient electrophoresis

Protein folding and unfolding mechanisms proceed through highly unstable intermediates. Lower temperatures reduce reaction rates, stabilizing these intermediates but distorting secondary and tertiary protein structure. Instead, denaturants like urea can be used. The cation portion of the urea zwitterion adsorbs to proteins and enables ion exchange, causing protein denaturation. At low urea concentrations, the folded protein state dominates; at high concentrations, the unfolded state dominates [29]. Using this interaction, a urea gradient can be employed to observe protein unfolding and refolding.

While these results have been established in macroscale slab gels, this project applies urea-gradient denaturation to the microscale. We utilize a photopatterning process to fix a urea-gradient in cross-linked polyacrylamide gels housed in microfluidic channels. Gels of uniform 0M and 8M urea concentrations are fabricated as controls, using the PAGE protocol detailed in the "Methods" section above. Fluorescently-labeled protein is electrophoretically analyzed in each gel. Dynamic measurement of unfolding (and folding processes) can be made by observing protein migration through a gel housing a gradient in urea concentration (0M to 8M). Protein mobility is monitored as the protein passes through a linear gradient—making protein fold-state a function of location. From urea-dependent behavior, we can deduce unfolding mechanisms, subunit interactions and tertiary structure, invisible to the inaccurate macroscale denaturation methods [30].

In preliminary analysis of purified bovine serum albumin, we observe significant velocity dependence on surrounding urea concentration (Table 4.6 and Figure 4.9). Comparison of mobility in a 0M urea gel yields a mobility that is two to three times faster than the mobility of the same protein in an 8M urea gel.

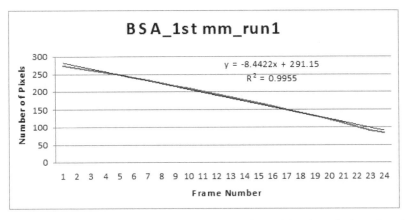

Figure 4.8 Plot of protein peak migration along separation channel (pixel number) as a function of time (frame number). Here migration of the protein (bovine serum albumin, BSA) is linear as a function of time.

Table 4.6 Summary of BSA Migration as a Function of Urea Concentration

	0M PAGE chip	8M PAGE chip
Average velocity (um/s)	289.9	124.5
Standard deviation (um/s)	43.5	11.4
Number of trials	18	12

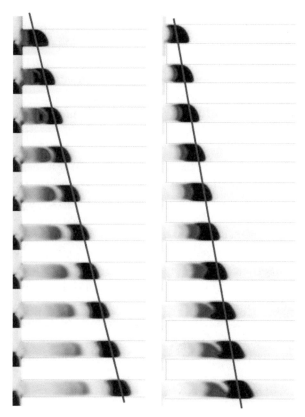

Figure 4.9 Protein migration shows dependence on urea concentration in sieving gel. (Left) Migration of BSA protein peak in a chip containing 0M urea. (Right) Migration of a BSA peak in a chip containing 8M urea. Images collected 0.3 seconds apart. An electric field of 151V/cm was applied across the separation channel in both chips.

Preliminary results for BSA establish the validity of the 0M and 8M controls, as well as the robustness of the uniform urea gel preparation process. Further work is focusing on analysis of well-characterized proteins in 0M to 8M linear urea gradients developed based on the methods reported here.

4.3.4.2 Analysis of isoforms: Isoelectric focusing

Analysis of three lectin isoforms was performed using the IEF protocols described here (Figure 4.10). Steady-state focusing was achieved in less than 10 minutes, with well-resolved species visible near the anticipated focusing location. Expected focusing loca-

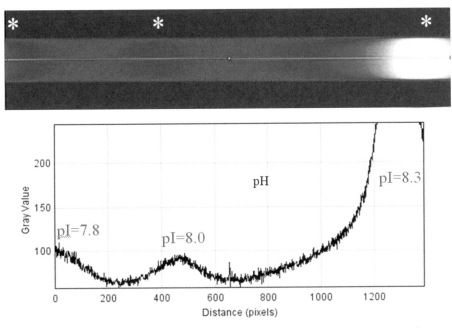

Figure 4.10   On chip PAGE IEF analysis of three lectin isoforms. Top: Fluorescence image of fluorescently labeled isoforms focused in microchannel (width 100 μm). Isoforms marked with a *. Bottom: Intensity plot of focused isoforms along separation axis.

tion is estimated from known channel length, IEF anolyte, and catholyte pH values, as well as known isoelectric point of protein species.

4.4   Discussion of pitfalls

A troubleshooting guide for PAGE on-chip electrophoresis is presented in Table 4.7. A similar troubleshooting guide for PAGE isoelectric focusing is presented in Table 4.8.

4.5   Summary notes

- Electrophoresis is a common procedure in basic science laboratories. By incorporating microfluidics into this familiar technique, we can achieve rapid and reproducible quantitation using small sample volumes and low applied voltage.
- Microfluidic gel electrophoresis allows for rapid assays, compatibility with a variety of diagnostic fluids, multianalyte detection, and consumption of very small quantities of samples and buffers.
- Proteins with different electrophoretic mobilities travel through the microfluidic channels at different velocities, allowing for clear resolution of analyte bands.
- Urea-gradient gel electrophoresis can be used to monitor protein mobility through a linear urea gradient, revealing information about protein unfolding mechanisms, subunit interactions and tertiary structure.

Table 4.7   Troubleshooting Guide for On-Chip Protein Electrophoresis Using PAGE Approach

Problem	Explanation	Potential Solutions
Fluid is not wicking through channels.	Debris is present.	Filtered DI water flushes.
		If persists, restart process with 10-minute filtered NaOH incubation.
		As a last resort, use methanol for three flushes.
There are bubbles in the gel after removing the chip from the UV oven.	Bubbles not completely removed from silane or acrylamide solutions.	Sonicate the acrylamide solution until no bubbles are present upon vigorous tapping (may take up to 60 minutes for beginners).
	Extra air introduced by centrifugation or pipetting.	Do not centrifuge to mix. Instead, carefully pipette with the tip in the middle of liquid volume when removing or depositing fluid.
Gel is liquid-like and not polymerized.	Missing photoinitiator.	Vacuum out gel. Restart the fabrication process from NaOH step.
Cloudy acrylamide solution after sonicating.	Too much waste buildup in degassing vacuum tube. Waste is mixing with the acrylamide solution.	Wearing goggles, gloves and lab coat, clean degassing vacuum tube with bleach and water. Use reagent alcohol if needed. Vacuum dry.
Protein not loading under applied electric field.	Electrodes are not touching the bottom of the wells.	Press down and tape to hold. Or use longer electrodes. If a big black dot is visible in the well, then electrode touches the bottom of well.
	The quick change from positive to negative can stimulate movement.	Try using negative voltages (i.e., A=B=D= -450V, C=0V) to load.
	Electrodes can be damaged or nonresponsive.	Change electrodes. Use E-H instead.
	Using wrong electrodes.	Confirm that the electrodes in the program for Step A-C are actually the electrodes being placed into the corresponding wells. Consult Figure 4.4 and Section 4.3.1.2 for directions.
Grayscale still saturating (Step 4 in Section 4.3.3.	Fluorescence of protein plug is still too bright.	Go to a lower exposure time. Alternatively, use a lower concentration of protein (for example 0.5 μM), depending upon microscope sensitivity.
I am loading my protein but bubbles are forming in parts of the gel.	Gel has failed due to current spike.	Save data that has been collected. This microchip device cannot be used any longer. Start gel fabrication process on the device on other side of NS12A chip.
	If no current spike, then existing bubbles have increased in size.	As long as these bubbles are not in the separation channel, continue use. If bubbles are blocking loading or separation, discard device.

Consult the PAGE isoelectric troubleshootingtable (Table 4.8) as well since many problems overlap between the two systems.

- Microfluidic multidimensional systems such as PAGE-based IEF achieve high peak resolution and afford automated operation.
- The assays discussed in this chapter can be used to validate disease biomarkers, predict onset of disease, indicate disease progression, guide treatment, and monitor its efficiency.

Table 4.8   Troubleshooting Guide for On-Chip Isoelectric Focusing Using PAGE Approach

Problem	Explanation	Potential Solutions
Debris builds up in the channel before and after silanization.	Even working under a hood, debris can often get caught inside the channels at any step.	If this happens before silanizing, rinse the channel with water, placing the vacuum at the well closest to the debris and loading water at all other channels. If this does not work, another NaOH incubation is recommended as described in the procedure above.
		If this happens after silanizing, rinse the channels once with methanol, and multiple times with water.
Sonicate acrylamide solution for the recommended time, but still see bubbles.	Either there is still air in the fluid or bubbles are created as you sonicate as the fluid splashes about in the tube.	If bubbles form only when the fluid inside splashes around too much, consider the sonication as complete. If the bubbles are still forming out of the solution, continue sonicating. It is acceptable to take breaks while sonicating as long as the fluid remains under tight vacuum conditions.
		Ensure a good vacuum seal while sonicating.
		Tap the tube under various frequencies, and in the location of the sonicator where sonication waves are propagated.
After cross-linking in the UV oven bubbles formed.	Sonication might not have worked properly and there is still air trapped in the solution, or after sonication, the solution was exposed to air for too long.	Sonicate the acrylamide solution for a longer period of time. Conduct all the steps after the sonication of acrylamide to placing the device in under UV rapidly to lessen the time of exposure of the solution to air.
	Debris in a device also causes location where bubbles form.	Confirm that there is no debris in device.
	Having the pipette tips not fit tightly enough can cause them to fall over, and not provide the solution at the wells. Having pipette tips jammed in may hinder the flow of fluids, especially if the fluid is highly viscous.	Have the pipette tips fit snugly enough that you can pick up the device holding the tip; at that point pull tip out ½ a millimeter, to ensure good flow of fluid.
Proteins not loading in the channel.	Connections are loose.	If the current is less than 2 μA, check the connections, and make sure the electrodes are contacting the fluid in the well.
	Polarity is off.	If the current is above 2 μA, and the protein is still not loading, check to see if the polarities are oriented correctly: Positively charged proteins should be loaded with the positive electrode location, and negatively charged proteins should be loaded in the well with the negative electrode.
	The protein may be loading but under the settings of the imaging software, it isn't evident.	Have the WinView software exposure time set to 500 ms. Adjust the auto-contrast and take intensity readings of channels and areas outside the channels. If the channel intensity is significantly higher, even if it's not visible, it's most likely that the protein has loaded. Besides, it is not desirable to have the protein concentration in the channels to be too high, because that may take away from a focused band.
Acid base loaded, but no current spike, no bands.	Connections are loose.	Check all connections: If the current readings are less than 2 μA, then connections are loose. There must be full fluidic connection.
	Acid and base not loaded in the right orientation in comparison to the electrodes.	Remove and replace acid and base, after checking orientation of acid/base in relation to electrode polarity.
		If connection is not loose: the current spike may come after a couple minutes of applying voltage.

Acknowledgments

The authors wish to thank the Department of Bioengineering at the University of California, Berkeley and the University of California Regents' Junior Faculty Fellowship (AEH).

References

[1] Tricoli, J. V., et al. (2004). "Detection of Prostate Cancer and Predicting Progression: Current and Future Diagnostic Markers." *Clin Cancer Res,* 10: 3943–3953.

[2] Folch, A., et al. (2000). "Microfabricated elastomeric stencils for micropatterning cell cultures." *Journal of Biomedical Materials Research,* 52(2): 346.

[3] Burgreen, D., et al. (1964). "Electrokinetic Flow in Ultrafine Capillary Slits." *Journal of Physical Chemistry,* 68: 1084–1091.

[4] Smithies, O., et al. (1956). "Two-dimensional electrophoresis of serum proteins." *Nature,* 177(4518): 1033.

[5] Giddings, J. C. (1991). *Unified Separation Science,* Wiley-Interscience.

[6] Moore, A. W., Jr., et al. (1995). "Comprehensive three-dimensional separation of peptides using size exclusion chromatography/reversed phase liquid chromatography/optically gated capillary zone electrophoresis." *Anal Chem,* 67(19): 3456–63.

[7] Tang, H. Y., et al. (2005). "A novel four-dimensional strategy combining protein and peptide separation methods enables detection of low-abundance proteins in human plasma and serum proteomes." *Proteomics,* 5(13): 3329–42.

[8] O'Farrell, P. H. (1975). "High-Resolution 2-Dimensional Electrophoresis of Proteins." *Journal of Biological Chemistry,* 250(10): 4007–4021.

[9] Bushey, M. M., et al. (1990). "Automated instrumentation for comprehensive two-dimensional high-performance liquid chromatography of proteins." *Anal Chem,* 62(2): 161–7.

[10] Mohan, D., et al. (2002). "On-line coupling of capillary isoelectric focusing with transient isotachophoresis-zone electrophoresis: a two-dimensional separation system for proteomics." *Electrophoresis,* 23(18): 3160–7.

[11] Chen, J., et al. (2003). "Capillary isoelectric focusing-based multidimensional concentration/separation platform for proteome analysis." *Anal Chem,* 75(13): 3145–52.

[12] Zhang, M., et al. (2006). "Two-dimensional microcolumn separation platform for proteomics consisting of on-line coupled capillary isoelectric focusing and capillary electrochromatography. 1. Evaluation of the capillary-based two-dimensional platform with proteins, peptides, and human serum." *J Proteome Res,* 5(8): 2001–8.

[13] Busnel, J. M., et al. (2007). "Capillary electrophoresis as a second dimension to isoelectric focusing for peptide separation." *Anal Chem,* 79(15): 5949–55.

[14] Cabrera, R., et al. (2008). "Tailoring orthogonal proteomic routines to understand protein separation during ion exchange chromatography." *J Sep Sci,* 31(13): 2500–10.

[15] Hagemeier, E., et al. (1983). "On-line high-performance liquid affinity chromatography-high-performance liquid chromatography analysis of monomeric ribonucleoside compounds in biological fluids." *J Chromatogr,* 282: 663–9.

[16] O'Connor, P., et al. (1997). "Combination of high-performance liquid chromatography and thin-layer chromatography separation of five adducted nucleotides isolated from liver resulting from intraperitoneal administration with 7H-dibenzo[c,g]carbazole to mice." *J Chromatogr B Biomed Sci Appl,* 700(1–2): 49–57.

[17] Morovjan, G., et al. (2002). "Metabolite analysis, isolation and purity assessment using various liquid chromatographic techniques combined with radioactivity detection." *J Chromatogr Sci,* 40(10): 603–8.

[18] Baranyi, M., et al. (2006). "Chromatographic analysis of dopamine metabolism in a Parkinsonian model." *J Chromatogr A,* 1120(1–2): 13–20.

[19] Ong, S. E., et al. (2001). "An evaluation of the use of two-dimensional gel electrophoresis in proteomics." *Biomol Eng,* 18(5): 195–205.

[20] Uttenweiler-Joseph, S., et al. (2008). "Toward a Full Characterization of the Human 20S Proteasome Subunits and Their Isoforms by a Combination of Proteomic Approaches." *Methods Mol Biol,* 484: 111–30.

[21] Cooper, J. W., et al. (2004). "Recent advances in capillary separations for proteomics." *Electrophoresis,* 25(23–24): 3913–26.

[22] Fournier, M. L., et al. (2007). "Multidimensional separations-based shotgun proteomics." *Chem Rev*, 107(8): 3654–86.

[23] Issaq, H. J., et al. (2005). "Multidimensional separation of peptides for effective proteomic analysis." *J Chromatogr B Analyt Technol Biomed Life Sci*, 817(1): 35–47.

[24] Neverova, I., et al. (2005). "Role of chromatographic techniques in proteomic analysis." *J Chromatogr B Analyt Technol Biomed Life Sci*, 815(1–2): 51–63.

[25] Giddings, J. C. (1984). "Two-Dimensional Separations : Concept and Promise." *Analytical Chemistry*, 56(12): 1258A–1270A.

[26] von Lode, P. (2005). "Point-of-care immunotesting: Approaching the analytical performance of central laboratory methods." *Clinical Biochemistry*, 38(7): 591–606.

[27] Yager, P., et al. (2006). "Microfluidic diagnostic technologies for global public health." *Nature*, 442(7101): 412–418.

[28] Chin, C. D., et al. (2007). "Lab-on-a-chip devices for global health: Past studies and future opportunities." *Lab on a Chip*, 7(1): 41–57.

[29] Creighton, T. E. (1986). "Detection of folding intermediates using urea-gradient electrophoresis." *Methods in Enzymology*, 131: 9–26.

[30] Carter, P., et al. (1986). "Construction of heterodimer tyrosyl-tRNA synthetase shows tRNA-Tyr interacts with both subunits." *Proc. Natl. Acad. Sci. USA*, 83: 1189–1192.

Electrowetting

Vijay Srinivasan*, Ramakrishna Sista, Michael Pollack, and Vamsee Pamula

Advanced Liquid Logic Inc., Research Triangle Park, NC

*Corresponding author: 615 Davis Dr., Ste 800, Morrisville, NC 27560, phone: 919-287-9010, fax: 919-287-9011, e-mail: vijay@liquid-logic.com.

Abstract

The emerging paradigm of the lab-on-a-chip powered by microfluidics is expected to revolutionize miniaturization, automation, and integration in the life science laboratory. In this chapter we describe methods to design, fabricate, and test a droplet-based microfluidic lab-on-a-chip, based on electrowetting actuation. Droplet-based protocols are developed from elementary operations to implement clinical chemistry assays and immunoassays. A colorimetric enzyme-kinetic assay for glucose is used as the model system to illustrate the use of the lab-on-a-chip for clinical chemistry applications. Methods to effectively manipulate magnetic beads are developed and used to implement heterogeneous immunoassays for IL-6 and insulin.

Key terms	electrowetting
	digital microfluidics
	clinical diagnostics
	droplet
	immunoassay
	ELISA
	magnetic beads

5.1 Introduction

The electrowetting phenomenon has recently emerged as a powerful method to manipulate liquids in microfluidic systems [1] The most common applications of electrowetting include lab-on-a-chip devices [2, 3], optical lenses [4], and electronic displays [5]. In this chapter we will describe methods of using electrowetting for lab-on-a-chip applications such as clinical chemistry [6] and immunoassays [7, 8]. Most of the methods are generally extensible to other application such as nucleic acid amplification [9], multiplexed proteomic sample preparation and analysis by MALDI-MS [10], and explosives detection [11].

The electrowetting-based lab-on-a-chip devices use discrete droplets as reactors instead of continuous flow of liquid in channels. There has been growing interest in droplet-based microfluidic systems in recent years as an alternative paradigm to continuous-flow channel-based systems [12]. Other methods for droplet manipulation reported in the literature include flow focusing [13], dielectrophoresis [14], surface acoustic waves [15], thermocapillarity [16], and magnetic methods [17]. Among these, flow focusing is the only method that has shown widespread applicability and progress beyond the proof-of-concept demonstrations. However, flow focusing based droplet devices still requires fixed channels and an externally pumped carrier fluid. Individual control of droplets and droplet operations is usually not possible. Electrowetting-based devices offer individual control of droplets and active control of droplet operations. Since there is no continuous flow of liquid, different regions of the device may be independently operated in parallel without affecting each other. Droplet operations are effected by simply switching voltages and therefore no valves or external pumps are required. The size of droplets and the devices themselves may be scaled arbitrarily within the limits of fabrication processes.

5.1.1 Electrowetting theory

The electrowetting phenomenon has its origins in electrocapillarity which was first described by the French physicist Gabriel Lippmann in 1875 [18]. Lippmann observed that the capillary depression of mercury in contact with electrolyte solutions could be varied by applying a voltage between the two liquids. More than 100 years later, Beni et al. [19] replaced the liquid mercury electrode with a solid electrode and introduced the electrowetting phenomenon, which he described as the change in solid-electrolyte contact angle by the application of a potential difference between the two. To prevent electrolysis of the liquid, Berge et al. [20] introduced a dielectric layer on top of the electrode and this configuration serves as the basis for almost all the electrowetting devices used today (shown in Figure 5.1).

The change in contact angle of a liquid as a function of applied voltage can be described by the electrowetting equation below:

$$\cos\theta_0 = \cos\theta_V + \frac{\varepsilon_o \varepsilon_r}{2 d\sigma_{lv}} V^2$$

θ_V = contact angle at voltage V, θ_0 = contact angle at voltage V=0, V = applied voltage, ε_o = permittivity of vacuum, ε_r = relative permittivity of the insulator, d = thickness of the insulator, and σ_{lv} = liquid-filler fluid interfacial tension. For a detailed theoretical

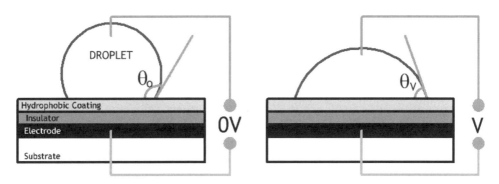

Figure 5.1 Classical electrowetting setup showing the reduction in contact angle due to an applied potential difference between the droplet and an underlying counter electrode. The counter electrode is insulated to prevent electrolysis. The insulator is made hydrophobic to maximize the contact angle change upon the application of voltage.

discussion of electrowetting the reader is referred to a recent review on the topic by Mugele et al. [1].

5.1.2 Droplet manipulation using electrowetting

The simplest configuration of an electrowetting device for droplet manipulation is shown in Figure 5.2 and consists of two parallel substrates separated by a gap [21, 22]. The first substrate (referred to as the electrowetting chip) consists of a patterned electrode array that is insulated and coated with a hydrophobic material. The hydrophobic coating is required to achieve a large and reversible contact angle change. The second substrate (referred to as top plate) consists of a large counter electrode and is also coated with a hydrophobic material. Droplets are sandwiched between the two substrates and are surrounded by an immiscible fluid such as air or oil.

The electrode array is typically connected to a common voltage source through a bank of switches and the top plate is connected to a reference potential. Turning an electrode ON applies a potential difference between the droplet and the electrode, causing the droplet to wet the electrode and align itself with it.

In the methods described in this chapter we use low viscosity silicone oil as the filler fluid. Silicone oil reduces the contact angle hysteresis of the system, which reduces the

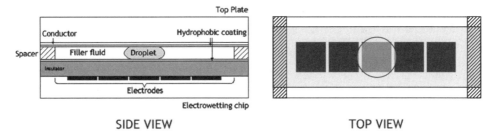

SIDE VIEW   TOP VIEW

Figure 5.2 Side and top view of the parallel plate electrowetting setup to manipulate droplets. The droplet is sandwiched between the electrowetting chip with control electrodes and a conductive top plate.

voltage required for operation and improves reliability and reproducibility of droplet operations. The oil also serves to contain evaporation of the droplet. Silicone oil forms a film between the droplet and the hydrophobic surfaces during electrowetting operation. This oil film isolates the droplet from the hydrophobic surfaces, minimizing adsorption of proteins and other biological molecules and facilitating transport. Further details on the use of silicone oil in the electrowetting system have been described elsewhere [2]. Surfactant additives in the aqueous droplets have been shown to reduce adsorption and enable manipulation of protein-rich droplets [23].

The electrowetting chip consists of a set of contiguous "unit" electrodes and a "unit" droplet is typically slightly larger than a unit electrode. Larger droplets may be integer multiples of the size of a unit droplet. The motion of a droplet in an electrowetting system is discrete in space (defined by the array of electrodes) and in time since it is typically operated by a single-system clock. Due to its similarity to the functioning of digital logic circuits this approach is also referred to as "digital" microfluidics.

The most basic droplet operation is the transport of a unit droplet from one electrode to its immediate neighbor. Repeating this process allows droplets to be transported across multiple electrodes. Basic droplets operations that involve or result in more than one unit droplet include merging and splitting. Figure 5.3 shows the switching sequences involved in the basic droplet operations.

Other important droplet operations include mixing and dispensing. Droplet mixing can be performed as a combination of merge, transport and split operations. Droplet dispensing is a special case of asymmetric splitting in which the dispensed droplet is typically a unit droplet. Washing is a key process in applications involving solid phases (typically magnetic beads) such as heterogeneous immunoassays. During washing, the solid phase is immobilized and unbound molecules can be removed by serial dilution, which involves repeated addition of a wash buffer droplet and splitting away the excess supernatant until the supernatant is free of unbound molecules.

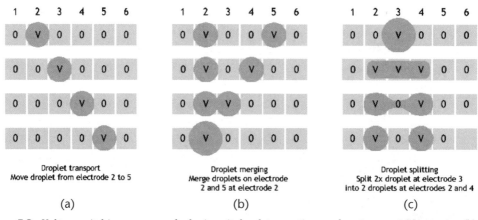

Figure 5.3 Voltage switching sequences for basic unit droplet operations such as transport (a), merging (b), and splitting (c). A "V" over the electrode indicated that a potential difference is applied between that electrode and the top plate. A "0" indicates that there is no potential difference between the electrode and the top plate. The images show different time instances from top to bottom.

5.1.3   Digital microfluidic lab-on-a-chip for clinical diagnostics

Clinical diagnostics is one of the most promising applications for microfluidic lab-on-chip systems, especially in point-of-care or near-patient setting [24]. All of the benefits of miniaturization such as reduced reagent consumption, reduced sample requirement, decreased analysis time, and higher levels of throughput and automation, are realized in this application.

Clinical chemistry   An enzymatic glucose assay used a model system to illustrate the use of the lab-on-a-chip for clinical chemistry applications [25]. A modified Trinder's reaction [26], which is a colorimetric enzyme based method, is used for glucose detection in the electrowetting system. Glucose is enzymatically oxidized in the presence of glucose oxidase to produce hydrogen peroxide, which reacts with 4-aminoantipyrine (4-AAP) and N-ethyl-N-sulfopropyl-m-toluidine (TOPS), to form violet colored quinoneimine, which has an absorbance peak at 545 nm. The glucose concentration is directly proportional to the initial rate of the reaction.

Immunoassays   Assays for human insulin and IL-6 are used to illustrate the implementation of a magnetic bead based enzyme linked immunosorbent assay (ELISA) on the digital microfluidic platform. The droplet-based magnetic bead immunoassay protocol consists generally of three steps: incubation, washing, and detection. During incubation a sample droplet (containing the antigen of interest) is mixed with a droplet containing primary (I^0Ab) capture antibodies immobilized on magnetic beads and a droplet containing enzyme labeled secondary (II^0Ab) antibodies. After the formation of the bead-I^0Ab-antigen-II^0Ab-enzyme complex, the magnetic beads are immobilized and washed to remove unbound secondary antibody. The washed bead droplet is then mixed with a substrate droplet to produce a detectable signal.

5.2   Digital microfluidic lab-on-a-chip design

The physical components of an electrowetting-based lab-on-a-chip are fluidic inputs, on-chip reservoirs, droplet generation units, and droplet pathways for transport, mixing, incubation, washing, and detection.

5.2.1   Fluidic input port

The fluidic input port is the interface between the external world and the lab-on-a-chip. The design of the fluidic input port in microfluidic systems is challenging due to the discrepancy in the scales of real-world samples (microliters-milliliters) and the lab-on-a-chip (submicroliter).

The fluidic input port is designed for manual loading of the reservoirs through a loading hole in the top plate, designed to fit a small volume ($<2\,\mu$L) pipette tip. The loading hole is connected to the reservoir by a narrow channel of width w, patterned in the spacer. Choosing w to be less than or equal height of the droplet causes the fluid pressure in reservoir to be lower compared to the loading channel. This prevents liquid from spontaneously flowing back into the loading hole. This pressure difference is initially overcome by the positive displacement pipetting action, to fill the reservoir with the liquid.

5.2.2 Liquid reservoirs

Droplet dispensing from an on-chip reservoir occurs in the following three steps. A liquid column is extruded from the reservoir by activating a series of electrodes adjacent to it. Once the column overlaps the electrode on which the droplet is to be formed, all the remaining electrodes are deactivated to form a neck in the column. The electrode in the reservoir is then activated to pull the liquid back causing the neck to break completely and form a droplet.

The reliability and repeatability of the dispensing process is affected by several design and experimental parameters, the most important being the aspect ratio of the droplet. Previous results [27, 28] indicate that droplet dispensing in a water-silicone oil system requires a droplet aspect ratio (diameter/height) greater than 5 and a water-air system requires an aspect ratio greater than 10. Larger aspect ratios cause droplets to split easily even while transporting. As a rule of thumb an aspect ratio between 4 and 6 is most optimal for droplet transport, dispensing, and splitting for an electrowetting system in silicone oil. A tapering pull-back electrode (wider at the dispensing end) ensures that the liquid is retained at the dispensing end of the reservoir as the reservoir is depleted.

5.2.3 Droplet pathways

Droplet pathways consist of contiguous electrodes, which connect different areas of the chip. These electrodes can be used either simply for transport or for other more complex operations such as mixing and splitting. In order to minimize the number of electrical contacts, a multiphase transporter design is used for the fluidic pathways. In an n-phase transporter every nth electrode is electrically connected, and droplets are always spaced apart by $k \times n$-1 electrodes, where k is any integer.

Mixing

Mixing of two droplets on the lab-on-a-chip is simply achieved by merging them and moving the merged droplet along a linear path. Droplet motion in an electrowetting system is known to create internal circulating flow patterns that enhance the mixing process. Since fluid flow in microfluidic systems is almost fully laminar, these flow patterns show a high degree of reversibility [29] when the direction of the droplet motion is reversed. A linear path in one direction is therefore preferable over repeated back and forth movement to avoid reversal of flow.

Incubation and detection

Chemical and biochemical reactions can be performed by merging and mixing reaction droplets. Mixing of more than 3 or 4 droplets at a time can be a problem due to chip real-estate concerns, which means dilution factors greater then 3 would require novel strategies. One way to do this would be to use a serial-dilution approach, which consists of repeated mixing and splitting operations [28].

For magnetic bead based immunoassays, it is essential to keep the magnetic beads resuspended to accelerate binding—to reduce the time to result and improve sensitivity. Magnetic beads have a tendency to settle down due to gravity and aggregate if they are exposed to strong magnetic fields for a long time. In the absence of a magnetic field beads may be resuspended by simply transporting the droplet back and forth owing to the internal circulation generated by droplet transport. The magnetic field may be effectively removed by moving the droplet away from the magnet or by moving the magnet away from the chip. The first method is preferable since moving parts add complexity and cost to the instrument.

Bead washing

The purpose of washing beads is to remove unbound molecules from those bound to beads. Washing is performed by serial dilution, which involves repeated addition of a wash buffer droplet and removal of the supernatant by splitting the droplet. During this process the magnetic beads are immobilized using a permanent magnet to prevent the beads from being washed away. Several parameters influence the attraction of magnetic beads during immobilization, including the buffer in which the beads are suspended, the pull force of the magnet, and the position and orientation of the magnet relative to the bead-containing droplet. Bead resuspension is also important during washing to remove unbound material trapped in bead interstices. Figure 5.4 shows the steps involved in one wash cycle.

5.3 Materials

5.3.1 Chemicals

Glucose assay   Glucose oxidase (G-6125), peroxidase (P-8125), 4-aminoantipyrine (A-4382), and TOPS (E-8506) were purchased from Sigma Chemicals (St Louis, MO). Glucose standards (100, 300, and 800 mg/dL) were obtained from Sigma and diluted in deionized water to give stock solutions with concentrations ranging from 25 to 800 mg/dL. The glucose reagent consisted of glucose oxidase (6 U/mL), peroxidase (6 U/mL), 4-aminoantipyrine (6 mM), and TOPS (6 mM) constituted in 0.1M pH=7.4 phosphate buffered saline. The reagent was stable for 2 days when stored at 4°C.

Immunoassay reagents   Dynal MyOne™ Streptavidin magnetic beads (1.05 μm diameter and 10 mg/mL stock concentration) and Amplex Ultra red reagent (A36006) were obtained from Invitrogen™ (Carlsbad, CA). Biotinylated horseradish peroxidase (HRP) was obtained from EY laboratories (San Mateo, CA). Lumigen PS-Atto and APS-5 were obtained from Lumigen Inc. (Southfield, MI). Tween 20 and APS-5 were obtained from Pierce Chemicals (Rockford, IL).

Chemiluminescence substrate for HRP (Lumigen PS-Atto) was prepared by mixing equal volumes of PS-Atto A and B solutions. The colorimetric substrate for HRP (Amplex® Ultra red) was prepared by mixing equal volumes of 0.1 mM Amplex Ultra red reagent in DMSO solution and 2 μM hydrogen peroxide. The Dynal MyOne Streptavidin beads were labeled with HRP by incubating 50 μL of 2-mg/mL magnetic beads (1/5x stock) with 10 μL of 10-μg/mL biotinylated HRP for 30 minutes in a microcentrifuge

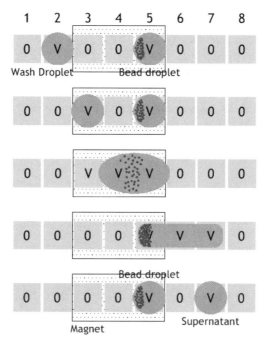

Figure 5.4 Droplet operations involved in one wash cycle. The process is repeated until the supernatant is free of unbound material. The shaded rectangle represents the magnet underneath the chip.

tube, followed by washing five times with TRIS wash buffer (0.05M TRIS-HCl, 0.1M NaCl, 0.1 mg/mL BSA, 0.01% Tween 20). The beads were finally resuspended in 50 μL of TRIS wash buffer.

Insulin and IL-6 immunoassay reagents   Insulin antibodies (Cat No 33410) and calibrators (Cat No 33415), and IL-6 antibodies (Cat No A30945) and calibrators (Cat No A30944) were obtained from from Beckman Coulter (Fullerton, CA).

Filler fluid   1 cSt silicone oil (DMS-T01) and 1.5 cSt silicone oil (DMS-T01.5) were obtained from Gelest, Inc. (Morrisville, PA).

5.3.2 Fabrication materials

Hydrophobic coating   Teflon AF-1600 (6% w/w in FC-75) was obtained from DuPont, (Wilmington, DE). The 6% stock was diluted to a final concentration of 1% w/w in FC-75 (3M, United States).

ITO glass   ITO glass was obtained from Delta Technologies (Stillwater, MN). The sheet resistance was 5 ohm/sq or 100 ohm/sq.

Magnets   0.5 Tesla and 1 Tesla Neodymium magnets (ND 42) with different pull forces (1.25, 5, and 10 lb) were obtained from KJ Magnetics (Jamison, PA).

5.4 Device Fabrication

5.4.1 Fabrication of single layer electrowetting chips

Single-layer electrowetting chips were fabricated by photolithographically patterning high reflective chromium on 0.060" thick glass wafer. The thickness of the conductive layer was 840Å. The devices consisted of four linear arrays of square electrodes. The electrode pitch (center-to-center) within an array was 1.27 mm. The gap between adjacent electrodes was 25.4 μm. The single-layer electrowetting chips were fabricated by Photosciences, Inc. (Torrance, CA). Multiple devices were arrayed on a 6"x6" wafer and diced into individual chips before use.

Alternative methods   Electrowetting chips have also been fabricated using other conductive materials such as ITO, gold, and platinum, and substrates such as silicon [30]. Features smaller than 10 μm can be achieved using a lift-off process instead of an etching process.

5.4.2 Fabrication of two layer electrowetting chips

To enable more complex electrode arrangements, a two-layer metal process was used to fabricate the chips. The process steps involved in the fabrication of two-layer electrowetting chips are described below.

1. A 2,000Å chromium or indium tin oxide coated glass wafer (0.040" thick) was photolithographically patterned to define the electrical contacts and interconnections to the electrodes.
2. A 6-μm thick photoimageable polyimide layer was spin-coated to act as the interlevel dielectric and photolithographically patterned to define 40-μm diameter vias.
3. Indium-tin oxide was sputtered to metallize the vias and define the electrodes that were to be used for optical detection.
4. A second layer of 1000Å chrome was sputtered and patterned to define the other electrodes.
5. A 4-mil (~100 μm) thick photoimageable polymer (~90 μm after curing) was laminated onto the wafer and patterned to serve as the spacer material. The polymer pattern also defined the reservoirs, waste area, and the loading channels.

The two layer electrowetting chips were fabricated by the Biomedical Microdevices and Sensors Laboratory at North Carolina State University, (Raleigh, NC). The minimum feature size was 10 μm. Chips consisted of sample injection elements, reservoirs (and waste), droplet formation structures, fluidic pathways, mixing areas, and optical detection sites. The electrode pitch was 500 μm with 10-μm gaps between electrodes.

The two-layer chip consisted of eight reservoirs and two waste areas (Figure 5.5). The fluidic pathways consisted of two 4-phase transporters and one optical detection site. The reservoirs were 4 μm in diameter, the electrode pitch 500 μm, and the spacer thickness was 90 μm. A 100-μm wide channel was defined on the spacer material to connect the fluid input port and the reservoir. The chip was designed to be controlled using 36 electrical inputs with 18 contact pads provided on each side of the chip.

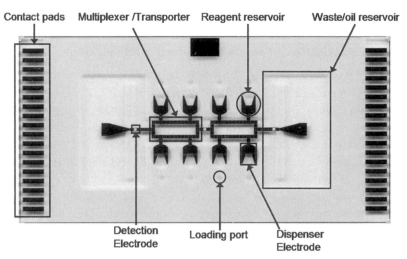

Figure 5.5 Two-layer electrowetting chip with **eight** reservoirs and **two** waste regions. The electrode pitch is 500 μm and the spacer thickness is 90 μm.

Alternative methods   Two-layer devices are not very common in the electrowetting research community. However, several groups have recently reported the use of printed circuit board manufacturing processes to manufacture multilayer electrowetting chips [31–33]. This manufacturing method shows a lot of promise for commercialization due to its widespread availability and low cost.

5.4.3 Dielectric deposition

Parylene C layer served as the dielectric for both the single-layer and two-layer devices. The parylene thickness was nominally 1 μm for clinical chemistry experiments (two-layer devices), 5 μm for bead wash optimization experiments, and 12 μm for insulin and IL-6 immunoassay experiments. The parylene C coating was performed by Paratronix, Inc. (Attleboro, MA) or at the Shared Materials Instrumentation Facility (Duke University, NC), using a vapor deposition process. The electrical contact pads were masked with tape before the parylene deposition.

Parylene C is known to have poor adhesion to glass and a silane adhesion promoter is typically recommended. For thin coatings (~1 μm) we found that the adhesion promoter is not absolutely required. For thicker coating or if adhesion is a problem, it is recommended that an adhesion promoter be used since there is a larger tendency for thicker coatings to peel off. Thicker coatings of parylene (5 and 12 μm) are more reliable than a 1-μm coating. The thickness of the dielectric has a direct impact on the voltage required for operation. For a 1-μm thick Parylene C layer, 30 to 50V DC is optimal for droplet operations while lower voltages will suffice for liquids with low surface tension. This voltage scales with the square root of the dielectric thickness. Devices with 5-μm thick parylene require 60 to 100V and those with 12-μm thick parylene require 90 to 150V.

Alternative methods   Several other dielectrics have been used successfully for electrowetting including silicon dioxide [29], silicon nitride[40], and Teflon AF [34]. A more comprehensive list may be found in this recent review [1].

5.4.4   Fabrication of the top plate

An indium-tin oxide coated glass slide served as the conductive top plate. For the single-layer chips with no reservoirs the ITO slide was used as is without any modification. For the two-layer chips with reservoirs, holes were drilled on the ITO slide and acted as the fluidic input ports. The holes were aligned to the loading channels defined in the spacer layer. The diameter of the holes was chosen to be 1 mm in order to fit a small volume pipette tip.

5.4.5   Hydrophobic coating

The parylene C surface was made hydrophobic by spin-coating or dip-coating a layer of Teflon AF. The hydrophobic layer was ~100-nm thick for clinical chemistry, and bead wash optimization experiments. The hydrophobic layer was ~1000 nm for insulin and IL-6 immunoassay experiments. A 50-nm thick coating on parylene was sufficient for a low contact angle hysteresis surface (<10°) and the contact angle hysteresis is even lower if oil is used as the filler fluid. However, thicker coatings slowed down the degradation of the dielectric layer.

A 1% Teflon AF solution was used to produce the 100-nm coating and a 6% Teflon AF solution was used for the 1000-nm coating. The coating was cured using a hot plate or a convection oven at 170°C for 20 minutes to remove all the solvent. The ITO top plate was also made hydrophobic using the same process with a coating thickness of ~100 nm. Teflon AF has poor adhesion to most substrates and a coated device must be handled carefully to avoid damage to the coating.

Alternative methods   Cytop is another amorphous fluoropolymer that has been used as a hydrophobic coating for electrowetting [35]. Cytop has a slightly higher surface energy than Teflon AF. However, it is a much more robust coating material than Teflon AF with significantly better adhesion to all materials. Special grades of Cytop are also available depending on the substrate that needs to be coated.

5.5   Instrumentation and system assembly

5.5.1   Detection setup

Detection in the electrowetting system is most conveniently performed using optical techniques such as absorbance, fluorescence or chemiluminescence. This is due to the transparent nature of the materials involved in an electrowetting chip, including the electrodes which can be made out of transparent indium tin oxide. Optical detection is performed in a plane perpendicular to that of the device.

Colorimetry for clinical chemistry

The absorbance measurement setup consisted of a green LED (545 nm, RadioShack, USA) and a light-to-voltage converter (TSL257, Texas Advanced Optoelectronic Solutions, Plano, TX). The output voltage of the photodiode $V(t)$ is directly proportional to the light intensity $I(t)$ incident on it. The voltage output is logged on a computer using a data acquisition card (PCI-DAS08, Measurement Computing, Middleboro, MA) with a

12-bit A/D converter with a measuring range of −5V to +5V. All the data collected was analyzed in Microsoft Excel.

The absorbance $A(t)$ is related to the intensity by the following equation.

$$A(t) = ln\left(\frac{I(0)}{I(t)}\right) = ln\left(\frac{V(0) - V_{dark}}{V(t) - V_{dark}}\right)$$

$I(0)$ and $V(0)$ correspond to zero absorbance (or 100% transmittance), and V_{dark} is the dark voltage of the photodiode.

The absorbance equation assumes that all the light that is incident on the photodiode passes through the liquid. Since the diameter of the LED (5 mm) is much larger than the droplet (500 μm) the aperture of the light emitted by the LED was reduced by masking it using an opaque aluminum block with a 400-μm hole in the middle.

Chemiluminescence for immunoassays

Chemiluminescence was used as the detection mechanism for immunoassays due to its extremely high sensitivity. Chemiluminescence measurements were obtained in a plane perpendicular to the digital microfluidic chip using a photo multiplier tube (PMT) obtained from Hamamatsu (H9858, Bridgewater, NJ). The output of the PMT is a current signal proportional the light incident on its photocathode. The current is converted to a voltage signal using a transimpedance amplifier and digitized using an analog-to-digital converter. The digital signal was logged on a computer through an external microcontroller. Data was collected using custom software and analyzed using Microsoft Excel.

The PMT has an 8-mm diameter window for light collection, which is much larger than the 1.5-mm diameter single droplet. The PMT was placed at a distance of 5 mm from the top surface of the droplet to maximize the light collected from the droplet.

5.5.2 System assembly

The chip-top plate assembly was used with single layer devices. The spacer was constructed using polyester shims. The top plate was held down using microscope stage clips. Electrical connection to the top plate (and the droplet) is made by directly connecting the top plate to the reference potential using a flat crocodile clamp.

Two-layer devices were assembled such that the holes on top plate were aligned with the loading ports on the chip. The chip and top plate were clamped and held down by clips on the sides. Electrical connection was made between the ground pad and the top plate using conductive silver paint.

Electrical connections to the contacts on the chip were made using a 44-pin SOIC test clip (Model 6109, Pomona Electronics, Everett, WA). Voltages were applied to the test clip using an electronic controller that was essentially an array of high-voltage switches. The state of the switches (ON or OFF) was controlled by custom software through the parallel port or USB port of a computer. Magnets were placed underneath the chip at appropriate locations. A CCD camera with a zoom lens attachment was used to capture videos of droplet operations for qualitative analysis.

For experiments requiring chemiluminescent detection, the entire setup (chips, electrical control, PMT, and video camera) was enclosed in a dark box constructed out of opaque black foam board and opaque aluminum tape. The darkness of the box was checked using the PMT such that the background signal obtained was similar to the signal obtained with opaque aluminum tape directly covering the PMT photocathode.

5.6   Methods

5.6.1   Automated glucose assays on-chip

Two layer electrowetting devices were used to perform fully automated and integrated glucose assays. 1-cSt silicone oil was used as the filler fluid. The voltage used for droplet actuation was 50V and the droplet switching frequency was 1 Hz.

Three different concentrations of glucose (40, 80, and 120 mg/dL) were assayed in a serial fashion on the chip shown in Figure 5.5. The samples were loaded in three reservoirs and the reagent mixture was loaded in the fourth reservoir on one half of the chip. Droplets of sample and reagent were formed in succession from the on-chip reservoirs. The droplets were merged and transported to the detector location. There was no separate mixing step since just moving the droplet on the bus resulted in the droplets being mixed completely. The absorbance was monitored for 60 seconds using the LED-photodiode setup described previously and the droplet was discarded to a waste area on-chip after detection. Three assays were performed on each sample for a total of nine assays. A 120-mg/dL glucose sample was analyzed nine times serially to further evaluate the within-run variation. Droplets of the sample and the reagent were dispensed, mixed and the absorbance change monitored for 30 seconds.

5.6.2   Magnetic bead manipulation on-chip

Bead washing, retention, and resuspension experiments were conducted on chips fabricated using the single layer process. 1.5-cSt silicone oil was used as the filler fluid. The voltage used for droplet actuation was 100V and droplet switching frequency was 1 Hz unless otherwise specified.

Magnetic bead attraction

The effect of the parameters listed in Table 5.1 on the attraction of the 10-mg/mL Dynal magnetic beads was observed qualitatively using a microscope with images collected at regular intervals.

A 1-μL droplet containing magnetic beads was manually dispensed and sandwiched between two Teflon AF coated glass plates separated by a gap of ~200 μm (set using a polyester shim) in order to simulate the conditions on an electrowetting chip. Silicone oil (1.5 cSt) was used as the filler fluid around the droplet.

Magnetic bead wash efficiency

Dynal MyOne streptavidin coated magnetic beads (2.5 mg/mL) were resuspended in 10 mg/mL of unmodified HRP in TRIS wash buffer. A 1-μL unit droplet of this bead suspen-

Table 5.1 Parameters That Were Evaluated for Effect on Magnetic Bead Attraction

Parameter	Values
Buffer	Phosphase buffered saline (PBS)
	TRIS buffered saline (TBS)
	PBS/TBS with 0.005% Tween 20
	PBS/TBS with 0.01% Tween 20
Magnet pull force	1.25 lb
(field strength 0.5 Tesla)	5 lb
	10 lb
Position of magnet	Varied with respect to droplet position

sion was immobilized using a neodymium permanent magnet (ND-42, 1 Tesla field strength, 1.25-lb pull force) and washed 5 times with 5-μL droplets (5× droplet) of wash buffer per cycle. The 5× supernatant droplet (5 μL) from each cycle was collected and mixed with 50-μL Amplex® Ultra red substrate. The amount of HRP present in the supernatant was measured by reading the absorbance after 8 minutes at a wavelength of 570 nm using a BioTEK Synergy® plate reader. The same experiment was performed in microcentrifuge tubes on the bench for comparison to on-chip data.

Magnetic bead retention

Dynal streptavidin coated magnetic beads (1.05-μm diameter) were labeled with biotinylated HRP as described previously and resuspended in 50 μL of TBS with 0.01% Tween 20. A 1-μL unit droplet of biotinylated-HRP labeled magnetic beads was washed 5 times with 5-μL droplets of wash buffer. Each wash results in a 6x dilution and after 5 wash cycles the bead droplets is expected to be diluted 6^5 = 7776-fold. The supernatant (5 μL) after each wash was collected and pooled together and mixed with 50 μL of chemiluminescence substrate. The chemiluminescence signal was measured on a plate reader (BioTEK Synergy) to quantify the bead loss that may have occurred during the entire wash protocol. Measurements were taken for a total of 8 minutes and a reading was taken every 20 seconds. To quantify the loss of beads, a standard curve of HRP-labeled magnetic beads was generated for bead concentrations ranging from 2 mg/mL to 2×10^9 mg/mL. TRIS wash buffer was used as the diluent and as the negative control. The standards were assayed in same manner as the supernatant from the chip.

Magnetic bead resuspension

A 1-μL droplet of Dynal streptavidin coated magnetic beads labeled with biotinylated HRP was pipetted onto the chip as described previously. The droplet of HRP magnetic beads was shuttled back and forth across a set of six electrodes for 30 seconds at a switching rate of 4 Hz and an actuation voltage of 100V. The bead-containing droplet was shuttled both in the presence and absence of an external permanent magnet. The resuspension of magnetic beads within the droplet and the cycle number at which it occurred was determined by visual inspection.

5.6.3   Droplet-based immunoassay

Droplet-based immunoassays were performed on chips fabricated using the single-layer process. Silicone oil (1.5 cSt) was used as the filler fluid. The voltage used for droplet actuation was 100V. The droplet switching frequency was 1 Hz.

Equal volumes of magnetic beads coated with the primary antibody, secondary antibody labeled with alkaline phosphatase, and the blocking proteins were mixed on the bench and a 3-μL droplet of the reagent mixture was pipetted onto the chip. A 1-μL droplet of the sample (insulin or IL-6 standard) was pipetted onto another electrode on the chip. The reagent droplet and the sample droplet were merged to result in a 4-μL (4×) droplet that was incubated by shuttling on six electrodes for a total of 2 minutes. The shuttling of the droplet ensured dispersion of the magnetic beads and therefore mixing even in the presence of a permanent magnet placed underneath the chip. After incubation, the magnetic beads were immobilized using the permanent magnet and a 3× supernatant droplet was split off and removed, leaving behind a 1× bead droplet. The entire remaining unbound secondary antibody was removed through five cycles of the wash protocol described previously. After washing, a 1-μL droplet of Lumigen APS-5 was pipetted onto the digital microfluidic chip through the hole in the top plate and it was mixed with the droplet containing the washed magnetic beads. The light output was collected for 4 minutes using the PMT placed over the chemiluminescent droplet and the area under the curve was calculated. Immunoassays were performed in duplicate for insulin and IL-6 standards (S0 through S4) using a new chip each time. Standard curves were generated using SigmaPlot without using weighting parameters in the fit.

Identical experiments were performed on the bench in tubes keeping the sample and reagent volumes the same. Both incubation and washing were performed in tubes and the washed magnetic beads with the I^{0}Ab-antigen-II^{0}Ab-enzyme sandwich were pipetted onto the chip. The detection reaction was performed on chip as described previously. The bench and chip values were then compared.

5.7   Results and discussion

5.7.1   Testing of two layer electrowetting device

Chip loading

Silicone oil was injected first before the samples are loaded in order to avoid trapping of air in the reservoirs. Some practice was required to consistently load multiple reservoirs without any air bubbles. 1-cSt silicone oil evaporated through the loading holes over an extended period of time since they were not sealed after loading. This problem was alleviated by using silicone oil with higher boiling point such as 1.5-cSt silicone oil, which was used in the immunoassay experiments.

Droplet dispensing

Droplet dispensing was reliable for serum, plasma, and other reagents using the standard operating voltage of 50V DC. Three different serum samples with different total protein contents were tested and the dispensing reliability was identical for all of them. The maximum number of droplets that could be generated from a reservoir was 10. The volume variation was much higher for the last few droplets generated from a reservoir (i.e., when the reservoir is close to being empty). The volume variation was also larger when the droplet was formed farther away from the reservoir.

Alternative methods   Other methods have also been used for dispensing droplets. Dispensing of 700-nL droplets of 0.1M potassium chloride (KCl) solution from a larger initial droplet has been demonstrated [21]. The droplets were formed by extending a finger of liquid away from the reservoir till the droplet breaks off. Droplets were also dispensed from an external source using a pipette to inject liquid onto an energized electrode and retract the liquid to form a droplet on the energized electrode. Dispensing of microliter droplets using external pressure from an off-chip reservoir by capacitance metering has also been demonstrated [36]. The volume variation measured by capturing images of the dispensed droplets was measured to be less than 2%.

Droplet pathways

Droplets were transportable in a loop around the 4-phased transporter and the 3-phased transporter at a maximum frequency of 50 Hz. This corresponds to a speed of 2.5 cm/sec. To ensure reliable operation, the maximum clock frequency that was used in experiments was 10 Hz. Droplet transport was reliable and repeatable for a variety of liquids including physiological samples such as whole blood, plasma, serum, and reagents for the glucose assay and immunoassay.

Droplet mixing

The mixing times were evaluated by analyzing the mixing process between a fluorescein droplet and a water droplet. The two droplets were merged and transported in one direction (not shuttled) along a transporter at 10 Hz. Images of the merged droplet were captured and the time required for complete mixing was estimated to be ~0.8 seconds. A variety of other mixing strategies may be used incorporating shuttling, splitting, and merging operations.

5.7.2   Automated glucose assay on-chip

The absorbance as a function of time for generating three 3-point standard curves (total of nine serial assays) is shown in Figure 5.6. The rate of change of absorbance for each assay was estimated by linear curve fitting in Excel and the calibration curve as a function of concentration is shown in Figure 5.7.

The percentage coefficient of variation for the three different concentrations is 1.8% for 40 and 80 mg/dL and 0.72% for 120 mg/dL. The excellent reproducibility of the results indicates minimal variability in the volume of on-chip dispensed droplets and negligible cross-contamination.

A 120-mg/dL glucose sample was analyzed nine times serially to further evaluate the within-run variation. Droplets of the sample and the reagent were dispensed, mixed, and the absorbance change monitored for 30 seconds. The coefficient of variation for the entire experiment is 5.7%. The CV drops to 2.2% if only the first six data points are considered. The reason for larger error in the last three runs is due to wider variation in droplet volumes as the reservoir becomes empty.

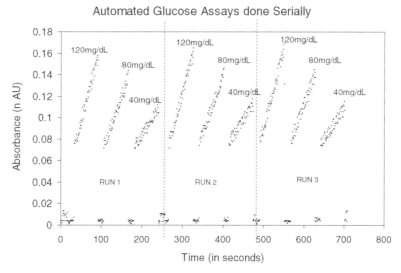

Figure 5.6 Absorbance as a function of time for **three** serial 3-point standard curves for glucose.

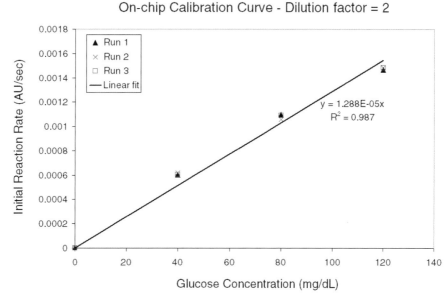

Figure 5.7 Calibration curve for serial glucose assays performed in triplicate.

5.7.3 Optimization of magnetic bead washing

Magnetic bead attraction

The composition of the buffer has a significant impact on attraction and resuspension of magnetic beads using a permanent magnet. Using PBS and TBS alone without any additional surfactants resulted in sufficient attraction but the beads aggregated irreversibly and complete resuspension was not possible. The addition of 0.01% Tween 20 in both these buffers resulted in no aggregation and the beads were readily resuspended upon removal of the magnet. 0.1mg/mL BSA was added to the buffer to further improve bead attraction and act as an additional blocking agent for immunoassays. The time taken for attraction of the magnetic beads was same (20 seconds) for magnets with different pull forces. Magnets with a lower pull force (1.25 lb) prevented irreversible aggregation of the beads while beads irreversibly aggregate at pull forces of 5 and 10 lb. A single magnet with field strength of 1 Tesla and a pull force of 1.25 lb was placed underneath the droplet for ease of implementation.

Wash efficiency   Figure 5.8 shows the end point absorbance (at 8 minutes) of the supernatant after each wash cycle obtained from experiments performed both on bench and on chip. The supernatant contains a decreasing amount of unbound horseradish peroxidase labeled secondary antibody with each wash cycle, therefore the absorbance is expected to reduce with each wash cycle. A negative control (with no labeled secondary antibody in the wash buffer) was mixed with the substrate and measured on the same plate reader. Although there is a significant difference in the washing protocol between the chip and bench, the absorbance value reached that of negative control in 6 wash cycles in both the cases.

Magnetic bead retention

The presence of HRP in the supernatant droplet would be an indication of bead loss during the wash protocol—the greater the signal, the more substantial the loss of beads. The chemiluminescent signal obtained from the supernatant after 7,776-fold serial dilution-based washing was 1,350 milli Lum/min on the chip, whereas the signal obtained for the negative control (TRIS wash buffer) was 900 milli Lum/min. The signal for the supernatant after washing is only marginally higher than the negative control. Considering the large variation in the starting quantity of beads ($7\text{-}12\times10^{9}$ beads/mg) the bead retention can be considered 100% for practical purposes.

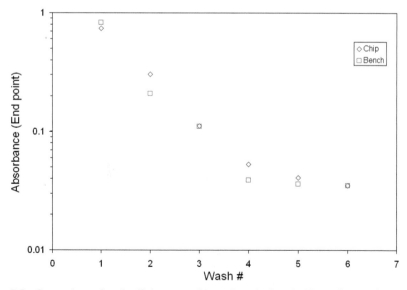

Figure 5.8 Comparison of wash efficiency on-chip and on the bench. The end point absorbance of the supernatant at the end of each cycle is plotted against the cycle number.

Magnetic bead resuspension

The parameters that affect bead attraction also affect bead resuspension. The addition of 0.01% Tween 20 was therefore critical for resuspension of the magnetic beads. 9 shows the process of resuspending beads in a droplet. Frame 1 of the Figure 5.9 shows a droplet above a permanent magnet with the beads aggregated. The droplet was then shuttled across 6 electrodes (right to left to right within the frame). Resuspension of magnetic beads within the droplet was obtained after 1 cycle even in the presence of an external permanent magnet as shown frame 2.

Immunoassays on human insulin and interleukin-6

0 shows the 4-parameter logistic fits to standard curves for IL-6 and insulin, respectively, obtained on-chip. Each data point is the average of two measurements performed on different chips. The chemiluminescence readings obtained from the bench were compared to that obtained from the chip by analyzing the linear correlation (forced through origin) between the bench values and the chip readings. For IL-6, the slope was 1.0572 and R^2 was 0.9912. For insulin, the slope was 0.9781 and R^2 was 0.9812. The near-unity values for slope and R^2 indicate excellent correlation between the bench assay and the on-chip assay.

5.8   Method challenges

Electrowetting is a powerful method to manipulate liquids in a microfluidic lab-on-a-chip. The method is useful for a wide range of life science applications including clinical chemistry and immunoassays. However, the method is not without its challenges, and some of them are discussed below.

Dilution factor   The main challenge with the electrowetting platform is the translation of bench protocols to chip protocols. Though this process is straightforward in many cases it is challenging in certain scenarios. For example, most bench protocols require reagents and substrates to be in vast excess to the sample. This is not easily imple-

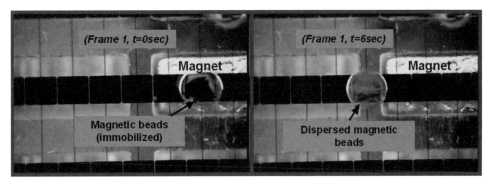

Figure 5.9   Resuspension of magnetic beads in a droplet. Frame 1 shows the droplet at the beginning of the first oscillation cycle. Frame 2 shows the droplet towards the end of the first cycle. The beads are dispersed after just 1 cycle.

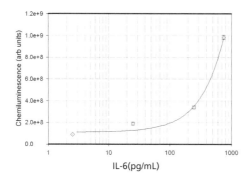
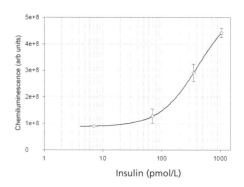

Figure 5.10 Left: IL-6 calibration curve obtained on-chip. Each data point is the average of two measurements. Right: Insulin calibration curve obtained on-chip. Each data point is the average of two measurements.

mented on an electrowetting-based lab-on-a-chip and concentrations of reagents need to be adjusted in such cases.

Oil as filler fluid   The use of oil as a filler fluid excludes certain lipophilic molecules from being compatible the electrowetting platform. In such cases the molecules need to be solubilized in the aqueous phase using surfactants. Droplets may be manipulated using air as the filler fluid; however the droplet operations are not very reliable and repeatable.

Low interfacial tension liquids   Electrowetting is based on reducing the effective interfacial tension between the liquid phase and a solid electrode. Liquid that already have a very low interfacial tension can be a challenge to manipulate on the platform since the contact angle change achievable is small.

System integration   The lack of good sample preparation methods is currently the biggest impediment to the commercial acceptability of microfluidic technologies and considerable research effort is required in this area to integrate this critical operation on-chip.

Troubleshooting Table

Problem	Possible Cause	Potential Solutions
Electrolysis of droplets after manipulation for certain cycles.	Degradation of insulator over time.	Use thicker dielectric and a thicker hydrophobic coating.
Droplet moves sluggishly.	Poor hydrophobic coating.	Increase thickness of hydrophobic coating and thermally cure for longer time.
Droplet is stuck to electrodes after manipulation for some cycles.	Dielectric charging.	Use low frequency AC voltages for droplet manipulation.
Air bubble is injected while loading liquid.	No oil in loading port.	Add a drop of oil to loading port, expunge air from the end of the pipette tip and then form tight fit with loading port.
Droplet splits during transport.	Droplet aspect ratio is incorrect.	Increase the gap between chip and top plate.

5.9 Summary points

1. Electrowetting is a powerful method to manipulate droplets to implement biological assays.
2. A two layer fabrication process is required to manufacture devices to perform fairly complex fluidic protocols.
3. Silicone oil or other low viscosity oil is necessary to reduce biofouling and improve reliability of droplet manipulation.
4. Using magnetic beads as the solid phase heterogeneous immunoassays can be performed without device modification.
5. Beads should be suspended in a buffer with surfactant to enable good immobilization and resuspension.

Acknowledgments

The authors would like to acknowledge Dr. Richard Fair, Dr. Hugh Crenshaw, Dr. Srinivas Palanki, and Dr. Philip Paik. The authors would also like to acknowledge contributions from Dr. Allen Eckhardt, Ryan Sturmer, Greg Smith, and Dr. Zhishan Hua of Advanced Liquid Logic, Inc. Portions of the work reported in this chapter were partially supported by grants (R43 DK066956 & R43 CA114993) from the National Institutes of Health, Bethesda, MD.

References

[1] Mugele, F., and J-C. Baret, "Electrowetting: From Basics to Applications," *J. Phys. Condens. Matter,* Vol. 17, 2005, pp. R705–R774.
   This paper presents a comprehensive review of electrowetting literature up until 2005. The paper discusses both theoretical and practical aspects of electrowetting.

[2] Srinivasan, V., V. K. Pamula, and R. B. Fair, "An Integrated Digital Microfluidic Lab-on-a-Chip for Clinical Diagnostics on Human Physiological Fluids," *Lab Chip,* Vol. 4, 2004, pp. 310–315.

[3] Fair, R. B. "Digital Microfluidics: Is a True Lab-on-a-Chip Possible?" *Microfluid Nanofluid,* Vol. 3, 2007, pp. 245–281.

[4] Berge, B., and J. Peseux, "Variable Focal Lens Controlled By an External Voltage: An Application of Electrowetting," *Eur. Phys. J.,* Vol. E 3, 2000, p. 159.

[5] Hayes, R. A., and B. J. Feenstra, "Video-Speed Electronic Paper Based on Electrowetting," *Nature,* Vol. 425, 2003, p. 383.

[6] Srinivasan,V., "A Digital Microfluidic Lab-on-a-Chip for Clinical Diagnostic Applications," Ph.D. thesis, Duke University, 2005.

[7] Sista, R., "Development of a Digital Microfluidic Lab-on-a-Chip for Automated Immunoassay with Magnetically Responsive Beads," Ph.D. thesis, Florida State University, 2007.

[8] Sista, R. S., A. E. Eckhardt, and V. Srinivasan, et al., "Heterogeneous Immunoassays Using Magnetic Beads on a Digital Microfluidic Platform," *Lab Chip,* 2008.

[9] Paik, P. Y., D. J. Allen, and A.E. Eckhardt, et al., "Programmable Flow-Through Real Time Pcr Using Digital Microfluidics," *Proc. μTAS,* 2007.

[10] Srinivasan, V., V. K. Pamula, and P. Paik, et al., "Protein Stamping for MALDI Mass Spectrometry Using an Electrowetting-Based Microfluidic Platform," In *SPIE Optics East, Lab-on-a-Chip: Platforms, Devices and Applications,* 2004.

[11] Pamula, V. K., V. Srinivasan, and H. Chakrapani, et al., "A Droplet Based Lab-on-a-Chip for Colorimetric Detection of Nitroaromatic Explosives," *Proceedings of IEEE MEMS 2005 Conference,* 2005.

[12] Teh, S., R. Lin, and L. Hung,et al., "Droplet Microfluidics," *Lab Chip,* Vol. 8, 2008, pp. 198–220.

[13]   Song, H., J. Tice, and R. Ismagilov, "A Microfluidic System for Controlling Reaction Networks in Time," *Angew. Chem. Int. Ed.*, Vol. 42, 2003, pp. 767–772.

[14]   Gascoyne, P. R. C., J. V. Vykoukal, and J. A. Schwartz, et al., "Dielectrophoresis-Based Programmable Fluidic Processors," *Lab Chip*, Vol. 4, 2004, p. 299.

[15]   Guttenberg, Z., H. Müller, and H. Habermüller, et al., "Planar Chip Device for PCR and Hybridization with Surface Acoustic Wave Pump," *Lab Chip*, Vol. 5, 2005, pp. 308–317.

[16]   Darhuber, A. A., and S. M. Troian, *Annual Review of Fluid Mechanics*, Vol. 37, 2005, pp. 425–455.

[17]   Lehmann, U., S. Hadjidj, and V. K. Parashar, et al., "Two Dimensional Magnetic Manipulation of Microdroplets on a Chip as a Platform for Bioanalytical Applications," *Sensors and Actuators B*, Vol. 117, 2006, pp. 457–463.

[18]   Lippmann, G., "Relations Entre les Phénomènes Electriques et Capillaries," *Ann. Chim. Phy*, Vol. 5, 1875, p. 494.

[19]   Beni, G., and S. Hackwood, "Electro-Wetting Displays," *Appl. Phys. Lett.*, Vol. 38, No. 4, 1981, pp.207–209.

[20]   Vallet, M., B. Berge, and L. Vovelle, "Electrowetting of Water and Aqueous Solutions on Poly(Ethylene Terephthalate) Insulating Films," *Polymer*, Vol. 37, No. 12, pp. 19962465–19962470.

[21]   Pollack, M. G., "Electrowetting-Based Microactuation of Droplets for Digital Microfluidics," Ph.D. thesis, Duke University, 2001.
        The concept of digital microfluidics was first introduced in this seminal thesis. The work presented in this thesis serves as the basis for almost all lab-on-a-chip devices based on electrowetting.

[22]   Pollack, M. G., A. D. Shenderov, and R. B Fair, "Electrowetting-Based Actuation of Droplets for Integrated Microfluidics," *Lab Chip*, Vol. 2, No. 1, 2002, pp. 96–101.

[23]   Luk, V.N., G. C. Mo, and A. R. Wheeler, "Pluronic Additives: A Solution to Sticky Problems in Digital Microfluidics," *Langmuir,* Vol. 24, 2008, pp. 6382–6389.

[24]   Park, J. Y., and L. J. Kricka, "Prospects for Nano- and Microtechnologies in Clinical Point-of-Care Testing," *Lab Chip*, Vol. 7, 2007, pp. 547–549.

[25]   Srinivasan, V., V. Pamula, and R. B. Fair, "A Droplet-based Microfluidic Lab-on-a-Chip for Glucose Detection," *Analytica Chimica Acta*, Vol. 507, 2004, pp. 145–150.

[26]   P. Trinder, "Determination of Glucose in Blood Using Glucose Oxidase with and Alter-Native Oxygen Acceptor," *Annals in Clinical Biochemistry*, Vol. 6, 1969, pp. 24–27.

[27]   Ren, H., V. Srinivasan, and R. B. Fair, "Design and Testing of an Interpolating Mixing Architecture for Electrowetting-Based Droplet-on-Chip Chemical Dilution," *Transducers Conference*, 2003.

[28]   Cho, S. K., H. Moon, and C. J. Kim, "Creating, Transporting, Cutting and Merging Liquid Droplets by Electrowetting-Based Actuation for Digital Microfluidic Circuits," *Journal of Microelectromechanical Systems*, Vol. 12, No. 1, 2003, pp. 70–80.

[29]   Paik, P., V. K. Pamula, and M. G. Pollack, et al., "Electrowetting-Based Droplet Mixers for Microfluidic Systems," *Lab Chip*. Vol. 3, No. 1, 2003, pp. 28–33.

[30]   Fouillet, Y., D. Jary, and C. Chabrol, et al., "Digital Microfluidic Design and Optimization of Classic and New Fluidic Functions for Lab on a Chip Systems," *Microfluidics Nanofluidics*, Vol. 4, 2008, pp. 159–165.

[31]   Paik, P.Y., V. K. Pamula, M. G. Pollack, and K. Chakrabarty, "Coplanar Digital Microfluidics Using Standard Printed Circuit Board Processes," *Proceedings of MicroTAS 2005*, 2005, pp 566–568.

[32]   Gong, J., and C. -J. Kim, "Direct-Referencing Two-Dimensional-Array Digital Microfluidics Using Multilayer Printed Circuit Board," *Journal of Microelectromechanical Systems*, Vol. 17, No.2, 2008, pp. 257–264.

[33]   Abdelgawad, M., and A.R. Wheeler, "Low-Cost, Rapid-Prototyping of Digital Microfluidics Devices," *Microfluidics and Nanofluidics*, Vol. 4, 2008, pp. 349–355.

[34]   Seyrat, E., and R. A. Hayes, "Amorphous Fluoropolymers as Insulators for Reversible Low-Voltage Electrowetting," *J. Appl. Phys.*, Vol. 90, 2001, pp. 1383.

[35]   Berry, S., J. Kedzierski,, and B. Abedian, "Low Voltage Electrowetting Using Thin Fluoroploymer Films," *J Colloid Interface Sci.*, Vol. 303, No. 2, 2006, pp. 517–524.

[36]   Ren, H., "Electrowetting-Based Droplet Formation," PhD thesis, Duke University, Durham, NC, 2004.

Dielectrophoresis for Particle and Cell Manipulations

Benjamin G. Hawkins[1], Jason P. Gleghorn[2], and Brian J. Kirby[*]

[1]Department of Biomedical Engineering, [2] Sibley School of Mechanical & Aerospace Engineering, Cornell University
*Corresponding author: 238 Upson Hall, Ithaca, NY 14853,
phone: 607-255-4379, fax: 607-255-1222, e-mail: bk88@cornell.edu

Abstract

In this chapter, we will explore the use of dielectrophoresis (DEP) for particle and cell manipulation. This is a broad field, and so our aim will be to present the theory behind several dielectrophoresis techniques, as well as specific experimental recipes. Initially, we will present the concept and theoretical underpinnings of DEP and build a framework of approximations that connect our theoretical development to engineering design and experimental implementation. First, from a theoretical perspective, we will approach DEP techniques based on the properties of the driving electric field and introduce experimental techniques for manipulating these field properties. Next, from an experimental perspective, we will approach DEP techniques based on device geometries used to generate electric field nonuniformities and report specific methodologies for reproducing experimental results. Finally, from a design perspective, we will present DEP techniques based on the resulting motion of particles in the system and examine a variety of demonstrated applications.

Key terms   dielectrophoresis
insulating dielectrophoresis, iDEP
negative dielectrophoresis, nDEP
positive dielectrophoresis, pDEP

6.1 Introduction: physical origins of DEP

The term "dielectrophoresis" was used by Herbert A. Pohl as early as 1951 in describing the motion of particles in response to a nonuniform electric field [1]. In that work, "dielectrophoresis" describes the force—exclusive of electrophoresis—exerted on polarizable particles, as a function of their complex permittivity, in the presence of an externally applied, nonuniform electric field. In this work, we additionally include the torque experienced by a particle in a rotating electric field. We reason that these phenomena are related in that each is dependent, in some manner, on the relative complex permittivities of a particle and its surrounding media.

Dielectrophoresis originates from the response of matter to an electric field, and more specifically, from differences in this response across an interface. In an electric field, a perfectly conducting material will transport electrons instantaneously along the field, effectively making the electric potential uniform within the material. For a material that is perfectly insulating, electrons are immobile, and the electric current within is zero; thus the electric potential is defined by the charge distribution according to Gauss' law. For most materials subjected to dielectrophoresis, the response to an externally applied electric field is neither that of a perfect conductor nor that of a perfect insulator, but rather behaves as a "leaky dielectric" or "lossy dielectric." For harmonic fields, this property of matter is described using a complex, frequency-dependent permittivity, $\tilde{\varepsilon}$ (6.1). A material subjected to an electric field will respond, or polarize, in a manner that is dependent on its complex permittivity as well as the strength and frequency of the local electric field.

$$\tilde{\varepsilon} = \varepsilon - i\frac{\sigma}{\omega} \qquad (6.1)$$

Here, i is $\sqrt{-1}$, σ is the conductivity, and ω is the radial frequency of the electric field. The complex permittivity $\tilde{\varepsilon}$ is a function of electric field parameters (magnitude and frequency), thermodynamic parameters (temperature and pressure), and material parameters (composition) [2]. The real and imaginary components of $\tilde{\varepsilon}$ correspond to displacement and conduction current, respectively, and relate respectively to the localization of bound charge and the motion of free charge. The "leaky-dielectric" model implies that when matter within the system reorients or charge redistributes in response to changes in the external electric field, there is a finite "lag" time, which is a function of frequency, between the electric field and the response of a material.

Fundamentally, dielectrophoretic forces (or torques) result from a nonuniform polarization along an interface between materials with different dielectric properties. This implies two things: (1) there is an interface between two materials that respond differently to an imposed electric field (described by different values of the complex permittivity, $\tilde{\varepsilon}$), and (2) the imposed electric field varies significantly along this interface[1]. The Maxwell-Wagner interfacial polarization takes the form of electric molecular dipoles within the material. The creation of dipoles manifests at the interface between materials of different complex permittivity, due to different relative dipole strengths, as a "bound charge." Bound charge is distinguished from "free charge" in that it manifests

---

[1] Interestingly, it is also possible to observe a force with a uniform electric field, if the relative complex permittivities vary significantly along the interface.

itself only in response to an external field (i.e., it is induced) and spatial variation in permittivity and is not free to move through conducting media.

The electric field exerts a Coulomb force on these bound charges, whose sum along the interface is nonzero only when the field is nonuniform. Thus DEP allows a force to be applied to particles as long as the electric field is nonuniform and $\tilde{\varepsilon}$ is different for the particle as compared to its surrounding medium. The frequency dependence of $\tilde{\varepsilon}$ gives experimenters an extraordinary amount of flexibility with regards to what biological particles can be manipulated with DEP, while the favorable scaling of DEP forces as length scales diminish motivates DEP's use in microfabricated systems.

6.2   Introduction: theory of dielectrophoresis

Now we wish to put the above description in mathematical terms that are relevant to the experimental researcher and recapitulate the canonical equations for the DEP force, F_{DEP}. In order to do this, we consider a particle submerged in an electrolyte, each described using complex permittivities ($\tilde{\varepsilon}_p$ and $\tilde{\varepsilon}_m$, for particle and media, respectively), under the influence of a harmonic electric field, $\vec{E}$. The electric field frequency is assumed to be low enough such that the permittivities, ε_p and ε_m, can be considered constant. The electric field is created by applying a potential at some point in the device, and, as expected, we describe the electric field in general as the gradient of this potential:

$$\vec{E} = -\nabla_v \tag{6.2}$$

We have made no assumptions yet as to the temporal or spatial properties of the field, except to say that it is harmonic. Owing to the mathematical simplicities stemming from the use of complex algebra in treating these sinusoidal functions, we represent the electric field as:

$$\vec{E}(\vec{r}, \omega, \phi) = \Re\left[\vec{E}_0(\vec{r}) e^{j(\omega t + \phi)} \right] \tag{6.3}$$

where $\vec{r}$ is a position vector, $\vec{E}_0(\vec{r})$ captures the spatial distribution of the electric field (still arbitrary), $\Re[\cdots]$ is the real component of the quantity in brackets, and ω and ϕ are the angular frequency and phase of the field, respectively. In this sort of complex analysis it is common practice to omit the harmonic portion in notation. As long as the system remains linear, and we consider only the steady state, we can continue to use principles of superposition, mesh analysis, and other analytical tools. Later, we can include the harmonic portion when we wish to calculate derivatives or time-averages of these harmonic quantities.

Moving forward, we present the basic theory of dielectrophoresis and discuss its limitations. We begin with a brief derivation of the dielectrophoretic force employing the commonly used dipole approximation for an isolated sphere in an infinite medium. Following this, we expand the discussion to illustrate the limitations of the electrodynamic and fluid-mechanical approximations.

In the most general case, the solution for the electromagnetic force (which includes DEP) on an arbitrary object is found by integrating Maxwell's stress tensor. The Maxwell stress tensor formulation is the most general and powerful, but in problems lacking the appropriate symmetry, solutions can only be obtained numerically or as close approximations. For those geometries with suitable symmetry, there are closed-form solutions and approximations available in literature. Relevant analytical [3–7] and numerical [5–11] references are summarized in Tables 6.1 and 6.2.

Consider a homogeneous, isotropic sphere in a semi-infinite, homogeneous, isotropic electrolyte, with an electric field applied along the z-axis. If the external field varies linearly over the characteristic particle dimensions, the polarization of a dielectric sphere in an electric field can be represented by replacing the particle with an equivalent, *effective* dipole located at the center of the particle. Such a dipole consists of two charges of equal magnitude and opposite sign at a position, $\vec{r}$, separated by a vector distance, $\vec{\delta d}$. If the electric field is nonuniform over the dipole length, δd, then the sum for electrostatic forces on the dipole is,

$$\vec{F} = q\vec{E}\left(\vec{r} - \vec{\delta d}\right) - q\vec{E}(\vec{r}) \tag{6.4}$$

where q is the dipole charge and $\vec{\delta d}$ is the vector from the negative dipole charge to the positive dipole charge, by convention. Expanding the first term in a Taylor series, we obtain:

$$\vec{E}\left(\vec{r} + \vec{\delta d}\right) = \vec{E}(\vec{r}) + \vec{\delta d} \cdot \nabla \vec{E}(\vec{r}) + \ldots \tag{6.5}$$

Table 6.1 A Brief Summary of Literature Sources That Deal with Maxwell Stress Tensor Solutions and Dielectrophoresis

No.	Author. Year	Title	Technique	Application
[8]	Al-Jarro, A., et al. 2007	Direct calculation of Maxwell stress tensor for accurate trajectory prediction during DEP for 2D and 3D structures.	Numeric	2D and 3D particle trajectories, cell models, cell-cell interactions.
[9]	Jones, T.B. and Wang, K.-L. 2004	Frequency-Dependent Electromechanics of Aqueous Liquids: Electrowetting and Dielectrophoresis.	Numeric	Deformation of air-water interface, use of DEP/EWOD for fluid transport.
[5]	Kang, K.H. and Li, D.Q. 2005	Force acting on a dielectric particle in a concentration gradient by ionic concentration polarization under an externally applied DC electric field.	Approximate Analytic, Numeric	Concentration polarization force.
[6]	Liu, H. and Bau, H.H. 2004	The dielectrophoresis of cylindrical and spherical particles submerged in shells and in semi-infinite media.	Analytic	DEP force on a spherical and cylindrical particle.
[10]	Liu, Y., et al. 2007	Immersed electrokinetic finite element method.	Numeric	Nonspherical particles, particle deformation, numeric technique DEP force in single-particle traps, nonspherical particles.
[7]	Rosales, C. and Lim, K.M. 2005	Numerical comparison between Maxwell stress method and equivalent multipole approach for calculation of the dielectrophoretic force in single-cell traps.	Numeric	
[11]	Singh, P. and Aubry, N. 2005	Trapping force on a finite-sized particle in a dielectrophoretic cage.	Numeric	DEP force on various size particles in a single particle trap.

Table 6.2 Table 6.1 continued.

No.	Comparison/ Validation	Relevant Conclusions
[8]	effective dipole	MST formulation is necessary when particle size is on the order of electrode size, and particle-particle interactions play a significant role on this length scale.
[9]	n/a	MST solution matches experimental behavior for air/water interface. Electrowetting on dielectric is dominant in this case.
[5]	numeric solution	MST calculation is used to determine electrical force on a sphere and combined with the hydrodynamic stress tensor to yield net force due to concentration polarization.
[6]	effective dipole, numeric solution	Dipole approximation fails when particle size is the same order of magnitude as the characteristic length scale of the electric field. The dipole approximation is worse for a cylinder.
[10]	effective dipole	Dipole approximation becomes increasingly inaccurate as a particle approaches an electrode. MST is used to effectively calculate DEP force and deformation of spheres, CNTs, bacteria, and viruses.
[7]	multipole	The results show that a small number of multipolar terms need to be considered in order to obtain accurate results for spheres. The full MST calculation is only required in the study of nonspherical particles.
[11]	effective dipole	Point dipole model overestimates DEP force by 40% for particles with 10% variation in permittivity and size 25% of electrode size (when particles are "close" to the electrode). As particles approach electrodes or approach electrode size, even the quadrupole term is insufficient.

If the characteristic length scale of the electric field non-uniformity is large compared to the particle size, then we can neglect terms of higher order in (6.5) and plug (6.5) into (6.4) to find:

$$\vec{F} = q\vec{\delta d} \cdot \nabla \vec{E}$$
$$= \vec{p}_{eff} \cdot \nabla \vec{E} \tag{6.6}$$

which gives us the force on some unspecified effective dipole moment, but does not intimate how the properties of particle and media contribute to this effective dipole moment. There are several approaches to determine these details; the most common is to examine a homogeneous dielectric sphere in a dielectric media, and solve Laplace's equation for the electric potential inside and outside the sphere, applying the appropriate boundary conditions at the interface. Solution by the separation of variables technique yields (6.7) and provides analytical results that are physically intuitive.

$$\vec{p}_{eff} = -4\pi\varepsilon_m a^3 E_0 \Re\left(\frac{\tilde{\varepsilon}_p - \tilde{\varepsilon}_m}{\tilde{\varepsilon}_p + 2\tilde{\varepsilon}_m}\right) \tag{6.7}$$

$$\langle F_{DEP} \rangle = \pi\varepsilon_m a^3 \Re[f_{CM}]\nabla\left(\vec{E}_0 \cdot \vec{E}_0\right) \tag{6.8}$$

$$f_{CM} = \frac{\tilde{\varepsilon}_p - \tilde{\varepsilon}_m}{\tilde{\varepsilon}_p + 2\tilde{\varepsilon}_m} \tag{6.9}$$

where the angle brackets denote time averaging, $\vec{p}_{eff}$ is the effective dipole moment for a sphere, a is the particle radius, ε_m is the permittivity of the media, $\tilde{\varepsilon}_p$ and $\tilde{\varepsilon}_m$ are the complex permittivities of the particle and media, respectively, and $\Re[f_{CM}]$ is the real part of the Clausius-Mossotti factor. This analytical expression for dielectrophoretic effects is intuitive, tractable, and is applicable under numerous experimental conditions (limitations on the applicability of this form of F_{DEP} will be discussed in Section 6.2.1). If we expand this analysis to a general electric field with spatially varying phase, we find an additional term that leads to electrorotation effects:

$$\left\langle \vec{F}_{DEP} \right\rangle = \pi a^3 \varepsilon_m \Re[f_{CM}] \nabla \left(\vec{E} \cdot \vec{E} \right) - 2\pi a^3 \varepsilon_m \Im[f_{CM}] \nabla \times \Re\left[\vec{E} \right] \times \Im\left[\vec{E} \right] \qquad (6.10)$$

The first term in this equation is the well-known result for the time-averaged dielectrophoretic force, and appears in the presence of any nonuniform electric field (of the form prescribed). The second term arises only in the presence of a spatially nonuniform electric field phase, $\phi \neq 0$, and is the driving term for traveling-wave dielectrophoresis (twDEP). In both terms, forces scale with particle radius cubed, making particle size an important factor in most DEP experiments.

6.2.1 Limiting assumptions and typical experimental conditions

Though seldom explicitly defined in the literature, there are clear boundaries for the applicability of the dielectrophoretic force equations presented above. Particle shape, characteristic particle dimensions and device length scale, particle concentration, and characteristic length scale of changes in the electric field must be considered[2].

We will examine each of these parameters as variations from our "baseline case" (6.10): a sparse concentration of homogeneous, spherically symmetric particles in an infinite domain, under the influence of an electric field that is well approximated by a first order linearization. We will consider, as they affect dielectrophoretic forces and torques:

1. The limits of the spherical approximation, as particle shape deviates from sphericity and the dipole approximation fails.
2. The limits of the dipole approximation as the length scale of variations in the electric field decreases—due to shrinking channel dimensions, shrinking interparticle spacing, or increasing particle size—and our linearization of the electric field fails.
3. The limits of the effective permittivity model, and the Maxwellian equivalent body, as particle composition becomes increasingly complex.

We will also consider, as they affect drag forces and terminal particle velocity:

1. The limits of the spherical approximation, as particle shape deviates from sphericity and the drag coefficient takes on a different form.
2. The limits of the isolated particle approximation, as particle concentration increases.
3. The limits of the infinite domain assumption, as channel dimensions decrease.

---

[2] Throughout this chapter, however, we will continue to make the assumption that the double-layer thickness (Debye length, λ_D) is small in comparison to channel and particle dimensions.

6.2.1.1   Assumptions and approximations for F_{DEP}

Spherical approximation   While many of the particles in DEP experiments are not spherical, the spherical approximation, and the simple multishell models associated with it, is often used. Spherical symmetry leads to analytical solutions that are intuitive and easy to apply. There are, however, situations where these approximations are inappropriate (e.g., cellular samples of rod-shaped bacteria). We will present analytical solutions for the dipole moment of an elliptical particle and compare this result to the spherical moment presented earlier (6.7). In all of these cases we will assume that the major axis of our ellipsoidal particle will be aligned with the applied electric field and that the field varies only along this direction; a reasonable assumption for isotropic particles in non-rotating fields[3].

The shape of a polarized particle will influence the electric field it creates. As particle shape deviates from spherical, the applicability of the equivalent dipole representation based on a spherical particle rapidly decreases. Consider a prolate ellipsoidal particle with a long axis, a_1, and equivalent minor axes, $a_2 = a_3$. In general, the effective dipole moment will have three components, one along each axis due to the variation of the electric field along that axis. We have assumed that the particle aligns instantaneously with the field, and that the field varies only along the major axis. In this case, the effective dipole moment will be:

$$\vec{p}_{effective,1} = 4\pi a_1 a_2 a_3 \varepsilon_m \tilde{f}_{CM,1} \tilde{E}_{0,1}$$

$$\tilde{f}_{CM,1} = \frac{\tilde{\varepsilon}_p - \tilde{\varepsilon}_m}{3\left[\tilde{\varepsilon}_m + \left(\tilde{\varepsilon}_p - \tilde{\varepsilon}_m\right)L_1\right]}$$

$$L_1 = \frac{a_2^2}{2a_1^2 e^3}\left[ln\left(\frac{1+e}{1-e}\right) - 2e\right]$$

$$e = \sqrt{1 - \frac{a_2^2}{a_1^2}}$$

(6.11)

where the subscript 1 denotes a component along the 1-axis[4]. $\vec{p}_{effective,1}$ is the effective dipole moment in the 1-direction. $f_{CM,1}$ is the Clausius-Mossotti factor from our previous representations, but now depends on the axis under consideration. L_1 is the "depolarization factor" along the 1-axis [12].

Under typical experimental conditions ($E_0 = 1V/\mu m$, $2\pi\omega = 1$ MHz, $f_{CM} = 0$ when $2\pi\omega = 0.1$ MHz, $a_2 = a_3 = 1$ μm), for a prolate ellipsoid with complex permittivity shown in Table 6.3, a 30% variation in major axis length leads to 10% error in the effective dipole moment. This error quickly rises to 50% as the aspect ratio of the ellipse approaches 2:1. In the limiting case of a long, thin ellipsoid ($e \rightarrow 1$), the relative error between the real f_{CM} and the spherical approximation thereof can be shown to be equal to the spherical approximation of f_{CM}, so the relative errors of the spherical approximation approach zero near crossover points, and can range from 1 to –0.5.

---

[3] A particle consisting of an isotropic material (or an effectively isotropic material as will be discussed later in Section 6.2.1.1) will obtain three effective moments along its major and semimajor axes and experience a torque on each moment. For isotropic materials, the major axis has the largest moment, leading to a torque that will align the major axis with the external field.

[4] For Cartesian coordinates, {1, 2, 3} corresponds to {x,y, z}.

Table 6.3 Typical Experimental Conditions Used To Calculate Errors Associated with the Spherical Approximation

Material Property	Value
ε_m	$80\varepsilon_0$
ε_p	$2.6\varepsilon_0$
σ_m	$5.5\,\mu S/m$
σ_p	$100\,\mu S/m$

While the errors associated with the spherical approximation can be large, it remains a valuable tool for a first-pass at device design, especially in applications of sorting, screening, or trapping where subtle variations in the magnitude of polarization response are unimportant. In particular, the addition of multipoles and ellipsoidal adjustments does not alter the behavior of the *sign* of f_{CM}.

When performing ROT or DEP spectra experiments in an attempt to draw conclusions about particle structure or composition, a practice common in biological assays, spherical models are generally insufficient—ellipsoidal, cylindrical, or spheroidal models are available and allow for more accurate structural inferences.

Dipole approximation   When considering the electric field produced by a polarized particle of spherical shape, the characteristic length scale of the system is important. Since higher-order multipole effects drop off more rapidly than dipole fields, the field produced by a nonspherical particle becomes equal to that produced by a spherical particle as the distance from the particle center increases. Close to the particle, however, the effects of higher-order multipoles are apparent. As the characteristic length scale of the device shrinks, owing to reductions in channel dimensions or increases in particle concentration, the contributions of higher-order multipoles becomes significant. Additional multipoles are also important when considering rapidly varying electric fields and nonspherical particles. Several different geometries have been considered in detail in the literature [6, 7, 13–19]. A selection of these references is summarized in Tables 6.4 and 6.5.

The characteristic length scale of the particle and electric field under consideration must be examined to evaluate the applicability of classical DEP representations such as that shown in (6.7). When the electric field varies significantly over the length scale of the particle, then, in order to model the induced moment, we must replace the particle with an equivalent multipole. Green and Jones [17] compare the results from up to 9 moments to analytical solutions for spheres and ellipsoids, and show that higher-order moments are necessary to accurately obtain equivalent potential solutions for nonspherical particles.

While multipolar solutions are necessary when attempting to determine the effective moment or potential field surrounding a particle, the spherical approximation retains significant utility in estimating particle behavior in many experimental settings. The application of these approximations should be undertaken with care in light of the ultimate goal of the experiment. When using DEP for transport, trapping, or screening, the spherical approximation often leads to uniform errors (which are obfuscated by experimental uncertainties) and the qualitative correctness of the spherical approximation often suffices. In screening or electrorotation studies that depend critically on the

Table 6.4   A Brief Summary of Literature Sources That Deal with Multipole Moment Solutions and Particles of Various Geometries in Dielectrophoresis

No.	Author. Year	Title	Technique
[6]	Liu, R.M. and J.P. Huang. 2004	Theory of the dielectrophoretic behavior of clustered colloidal particles in two dimensions	Multiple image method, Maxwell-Gannet approx, 2D
[13]	Castellarnau, M. et al. 2006	Dielectrophoresis as a tool to characterize and differentiate isogenic mutants of *Escherichia coli*	Multishell, prolate spheroid
[14]	Ehe, A.Z. et al. 2005	Bioimpedence spectra of small particles in liquid solutions: Mathematical modeling of erythrocyte rouleaux in human blood	Approximate
[15]	Gimsa, J. 2001	A comprehensive approach to electro-orientation, electrode formation, dielectrophoresis, and electrorotation of ellipsoidal particles and biological cells	Analytic, numeric
[16]	Gimsa, J. et al. 1994	Dielectric-Spectroscopy of Human Erythrocytes—Investigations under the Influence of Nystatin	Analytic, experimental
[17]	Green, N.G. and T.B. Jones. 2007	Numerical determination of the effective moments of nonspherical particles	Analytic, numeric

Table 6.5   Table 6.4 continued

No.	Application	Relevant Conclusions
[6]	The effective f_{CM} of isolated and randomly clustered spherical and cylindrical particles.	For closely spaced particles (clusters) an effective dipole factor can be defined based on the MIM and MGA.
[13]	An ellipsoidal, multishell model of *E. coli* was used to determine the differences between isogenic mutants.	Experimentally distinguished DEP characteristics of isogenic mutants. Presented a confocal, multishell, ellipsoid model for f_{CM}.
[14]	Erythrocyte dipole moments are modeled using a short-cylinder model as an approximation.	Dielectrophoresis of erythrocytes is predicted using the short-cylinder model for field frequencies between 10^7 and 10^8 Hz.
[15]	Traditional formulations for DEP and ER with numeric techniques allow for consideration of the interaction of these phenomena.	Predicts a discontinuous electrorotation spectra due to the influence of dipole moments along the three coordinate axes.
[16]	Ellipsoidal dielectrophoretic and electrorotation spectra more accurately explained experimental data.	The use of ellipsoidal dielectric models of DEP and ER resulted in more accurate interpretation of results. Changes in cell dielectric properties was observed after 5 min of experimentation, a fact that should be considered by other researchers.
[17]	Higher order moments for various particle geometries are calculated. Particle shapes include sphere, oblate-and prolate-ellipsoids, cylinder, and erythrocyte.	Higher order moments can be used to more accurately determine the DEP force, and are particularly important in the case of non-spherical particles. In addition, the approximation of short-cylinders using prolate spheroids in particularly poor in light of the multipolar analysis.

magnitude of induced polarization, however, the spherical approximation should be examined critically for accuracy.

Maxwellian equivalent body   An axisymmetric, anisotropic particle can be well-approximated by replacement with a particle with an effective, isotropic, complex permittivity: the Maxwellian equivalent body [20]. Many models assume a homogenous, isotropic (equivalent) material, and in so doing, also assume that particle structure or isotropy is unimportant. For certain applications, this is true (e.g., separations and trapping applications where these components are not of interest), but when attempting to predict behavior (rather than measure) or gain insight into changes in internal structure, particle anisotropy can affect particle response through changes in $\tilde{\varepsilon}_{eff}$ [16, 21–29, 30, 31][5]. The complex permittivity, as we have defined previously, applies for isotropic materials only. In general, the permittivity takes on a tensorial character, $\vec{\vec{\varepsilon}}$. We can define our coordinate axes to coincide with the principal axes of $\vec{\vec{\varepsilon}}$, and in so doing, obtain a diagonal matrix. Furthermore, the work of Simeonova et al. [20] showed that the Maxwellian equivalent body can be used in cases where the anisotropy is axisymmetric, yielding:

$$\tilde{\varepsilon}_{eff} = \tilde{\varepsilon}_{1,effective}\frac{E_1}{|E|} + \tilde{\varepsilon}_{2,effective}\frac{E_2}{|E|} + \tilde{\varepsilon}_{3,effective}\frac{E_3}{|E|} \tag{6.12}$$

In the specific case where the electric field is linear and along the direction of the major axis, the permittivity tensor will reduce to a scalar component along the direction of the field.

6.2.1.2 Assumptions and approximations for F_{drag}

Spherical approximation

In order to determine the velocity of a particle undergoing DEP, we consider equilibrium between the DEP force and the viscous drag. For low Reynolds number, the viscous drag is given by

$$F_{drag} = \frac{1}{2}\rho u^2 A_p C_D \tag{6.13}$$

$$C_{D,sphere} = \frac{24}{Re} \tag{6.14}$$

$$F_{drag} = 6\pi\mu u a \tag{6.15}$$

where $A_p = \pi d^2/4$ is the particle cross-sectional area perpendicular to the direction of flow, $d = 2a$ is the particle diameter, $Re = \rho u d/\mu$ is the particle Reynolds number, and C_D is the drag coefficient. We also know the analytical form of the DEP force and can balance this against the drag force to give the velocity at equilibrium:

$$F_{DEP} = \pi\varepsilon_m a^3 \Re[f_{CM}]\nabla(\vec{E}\cdot\vec{E}) = F_{drag} \tag{6.16}$$

$$\pi\varepsilon_m a^3 \Re[f_{CM}]\nabla(\vec{E}\cdot\vec{E}) = \frac{1}{2}\rho u^2 A_p \frac{24\mu}{\rho u d} \tag{6.17}$$

---

[5] These references primarily consider electrorotation phenomena rather than transport by dielectrophoresis. However, electrorotation spectra and dielectrophoretic spectra are related by the Kramers-Krönig relationships:

$$\Re[f_{CM}(\omega)] = \frac{2}{\pi}\int_0^\infty \frac{x\Im[f_{CM}(x)]}{x^2 - \omega^2}dx + \Re[f_{CM,\infty}]$$

$$\Im[f_{CM}(\omega)] = \frac{2}{\pi}\int_0^\infty \frac{\Re[f_{CM}(x)] - \Re[f_{CM,\infty}]}{x^2 - \omega^2}dx + \Re[f_{CM,\infty}]$$

where $f_{CM,\infty}$ is the limit of the complex Clausius-Mossotti factor as $\omega \to \infty$ [12].

$$u = \frac{\varepsilon_m a^2}{6\mu} R[f_{CM}] \nabla \left(\vec{E} \cdot \vec{E} \right) \tag{6.18}$$

We use an equilibrium relation because the characteristic equilibration time (the time for particles to reach their terminal velocity) is very short compared to experimental time scales. The constant term that relates the dielectrophoretic force to the resulting particle velocity, then, is termed the "dielectrophoretic mobility":

$$\mu_{DEP} = \frac{\varepsilon_m a^2}{6\mu} \Re[f_{CM}] \tag{6.19}$$

When a particle is not spherical, however, the expression for drag coefficient changes as a function of particle shape and orientation. This variation is expressed through changes in both A_p, the cross-sectional area perpendicular to the flow, and C_D, the drag coefficient based on an *effective* particle diameter. The effective particle diameter is the diameter of a sphere of equivalent volume. For isotropic materials, particles will typically orient with their long axis parallel to the external electric field.

Consider an ellipsoid with axes a_1, a_2, and $a_3 (a_1 > a_2 > a_3)$. The dipole moment of such a particle can be broken up into components along each axis. Moments will be proportional to the effective permittivity (along a particular direction) and the electric field along a particular direction.

$$p_i \sim \vec{\varepsilon} \cdot a_i \vec{E} \cdot \hat{a}_i \tag{6.20}$$

p_i is the dipole moment vector along the axis $i = \{1,2,3\}$, $\vec{\varepsilon}$ is the permittivity, $\vec{E}$ is the electric field, and $\hat{a}_i$ is a unit vector along one of the axes $\{1,2,3\}$.

For a nonrotating electric field, the particle will experience torques along each moment, tending to align the particle to the sum of moments. In a rotating electric field, there is still a torque on the particle, however, and orientation is a function of the frequency of rotation in addition to the factors determining the magnitude of torque on each moment of the particle [15].

For the viscous drag on a particle, we present approximate analytic results from [32] for the special case of a prolate spheroid $(a_1 > a_2 = a_3)$—useful due to its similarity to rod-shaped particles:

$$C_{D, ellipse} = \frac{128}{Re} \frac{\left(1 - e^2\right)e^3}{2e + \left(3e^2 - 1\right)ln\left(\frac{1+e}{1-e}\right)} \tag{6.21}$$

where we have again used the eccentricity, $e = \sqrt{1 - a_2^2 / a_1^2}$. The resulting dielectrophoretic mobility for prolate spheroid with its long axis parallel to the direction of the applied electric field is found by combining (6.21), (6.11), and (6.6):

$$\mu_{DEP} = \left(\frac{2e + (3e^2 - 1)ln\left(\frac{1+e}{1-e}\right)}{e^3}\right) \frac{a_1 a_2}{24} \frac{\varepsilon_m}{\mu} \left(\frac{\tilde{\varepsilon}_p - \tilde{\varepsilon}_m}{\tilde{\varepsilon}_m + (\tilde{\varepsilon}_p - \tilde{\varepsilon}_m)L_3}\right)$$ (6.22)

$$L_3 = \frac{a_2^2}{2a_1^2 e^3}\left[ln\left(\frac{1+e}{1-e}\right) - 2e\right]$$

The first term in parentheses contains geometric corrections to the drag coefficient and in the second term, these corrections are made by including L_3.

For a rod-shaped bacterium with a 2:1 major/minor axis ratio, the electrodynamic correction is on the order of 20%–40% far from the crossover point, while the drag correction is roughly 60%.

Infinite domain assumption   In many fluid mechanics and electromagnetics problems, it is common to invoke "boundary conditions at infinity." For instance, it is useful to argue that as distance from the origin goes to infinity, a parameter of interest (such as velocity potential or electric potential) must converge to an externally imposed value. However, in applications where channel dimensions approach the particle diameter, this assumption is typically poor. The electrodynamics of the problem is influenced by the proximity of the boundaries on the Laplace equation, requiring more terms in the multipole expansion, while the drag relation is influenced by the proximity of the boundaries on the Stokes flow equations, requiring drag adjustments.

As a particle moves through the surrounding fluid, it will exert a pressure on the surrounding fluid. We will consider only incompressible fluids, and so, this pressure will be transduced to fluid velocity. Solutions to the fluid velocity field produced by a moving particle in an infinite fluid are readily available. However, as the moving particle approaches a wall, the fluid velocity field resulting from the moving particle will be retarded due to the no-slip boundary condition at the wall. When the force, $\vec{F}$, on a particle is directed parallel to the wall, the terminal particle velocity is modified by the scalar coefficient, B:

$$\vec{u} = \frac{B}{6\pi\eta a}\vec{F}$$ (6.23)

The scalar coefficient, B, decays to 0 as $x \rightarrow a$, and approaches 1 for $x > 10a$. A plot of the mobility coefficient shows the retarding effects of the wall as x increases [33]. At distances greater than one particle radius, the scalar coefficient, B, can be approximated as:

$$B = \left[1 - \frac{9}{8}\frac{a}{x} + \frac{1}{2}\left(\frac{a}{x}\right)^3 + ...O\left(\left(\frac{a}{x}\right)^5\right)\right]^{-1}$$ (6.24)

Isolated particle assumption   As with the infinite domain assumption discussed above, the assumption that particles are isolated in a media may break down at low interparticle spacing distances, caused by high particle concentrations or by particle localization due to hydrodynamic and/or electrokinetic forces. We examine briefly the results of Batchelor [34] and Goldman et al. [35], with regard to two interacting spheres. The equation of relative motion is:

$$\frac{d\vec{r}}{dt} = \left(\vec{\mu}_{2,1} - \vec{\mu}_{1,1}\right)\cdot \vec{F}_1 + \left(\vec{\mu}_{2,2} - \vec{\mu}_{1,2}\right)\cdot \vec{F}_2 \tag{6.25}$$

$$\vec{\mu}_{i,j} = \frac{1}{3\pi\eta\left(a_i + a_j\right)}\left[A_{i,j}\frac{\vec{r}\vec{r}}{r^2} + B_{i,j}\left(\vec{I} - \frac{\vec{r}\vec{r}}{r^2}\right)\right] \tag{6.26}$$

where $\vec{r}$ is the vector between particle centers, r is the magnitude of $\vec{r}$, $\vec{I}$ is the identity tensor, $A_{i,j}$ and $B_{i,j}$ are scalar coefficients that depend on r, and $\{i, j\}$ refer to sphere 1 or sphere 2. Batchelor found that $A_{1,1}$, $A_{2,2}$, $B_{1,1}$, and $B_{2,2}$ decrease sharply as the distance between particles decreases and increase to 1 as $r/a \rightarrow 10$. These coefficients correspond to mathematical contributions to the mobility of one particle or the other. The interaction terms, $A_{1,2}$, $A_{2,1}$, $B_{1,2}$, and $B_{2,1}$ decay slowly compared to the rate of increase of $A_{i,i}$ and $B_{i,i}$, but still decay to zero as r grows beyond 10 radii. The importance of these correction factors again comes into play when particle distances are less than 10 times the radius. Since the average particle spacing normalized by the radius is approximately $\sqrt[3]{4\pi / (3\rho)}$, where ρ is the volume fraction of monodisperse particles, these effects are safely neglected at local volume fractions below 0.4%.

6.3   Materials: equipment for generating electric field nonuniformities and DEP forces

There are myriad different DEP devices and techniques, so much so that dividing them cleanly into categories proves difficult. The dielectrophoretic effect is fundamentally tied to the electric fields, which can be defined by spatial and temporal characteristics, and so we choose to divide DEP approaches by spatial and temporal characteristics accordingly. We will consider DC or sinusoidal AC fields[6], with variations in temporal (frequency, phase) and spatial (geometry) character, as they relate to the generation of DEP forces.

6.3.1   Electric field frequency

The frequency dependence of the DEP force is perhaps its most alluring feature, and as a result, the majority of DEP techniques employ sinusoidally varying electric fields.

6.3.1.1   AC fields

AC electric fields are the primary field type employed in dielectrophoresis applications, owing to the availability of straightforward analysis and experimental implementation

---

[6]   It is entirely possible to create DEP forces using non-sinusoidal fields, but this is rarely done. A mathematical treatment of these fields is slightly more difficult, and is usually done via Fourier analysis—treating a nonsinusoidal field as a sum (or integral) of sinusoidal fields. We can therefore intuit the properties of non-sinusoidal fields with appropriate knowledge of the frequency response characteristics of sinusoidally driven DEP forces and the spectral composition of the nonsinusoidal field.

techniques. AC fields of high frequency ($>10^3$ Hz) can be safely applied to micro-fabricated electrodes in microfluidic channels without significant buildup of electrolytic products (hydrogen and oxygen bubble production at the anode and cathode, respectively), significant changes in temperature, or significant changes in pH.

Equipment and experimental setup

The most common and straightforward method for generating a sinusoidal AC electric field is to employ a function generator connected to electrodes. Function generator performance can range from low to high according to the experimental requirements. Higher quality instruments offer finer control over the characteristics of the output waveform (phase, frequency, amplitude) and offer a wider range of operating parameters, achieving higher frequency and power output. The power output of these devices is typically low, on the order of milliwatts for frequencies in the kilohertz range, and tends to decrease as frequency increases beyond one megahertz. These limitations are primarily the result of component slew rate limitations, which will be higher as performance (and cost) increase. In the kilohertz and megahertz range, it is common for generators to offer a DC-offset option, which will allow for the simultaneous use of AC and DC electric field components. This has been utilized by some experimenters to actuate DEP and electrophoresis/electrosmosis independently [36].

In general, the DEP frequency response characteristic extends beyond the megahertz range, and for experimental applications requiring field frequencies in the gigahertz range, an RF signal generator must be used. These signal generators work quite well for low-power applications, where micro- fabricated electrodes are used and can generate high electric field gradients due to their close proximity. There are situations where higher-voltage signals must be applied, such as in insulative dielectrophoretic applications, where electrode spacing is necessarily orders of magnitude larger, requiring concomitant increases in applied voltage. General amplification of arbitrary signals is possible for low frequency (~1 kHz) signals at a reasonable cost [36]. In order to amplify higher frequency, arbitrary signals (including a DC-bias) to the kilovolt range, more specialized equipment is currently required. Ampliflcation of high frequency sinusoidal waveforms can be accomplished using RF-amplification equipment.

In this discussion, we have concerned ourselves with only sinusoidal or DC signals, but this does not represent the limits of possible waveforms. Indeed, some researchers have shown that the addition of multiple frequency components (or the higher frequency spectral distributions present in sawtooth, square, or triangular waveforms) can be beneficial to a particular DEP application [37]. Most function generators can generate these waveforms; more complex waveforms, however, require an arbitrary waveform generator.

6.3.1.2 DC fields

While the majority of DEP approaches use sinusoidally varying electric fields with relatively high frequencies, recent advances in microfabrication techniques have made spatial variation of the electric field magnitude possible on relevant length scales (typically in insulative-DEP techniques; see Section 6.3.3.2) and opened the door to the applications of low-and zero-frequency electric fields. DC fields are interesting because they allow combination of linear and nonlinear EK effects and offer simplified analysis and straightforward implementation.

Equipment and experimental setup   DC electric fields offer simplicity of implementation, and combine transport of fluid and particles via electroosmosis and electrophoresis. DC electric fields drive electrophoresis and electroosmosis by the familiar mechanisms, and can also drive dielectrophoresis (see Section 6.3.3.2 for an explanation of the mechanism). This combination is convenient, as it allows analyte transport and manipulation with only one field.

DC fields will tend to drive larger currents and lead to the production of electrolytic products at electrode interfaces. For this reason, DC dielectrophoresis applications are typically carried out by placing insulating constrictions in the electric current path and electrodes in external reservoirs, allowing the electrolytic products to escape to the atmosphere [38, 36]. Brask et al., in a non-DEP application, employ palladium electrodes, taking advantage of the high hydrogen permeability of palladium to minimize hydrogen bubble formation at the electrode surface.

Driving DEP via DC fields from external reservoirs usually requires high voltages. These can be generated and controlled relatively easily with a DC high-voltage power supply, though the electrical resistance of the channel should be considered when planning for power and thermal management of the system. A short, wide channel filled with high conductivity buffer (such as 1M phosphate buffered saline) will have a low electrical resistance. This will lead to a larger current draw from the power supply, increased rates of electrolysis, and increased Joule heating. Lower conductivity solutions and long, narrow, shallow channels will ameliorate these difficulties. As electrophoresis and electroosmosis are both dependent on $\vec{E}$ and dielectrophoresis is dependent on $\nabla\left(\vec{E}\cdot\vec{E}\right)$, these effects can be tuned independently using the field magnitude and the geometry of the device (specific geometries and fabrication techniques will be discussed in detail in Section 6.3.3).

6.3.2 Electric field phase

In addition to the field frequency, the phase (and the spatial variation of phase) can be used in system design as well. If the electric field phase (ϕ) is spatially uniform, then it can be neglected when calculating steady-state dielectrophoretic effects. However, when the phase is spatially nonuniform, additional force and torque components exist, which open up a wider range of possible applications and measurements.

One particularly useful method for taking advantage of spatial variations in phase is what has become known as "traveling-wave" dielectrophoresis (twDEP), which uses a spatially-varying phase to propel particles down a channel. Recall from Section 6.2, (6.10), that the dielectrophoretic force contains two components: the irrotational, "traditional" dielectrophoresis component and the curl-dependent, "traveling-wave" dielectrophoresis component.

The canonical example of twDEP is to use an array of electrodes patterned in a microfluidic channel. The electrode array is composed of alternating, independently driven electrodes with different phase. The electrode array is aligned at an angle to the direction of flow. These signals—irrespective of phase—are used to levitate particles against gravity within the flow field due to irrotational, negative dielectrophoresis, and the varying phase is used to drive particles transverse to the direction of flow according to the imaginary component of the Clausius-Mossotti factor.

In much the same way that spatially varying phase leads to traveling wave dielectrophoresis, a rotating electric field will induce a particle "electrorotation" response. In this case, the dielectrophoretic effect is instead expressed as a torque and is dependent on a different aspect of the particle's electrical properties, namely the imaginary component of the Clausius-Mossotti factor ($\Im[F_{CM}]$).

$$\langle \tau \rangle = -4\pi\varepsilon_m a^3 \Im[f_{CM}]\left(\Re[\vec{E}] \times \Im[\vec{E}]\right) \tag{6.27}$$

The results of an electrorotation experiment typically consist of measured particle rotation rates as a function of electric field rotation frequency, or "ROT spectra." ROT spectra are related to DEP spectra by the Kramers-Krönig relation (see footnote 5), and thus ROT spectra have large magnitudes at the frequencies where DEP forces are changing.

Equipment and experimental setups   Creating spatially varying phase signals in microfluidic systems is most often accomplished using arrays of interdigitated electrodes. By applying a different phase to various electrodes in an array, a spatially varying field is produced. The electrode array itself is often fabricated at an angle to the microfluidic channel, because the twDEP forces that are created drive particles perpendicular to electrode orientation. This configuration leads to sorting as a function of the imaginary component of the Clausius-Mossotti factor transverse to the direction of flow. The electric signals applied to elements of the electrode array generally consist of four sinusoidal signals with phase magnitudes ϕ = 0, $\pi/2$, π, and $3\pi/2$, in order. This ABCD-configuration effectively generates an electric field maximum that propagates perpendicular to the array at a frequency equal to the driving frequency. Chang et al. showed that separating particles with similar (but distinct) $f_{CM}(\omega)$ functions via twDEP techniques can be improved by using two (or more) signals simultaneously, with different frequency. The first frequency is chosen to optimize the levitation of particles above the electrodes (i.e., $\Re[f_{CM}(\omega)] < 0$ for both particles), and the second is chosen to optimize the transport of particles along the array as a function of $\Im[f_{CM}(\omega)]$[37].

The design and fabrication of twDEP devices is slightly more challenging than simpler electrode configurations because of the need for four different signal paths. This requires multilayer fabrication techniques, but these are well understood and can be readily accomplished in appropriate clean-room facilities.

Rotating electric fields can be created by forming a quadrupolar configuration and driving each electrode at a different phase (again, ϕ = 0, $\pi/2$, π, and $3\pi/2$ form the components of the ABCD configuration). Fabrication of these devices is single-layer process, and can be accomplished using standard photolithographic techniques. However, the planar quadrupolar configuration can induce nDEP and force particles away from the electrodes (in the z-axis direction, if the electrodes are in the x-y plane). Particles will then be out of the range of influence of the electrodes or trapped against the microchannel wall, inducing a force that can confound a ROT spectra analysis. This difficulty has been overcome by using an octode cage: two planar, quadrupolar electrode arrays are assembled facing one another, displaced in the z-axis direction [39]. This configuration creates an electric field null in the center of the three-dimensional configuration. This is effective for trapping, but a field null yields no net torque. In order to conduct the electrorotation study, one quadrupolar array is offset by a few degrees (rotated about the z-axis in space or by adding a phase increment between the signals driving the quadrupolar arrays), creating a minimum, not a null, in the center of the octode cage [39, 40].

In short, electrorotation and twDEP are two useful techniques that depend on the phase of the electric field and the imaginary component of the Clausius-Mossotti factor, $\Im[f_{CM}]$. Electrorotation studies can be used to gain a large amount of information about an individual particle. ROT spectra can be used to determine $\Im[f_{CM}]$, and $\Re[f_{CM}]$ can be found via the Kramers-Krönig relationships (see footnote 5). Quadrupolar arrays and other ROT measurement techniques work well for single particle studies, but have low throughput, making them inappropriate for sorting applications. twDEP has been used to successfully separate particles transverse to the direction of fluid flow, making it an effective continuous-flow sorting technique. Combined with multiple frequencies, twDEP can be applied to a wide range of experimental conditions, with judicious choice of the driving waveform [37].

6.3.3   Geometry

The temporal aspects (frequency and phase) of electric fields in DEP systems are, of course, critical. However, the magnitude of the DEP force in micro- and nanoscale systems is central to DEP's burgeoning importance in the last 10 years. Micro- and nanofabrication techniques have opened the door to applications of DEP at the cellular length scale, where subtle variations in particle composition and changes in the dielectrophoretic response can be observed on a particle-by-particle basis. By altering the geometry of energized electrodes or the channel boundaries that define the electric current path, dielectrophoretic effects can be tuned for numerous applications.

6.3.3.1   Electrode configurations

As previously discussed, changing the shape and orientation of electrodes, in addition to modulating the frequency and phase applied, can give rise to dielectrophoretic particle trapping, dielectrophoretic sorting, electrorotation, and traveling wave dielectrophoresis effects. Specific geometries will be discussed in Section 6.4.

Castellanos et al. examined the case of two adjacent electrodes in detail in [41]. Their results detail the relative effects of DEP and hydrodynamics on the motion of particles adjacent to the electrode array (Figure 6.1). The electrodes were energized with a 5V potential and separated by a distance of 25 μm. Their findings concluded that, in general, motion due to gravity and dielectrophoresis vary with particle volume, and dielectrophoresis will dominate provided an appropriate electric potential is chosen (frequency and magnitude). Brownian motion decreases with increasing particle size, depending on 1/a. For particle sizes in the micron regime, this can be overcome by dielectrophoresis with a relatively low electric field. A similar argument can be made for the effects of buoyancy.

Depending on the depth of the channel, both positive- and negative-dielectrophoresis can be observed in electrode-based systems. Positive dielectrophoresis (pDEP) is observed as some fraction of passing particles become trapped on the electrode structures. Negative dielectrophoresis (nDEP) will only be observable if the distance between the electrode array and the opposite channel wall is short, and particles become trapped against the channel wall owing to nDEP and hydrodynamic forces. If the chan-

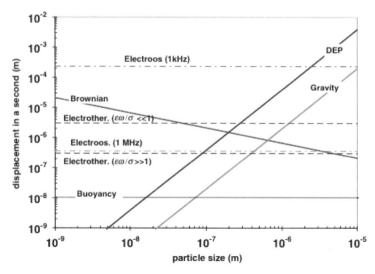

Figure 6.1 Scaling relationships developed in [41] show the relative effects of bouyancy, Brownian motion, electroosmosis, electrophoresis,and dielectrophoresis. Electrodes are separated by 25 μm and energized with a 5V potential. Copyright Institute of Physics. Used with permission.

nel is too deep, particles will be forced away from the array (by dielectrophoresis) and across (by hydrodynamic forces, either electrokinetic or pressure driven). In the absence of trapping (i.e., in a deep channel), it can be difficult to characterize the magnitude of nDEP forces, as deflection toward or away from the viewer is difficult to quantify in traditional microscopy.

Equipment and experimental setup   The fabrication of arbitrary electrode geometries has been an integral part of the development of dielectrophoresis as a viable technique for particle manipulation. The patterning techniques used are those of the microfabrication or MEMS industry, with electrode structures being fabricated via sputtering or electron-beam evaporation of metals, usually gold[7].

Devices with three-dimensional elements must be fabricated in planar configurations and then subsequently aligned using a mask aligner and bonded. The primary difficulty associated with octode cages (and other such three-dimensional structures) is that associated with visualization of trapped or sorted particles. Also, fabricating arrays of these structures can only be accomplished through complicated multilayer fabrication processes. Other geometries—point-lid, circle-dot, and grid-electrode—also require the use of multiple layers of electrodes leading to a concomitant increase in the complexity of device fabrication. In situ fabrication of three-dimensional dielectrophoretic traps in an array-able format has been achieved by electroplating gold onto an SU-8 mold [42]. The resulting post-trap configuration, allows for easier array fabrication and visualization of trapped particles.

Typically, electrodes are fabricated on glass substrates owing to the well-defined deposition process and visualization capabilities. Depositing metal onto the surface,

---

[7] Gold is chosen because it offers good conductivity, low production of electrolytic products at a reasonable cost (compared to platinum or palladium), and has well-established protocols for deposition.

however, limits the efficacy of traditional diffusion-based wafer-to-wafer bonding techniques, and so PDMS covers are generally used to define microfluidic channels [43–52]. Polymers other than PDMS have also been used, but adhesion remains a difficulty.

6.3.3.2 Insulative configurations

Insulative dielectrophoresis refers to a subset of DEP techniques that use insulating constrictions in the electric current path to create regions of high or low electric field magnitude. This is done by altering microchannel geometry in a glass, silicone, or plastic substrate. A simple analysis demonstrates the concept: consider a microchannel made of insulating material with two different regions filled with a conducting fluid. Treating each region separately as a one-dimensional problem, we can apply Ohm's law and find, for the electric fields in region 1 and 2:

$$\vec{E} = \nabla V \rightarrow \vec{E}_n = E_n \hat{x} = \frac{V_n}{L_n} \hat{x} \tag{6.28}$$

$$E_1 = \frac{V - V_2}{\dfrac{\sigma L_1}{w_1 d}} \tag{6.29}$$

$$E_2 = \frac{V_2}{\dfrac{\sigma L_2}{w_2 d}} \tag{6.30}$$

Solving for V_2 and substituting, we can determine a relationship between E_1 and E_2 as a function of the channel geometry:

$$E_2 = E_1 \frac{w_1}{w_2} \tag{6.31}$$

This is the fundamental principle that leads to the utility of insulative DEP (iDEP) configurations. In addition, there are practical advantages to iDEP; the lack of internal electrodes simplifies fabrication considerably, as electrodes are typically placed in reservoirs external to the microfluidic channel. External electrodes avoid complications due to electrolysis and electrolytic products at low frequencies (which can occur when electrodes are placed inside microfluidic channels), allowing for low-frequency operation.

Insulative or electrodeless dielectrophoresis has been primarily used to concentrate samples for later analysis, but we will also explore some designs that accomplish separation based on particle characteristics.

Equipment and experimental setup   The great appeal of iDEP devices is due to the simplicity of their operation and design. Their fundamental mode of operation does not involve integrated electrodes, inducing variation in electric field gradients via variations in channel geometry. Without integrated electrodes, devices can be fashioned from a single material, greatly simplifying the fabrication process. This process will vary depending on the substrate, but can often be accomplished on the benchtop, outside of a cleanroom.

Substrate choice will depend on the needs of the experiment. Glass devices are well characterized, with well-known physical and electrokinetic properties. Fabrication techniques for glass are well established, but are usually limited to quasi-two-dimensional configurations due to available wet etching techniques. Multilevel glass devices require careful alignment of photomasks to achieve multiple etch steps [53, 54]. Polymeric devices—polystyrene, Zeonor, TOPAS—have seen increasing use recently, owing to their low cost and greater flexibility. They achieve greater flexibility by leveraging existing silicon fabrication techniques and using the silicon device as a master for hot-embossing [36]. PDMS—a seemingly ubiquitous substrate for microfluidic devices—has not been used to a large degree in iDEP devices. This is due to its geometric variability due to high compliance and relatively high ion permeability (and resulting conductivity) making it a poor insulator. Glass devices, while more difficult to fabricate, can be reused consistently and will perform consistently, given their detailed characterization. Polymeric devices, on the other hand, offer simpler fabrication and lower cost, lending themselves to disposable applications.

6.4 Methods: data acquisition, anticipated results, and interpretation

In this section, we provide general recommendations for DEP techniques, followed by specific protocols for select applications. We have divided the protocols into two main classes based on device characteristics: (1) electrode-based, and (2) electrodeless or insulating devices. First, we will outline some general considerations when working with each type of device. Following that, we examine some of the more common, representative experiments and summarize the relevant experimental parameters for each.

6.4.1 General considerations for dielectrophoretic devices

Electrolysis

Metal electrodes in aqueous media can lead to the generation of gas bubbles at the electrode-media interface. In order to conduct current through the media between electrodes, electrons are donated at the anode, creating anions and/or eliminating cations; the reverse then occurs at the cathode to return electrons to the circuit. These Faradaic reactions (often electrolysis of water) generally produce gaseous products near this interface. In many cases, the amount of gas generated is small enough to remain in solution, but the characteristically small volumes used in most eDEP devices lead to decreased storage capacity for these products, and bubbles can form as a result. The specific chemical reaction that is responsible for gas generation depends on the media and the metal used in electrode fabrication. The relevant parameter for estimating the magnitude of these effects is typically current or current density per half cycle, so high-frequency techniques have little or no problems, while low frequency techniques can be challenging. iDEP devices have fewer problems with bubble generation due to electrolysis because electrodes are placed in relatively large, open, external reservoirs.

Media conductivity

Electrolyte concentration influences media conductivity (and therefore electric current and electrolysis), as well as the viability of biological samples. When working with cells or other biological samples, suspending solutions generally contain high concentrations of salts (~1M). High conductivity buffers have low electrical resistance, thus requiring more current to maintain an applied electric potential. Two major confounding effects are associated with large electric currents in dielectrophoresis devices: Faradaic reactions (discussed above) and Joule heating.

Joule heating is the result of ions flowing through the media, leading to a volumetric heat transfer rate that is proportional to the media conductivity and the square of the applied potential[8]:

$$q'' \sim (\nabla_v)^2 \sigma_m \tag{6.32}$$

where q'' is the volumetric heat transfer rate to the fluid as the result of the applied electric potential v. The total heat transfer increases with increasing conductivity, and the total integrated heat transfer will increase over time as the system heats up[9].

Electrode configurations in iDEP and eDEP devices vary; eDEP devices have closely spaced electrodes in a microfluidic channel, while iDEP devices have electrodes placed in external reservoirs. The large interelectrode distance in iDEP devices means that the applied electric potential may be orders of magnitude larger than in eDEP devices; however, the electric field magnitude will tend to be comparable in both types of devices, and Joule heating can occur in both cases.

Thermal effects

Variations in temperature, in both iDEP and eDEP devices, can influence a number of experimental parameters (e.g., conductivity, viscosity, and cell viability). We will briefly outline potential sources for temperature variation, possible outcomes, and strategies for maintaining isothermal conditions. Most researchers consider isothermal conditions within their devices—as we have above. Depending on experimental parameters, however, heating may become important, and the isothermal assumption may fail to explain certain results.

Given the electric field source term of Joule heating (6.32), any device region with high field magnitude is a potential source of thermal fluctuations. iDEP techniques can experience local changes in temperature in constricted regions, far from electrodes. Electrode-based DEP techniques can also generate temperature changes, especially near electrodes where current densities are typically highest.

Changes in temperature lead to myriad changes in experimental parameters, including fluid viscosity and conductivity. Changes in viscosity change the particle drag coeffi-

---

[8] It is important to note that "high conductivity" buffers are still poor conductors when compared to the conductivity of metal electrodes. Because of this, electric potential changes occur almost entirely across the media, driving the electric field source term (6.32) in the metal electrode to zero.

[9] This equation only models the heat transferred to the fluid media; a complete analysis will include the heat transferred from the fluid to its surroundings. This relationship is intended to illustrate the dependence of heating on potential and media conductivity. A full solution will also consider the dependence of conductivity on temperature.

cient, making measurements of dielectrophoretic or electrophoretic particle mobilities inaccurate. Local changes in viscosity, η, due to temperature variation, lead to changes in the local fluid Reynolds number and complicates the normally tractable low-Reynolds number Navier-Stokes equations. Even a 5°C temperature change will lead to a 10% difference in the viscosity of water [55]. Fluid conductivity is also a function of temperature; the conductivity of electrolyte solutions at 25°C increases approximately 2%/°C. Impurities tend to buffer these variations in conductivity, so variations should be verified experimentally in temperature-sensitive experiments. Local changes in conductivity create locally varying electric fields, which can complicate dielectrophoretic force predictions and lead to nonlinear fluid currents and vorticies [56, 57].

There are a few strategies for avoiding thermal variations in dielectrophoresis experiments. Lower conductivity buffer solutions can ameliorate the effects of high electric fields; many DEP experiments use deionized water as the running buffer for this reason. The dependence of dielectrophoretic effects on the gradient of the electric field also suggests that decreasing the characteristic length scale of a particular device may improve performance at lower electric field magnitudes [41]. Finally, choice of substrate can also improve thermal performance. A material that readily absorbs and dissipates heat (such as glass) will offer improved thermal performance over plastics (polycarbonate, cyclic-olefin polymers, etc.).

Particle adhesion

Depending on the substrate and particle, particle adhesion left unmanaged may be significant and, in some cases, prohibitive. Many techniques are available to ameliorate the effects of adhesion, such as surface coatings or surfactants in solution. An in-depth discussion of this topic is beyond the scope of this text; indeed there is a large body of literature on the subject [58], but we discuss a few common adhesion mechanisms, such as hydrophobic, electrostatic, and chemical interactions.

The hydrophobic properties of a particle-substrate system can affect particle adhesion. In an aqueous solution, hydrophobic substrates will promote adhesion of hydrophobic particles. Polystyrene particles are hydrophobic and, as a result, can adhere to hydrophobic substrates such as polycarbonate, cyclic-olefin polymers (TOPAS, Zeonor), and PDMS. Finally, biological samples present a complicated set of interactions that will depend on the specific analyte, which can be either hydrophobic or hydrophilic, positively or negatively charged, and have complicated chemical interactions with the substrate. Bioparticle adhesion is a broad topic that is covered extensively in the literature and remains an active area of research [59, 60].

Charge-carrying particles and surfaces will interact with one another electrostatically. Carboxylate-modified polystyrene particles carry a negative surface charge, which will repel them from negatively charged surfaces. The opposite is true for some amine-functionalized particles. Biological particles are invariably zwitterionic but usually net negative. Charge characteristics will also vary as a function of solvent pH, which will change the charge state and surface potential of both particle and substrate.

Certain applications, such as antibody assays and ELISA tests, purposely promote the adhesion of specific particles via chemical interactions. Such adhesion reactions depend on a number of conditions: pH, concentration of product and reactants, temperature, and interaction time.

Transport

In both iDEP and eDEP devices, analyte transport can impact device operation. A fundamental challenge when dealing with small planar electrodes in eDEP devices is how to bring particle suspensions into close enough proximity that the electric field variations caused by the electrode can manipulate the particles. Potential solutions are to use small channels or particle focusing techniques. Transport of fluid and analyte in iDEP systems is also of critical importance. Where pressure-driven flow is used, variations in channel geometry that create electric field gradients can break the similitude of electrical and fluidic potential field solutions, leading to difficulties in predicting particle paths. Alternatives are to use very low flow rates in pressure driven flow, or to use electro-osmotic/electropohretic effects for media and analyte transport.

Small channel dimensions are the most common solution to ensure that particles interact with DEP forces. However, as channel dimensions shrink, the point approximation for particles (meaning particle motion does not impact fluid behavior) becomes less applicable. A good rule of thumb is that if the bulk DEP relations presented in this chapter are to be used, channel dimensions should be at least ten times larger than particle diameter[10].

Particle-focusing techniques can be used, in both eDEP and iDEP devices, to ensure interaction between particles and DEP forces. Fluid flow focusing techniques use variable volumetric flow rates in multiple input channels to control the location of particles in the flow downstream of the intersection. Three (or more) channels converge near the dielectrophoretic section of the device. Two channels contain "sheath fluid" which does not contain any particles while the third channel contains the analyte particles. By setting the volumetric flow rates of the sheath fluid channels higher than the flow rate of the analyte channel, the fluid injected from the analyte channel will be focused by a factor of the ratio of volumetric flow rates [61].

Fabrication uncertainty

Insulative or electrodeless DEP techniques rely on variations in channel geometry defined by an electrically insulating substrate. Changes in channel geometry leading to changes in electric field magnitude can be estimated by defining a characteristic "constriction ratio" of bulk channel dimensions over constricted channel dimensions. The increase in electric field magnitude due to constrictions in channel geometry will be proportional to the constriction ratio [36]. Variations in channel geometry from manufacturing defects or mechanical strain before or during testing can lead to significant operational changes within the device. Due to the potential significance of these defects, substrate choices should include consideration of material properties such as hardness and modulus. A low modulus material, such as PDMS, might "sag" in wide, long channels (over 1 cm^2). The quality of channels fabricated in high modulus materials such as polycarbonate or Zeonor avoid this issue. Esch et al. examined the quality of devices fabricated with hot-embossing techniques—often used to fabricate devices in these high-modulus materials—as a function of master material choice (e.g., silicon, SU-8, copper) and device design [62]. They found that wet etching of the master material will yield nanometer-scale surface roughness that will transfer to the plastic, and is not recommended for dielectrophoresis applications. Also, high aspect ratio features and

---

[10] If trapping of particles becomes significant and agglomeration occurs, the effective diameter of agglomerates must be considered.

designs with high feature density will be prone to defects at corners and near features on the order of 5 μm. Similar confounding effects are observed in eDEP devices. Variations in channel geometry, due to channel "sag," or imperfections in electrode geometry can also lead to erratic electrokinetic effects.

6.4.2 Electrode-based dielectrophoresis

Given the general considerations above, we will now examine more specific applications. We will cover canonical electrode-based dielectrophoresis techniques and describe example experimental recipes. For each technique, we will summarize the general configuration of the experiment and then describe what the reader might expect to observe while conducting the experiment. Throughout this section, we will refer to Tables 6.6 and 6.8, which summarize experimental device parameters and solution parameters, respectively. Each experimental technique is identified with a number that corresponds to entries in Tables 6.6 and 6.8 at the end of this section.

6.4.2.1 Filtering/binary sorting

Binary sorting or filtering is one of the most common applications of dielectrophoresis. In these experiments, dielectrophoresis is used to separate one type of particle from a flowing solution based on size or material properties. In this way, a subset of particles (or all particles) are removed from a flowing solution. The goal of these experiments is either to remove a contaminant from solution or to capture particles for analysis.

Experiment ID #1: interdigitated electrode array

Discussion: The interdigitated electrode array is one of the most common electrode configurations used in DEP studies. The electrode array consists of two sets of electrodes, grounded and energized, that alternate spatially. This creates a nonuniform field in the region of the electrode array that can be used to trap particles against a flow.

Interdigitated electrode arrays have been used extensively to retain particles of interest from a microchannel flow, or to filter out unwanted particles from an analyte stream. Electrodes are typically gold and patterned on a glass substrate via a lift-off procedure. There are only a few relevant design parameters to optimize: electrode width, interelectrode distance, electrode length, and fluidic channel depth. With these parameters set, variations in applied electric field magnitude and frequency are left to vary in the experiment.

Electric fields and particle motion are easily modeled (and have been calculated analytically in [63]) in electrode systems and can lead to consistent results when accurate particle models are used [41, 64]. The advantages associated with planar, interdigitated electrode arrays are simplicity of fabrication and analysis. Disadvantages of the interdigitated electrode array are the potential for permanent particle adhesion during positive dielectrophoresis and the inherent "binary" separation achieved under a particular set of experimental conditions.

Trapping in these devices is usually due to pDEP, which attracts particles to electrodes. Occasionally, these devices will operate via nDEP, repelling particles from the array and trapping them against the opposing channel wall, but this regime is less common and more difficult to quantify. The interdigitated electrode array is enclosed in a channel of some sort to sustain pressure driven flow. A common and well-characterized channel can be fabricated using a silicon master and PDMS. Fabrication of the PDMS channel is covered briefly in Chapters 1, 2, and 3. Bonding the PDMS channel to the glass substrate is generally carried out by plasma cleaning, taking care to align the channel perpendicular to the electrode array. Once the channel is bonded to the glass device, particles can be introduced in a dilute suspension via a syringe pump.

Methods: For the device characteristics described in Table 6.6 and solution characteristics described in Table 6.8, we would typically observe the following. As particles flow past the interdigitated electrode array, application of an electric field at a frequency of 10 kHz will induce positive dielectrophoresis, attracting particles from the fluid to the electrodes. Particle accumulation may be observed via bright field or fluorescence microscopy. Changing the electric field frequency to 1 MHz will induce negative dielectrophoresis, reversing the direction of dielectrophoresis and forcing particles into the stream; accumulated particles will be effectively released.

6.4.2.2   Trapping

Several electrode configurations have been used by investigators to trap single particles or small populations for close observation or manipulation [39, 42, 65–68]. In general, these devices are used to examine the response of samples to changes in buffer solution or other external stimulus on a particle-by-particle basis.

Experimental ID #2: quadrupole traps

Discussion: Quadrupole traps consist of four electrodes placed in an "×" or "cross" configuration. Opposite electrodes are energized, with the remaining two grounded or energized 180° out of phase. This configuration creates an electric field minimum in the center of the array and a maximum at the region of closest interaction between adjacent electrodes. A device in this configuration will trap particles via positive dielectrophoresis in the high field regions. Particles repelled from the high field regions by negative dielectrophoresis may be observable near the central minimum, but typical channel dimensions are large such that particles will be forced away from the plane of the electrodes. A channel depth of approximately 10 μm would be necessary to observe particle aggregation via nDEP for micron-scale particles and the experimental parameters listed in Table 6.6. The quadrupole trap is typically used to trap particles via positive dielectrophoresis from a static suspension. Therefore, the channel geometry is largely irrelevant for pDEP, quadrupole trap configurations.

Methods: Quadrupolar traps tend to aggregate particles in the high field regions between adjacent electrodes due to positive dielectrophoresis. Particles are introduced using a syringe pump or by applying pressure by hand, but observation takes place at zero flow rate. For device parameters found in Table 6.6 and particle and solution details found in Table 6.8, particle aggregation will be observed between the electrodes for electric field frequencies between 10 kHz and 1 MHz. Below 8 kHz, no aggregation will be observed, due to dissipation from negative dielectrophoresis, unless channel depth is small, as discussed above.

Experiment ID #3: circle/dot traps

Discussion: Electrode-based DEP traps generally consist of geometries that tend to trap single particles. The goal is usually to create a system of addressable particle traps to observe individual particle responses to a stimulus or to study biological particle interactions as a function of distance. Methods used to achieve addressable trapping include quadrupole cages, circle-dot, and "DEP microwell" geometries (Figure 6.2) [69].

Common to all of these geometries is the goal of trapping an individual particle or a small number of particles in one location. To achieve this, electrode geometries are designed to create pointlike regions of high (or low) electric field magnitude. Individual addressing is possible in many cases, but usually limited by fabrication of control lines or addressing traces [70].

Circle/dot traps are a single-particle trapping technique that benefits from easy multiplexing and parallelization. Traps are formed by a two-layer lithographic technique and consist of a conducting "dot" exposed on one layer, surrounded by a circle fabricated on a second layer. This configuration traps particles via positive dielectrophoresis, drawing particles into the high field region in the center (dot). A key advantage of this technique is that particles of interest can be released into the flowing

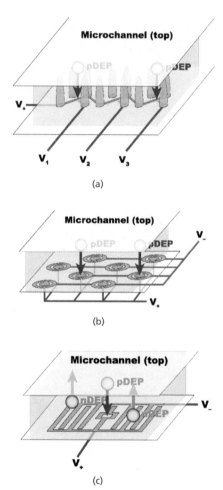

(a)

(b)

(c)

Figure 6.2   (a–c) Various electrode-based DEP trap geometries.

buffer by grounding a particular row and column, and collected in an output reservoir or via another collection scheme. This allows easier addressing of individual traps in large arrays. Typical addressing scales with the number of traps, but in this case, it scales with the square root of the number of traps [70].

Methods: Device operation involves sample loading, trapping, extra sample clearing, and release phases. Sample can be loaded by filling the device with particles and energizing circle electrodes (columns) to $+2V_{peak}$ and dot electrodes (rows) to $-2V_{peak}$ at a frequency of 1 MHz[11]. The dielectrophoretic trapping force can be modulated by changing the voltage on the circle and/or the dot. The strongest trapping forces will occur when both the dot and circle are energized. If either the circle or the dot is grounded, the dielectrophoretic force will be reduced, but not zero. The analyte suspension can be introduced until all traps are filled. After the loading phase, the excess analyte can be removed by flowing buffer solution over the array at a relatively low flow rate (0.06 mL/hr). This flow should continue through the release phase. If both circle and dot are grounded, then the dielectrophoretic force will be zero, and any particle in the trap will be released. This configuration will release only particles at the intersection of a grounded row and a grounded column. Observations can be made on a particle-by-particle basis in each of these traps while varying the buffer solution.

Experiment ID #4: castellated electrode traps

Discussion: Castellated electrode arrays have been used to align or trap particles. They consist of interdigitated electrodes with "castellations" along their length. This creates alternating regions of high and low electric field magnitude, corresponding to regions of close proximity between castellations and regions of separation in between castellations, respectively. Arrays can be formed by aligned or offset castellated electrodes placed adjacent to one another.

Castellated electrode arrays have been used to focus particles from a well-mixed solution into a line of particles for subsequent analysis. This technique has been used in DEP sorting devices and micro-scale dielectric spectroscopy experiments, where the electrical response of individual particles is examined [71, 72]. Both straight and castellated interdigitated electrode arrays can sustain a fluid flow and potentially trap particles against fluid drag forces using DEP. This characteristic has been used to measure not only the sign, but also the magnitude, of DEP forces as a function of frequency. This is accomplished by measuring the number of particles collected by the array at varying electric field frequencies. Castellated electrode arrays, however, are most typically used to concentrate samples (in the high electric field regions, under positive DEP) or to pattern particles at a specific location. The advantage of castellated interdigitated electrodes is the localization of high electric field regions.

In our experimental example, Sebastian et al. utilized aligned, castellated electrode arrays to create localized aggregates of mammalian cells. Such aggregates "could be used as artificial microniches for the study of interactions between cells" [73]. Experimenters trapped layers of stromal cells and Jurkat T-lymphocytes in the high field regions

---

[11] This is equivalent to a 180° phase differential between the circle and dot.

between castellations (Figure 6.3). Trapping fields were maintained until three-dimensional aggregates were formed containing both cell types.

Methods: Aggregation of mammalian cells (described in Table 6.6) using this array (described in Table 6.8) can be demonstrated in two phases: sample loading and sample aggregation. During the sample loading phase, cell suspensions can be introduced to a flow chamber via a syringe pump. In the aggregation phase, the electrodes can be energized at 1 MHz, to voltages ranging from 0 V_{peak} to 20 V_{peak}, and the cells can aggregate between castellations for 5 to 12 minutes. During this time, low conductivity, 480-mM D-sorbitol solution should be slowly circulated throughout the flow chamber to wash away ions released from the cells, preventing changes in the local conductivity that would confound dielectrophoretic trapping effects [73].

6.4.2.3   Sorting

Electrode-based designs used to sort particles based on size or frequency-response are less common than trapping techniques, but have still been well characterized in the literature [74–76]. These designs typically involve a spatially varying parameter such as electrode geometry or electric field phase that leads to a corresponding distribution of particles. We present two of the most conceptually illustrative techniques.

Experiment ID #5: angled or curved electrodes

Discussion: Angled electrodes can be used to separate particles based on DEP response or as a preconcentration system to create a localized stream of particles. Castellated electrodes have been used as "concentrators" as well, but the mechanism is slightly different. Castellated electrodes act on particles using negative dielectrophoresis (referring to the sign of f_{CM}, meaning particles are directed away from regions of high electric field) and focus by the cumulative action of a series of high field regions generated between two parallel castellated electrodes. Angled electrodes, in contrast, rely on negative dielectrophoresis to trap particles against fluidic drag forces. Dielectrophoretic forces and fluidic drag forces parallel to the direction of flow balance, and a net force parallel to the electrode results.

The parameters that determine particle behavior in this case are particle size (nearly always a significant contributor to DEP response; see (6.10) and Section 6.2.1.1), flow

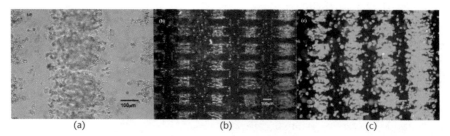

(a)   (b)   (c)

Figure 6.3   (a) Bright field image of cell aggregates formed between castellated electrodes. (b) Red fluorescent labeled Jurkat cells trapped close to the castellated electrode array. (c) Green fluorescent labeled stromal cells trapped on top of Jurkat cells. (From [73].) Copyright Institute of Physics. Used with permission.

Table 6.6 Experimental device characteristics for DEP applications described in Section 6.4. ID's correspond to data listed in Table 6.8.

ID	Application	Configuration	Substrate
1	Binary Sorting	Interdigitated Electrode Array	Gold electrodes on glass, channel in PDMS
2	Trapping	Quadrapolar Electrode Array	Gold electrodes on glass, channel in PDMS
3	Trapping	Circle-Dot Electrode Array	Gold electrodes on Si/SiO$_2$ (double layer fabrication); Channel in 3M tape/glass coverslip
4	Trapping	Aligned Castellated Electrode Array	ITO Electrodes on glass; Channel in tape/glass coverslip
5	Sorting	Paired Curved Electrodes	Gold electrodes on glass; channel in glass
6	Sorting	Traveling-wave dielectrophoresis	Gold electrodes on glass/Si$_3$N$_4$ insulator (double layer fabrication); channel in SU-8 and PDMS
7	Electrorotation	Polynomial quadrupolar electrodes	Gold electrodes on glass
8	Concentration	Insulating Post Array	Isotropically etched glass, channel in bonded glass coverslip
9	Sorting	Curved Insulating Constriction	Zeonor 1020R

velocity, particle position within flow (fluid velocity will vary with distance from channel walls in Poiseuille flow), electric potential, electric field frequency (through f_{CM}), and particle properties (again, through f_{CM}).

Angled electrodes have a few advantages that have led researchers to utilize them to preferentially guide particles or trap them. By angling the electrodes, with reference to the channel, it is possible to take particles trapped against a channel wall by nDEP (see Section 6.4.1.1) and displace them transverse to the direction of flow. In this manner, they have been used to preferentially direct particles to different outlets or focus particles into concentrated streams [77–80].

Schnelle et al. extended the concepts of angled electrodes and achieved a continuous flow separation by creating curved electrodes patterned on the top and bottom of a microfluidic channel. The curved electrodes were aligned with each other on top and bottom and shaped as shown in Figure 6.4 [39]. This curved electrode configuration balances negative dielectrophoresis against fluidic drag forces to achieve spatial separation of analytes in a continuous flow regime. The nDEP force generated at the electrodes is directed parallel to the radius of curvature of the electrode while the fluid drag force is directed parallel to the channel walls. The component of the nDEP force parallel to the direction of fluid flow will tend to trap or stop the forward motion of the particle. The component of the nDEP force perpendicular to the direction of flow will tend to move particles along the electrode. As the local angle of the electrode changes, the nDEP force opposing fluid drag decreases, eventually allowing the particle to pass the electrodes.

Methods: This configuration can be used to achieve continuous-flow separation of polystyrene particles as a function of size [74]. Sample experimental device and solution parameters are listed in Tables 6.6 and 6.8. A phosphate buffered saline solution with polystyrene particles of varying diameter can be introduced to the channel by constant pressure-driven flow. Electrodes with angled and curved geometries can be energized to

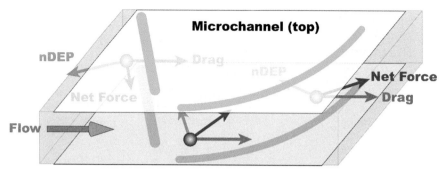

Figure 6.4 Schematic of aligning and sorting electrode pairs used to achieve size-based separation. Particle representations are not to scale. (From [74].) Copyright Elsevier. Used with permission.

5.3 V_{DC} to manipulate the particles. As the particles enter the device, they can be concentrated into a stream at the edge of the channel via a straight-angled electrode "aligner." The mechanism here is similar to that previously described for interdigitated electrodes. The "aligning" electrode pair can be designed to be wider than the sorting (curved) electrode pair, leading to higher nDEP forces at the same voltage compared to the narrower curved electrodes [74]. Once aligned at the edge of the channel, particles can then be deflected by the sorting (curved) electrode pair, resulting in a spatial separation that depends on the local fluid drag and dielectrophoretic forces. Depending on device design, output flow can be divided into a number of separate outlets. Schnelle et al. divided the output of their device into four separate channels.

Experiment ID #6: traveling-wave dielectrophoresis

Discussion: Travelling-wave dielectrophoresis (twDEP) is the result of an electric field with a spatially varying phase, and is typically created by using an interdigitated electrode array with four separate electric potentials of varying phase applied to individual electrodes (Section 6.3.2). This leads to a levitation force as well as a transverse force that can be used to sort particles according to (6.10). Cui and Morgan designed such a device and demonstrated motion of polystyrene particles via twDEP [81].

Methods: Sample experimental device and solution parameters are listed in Tables 6.6 and 6.8, respectively. In this example, the experiment proceeds in two phases: loading and sorting. Initially, a suspension of particles can be introduced to the channel via pressure-driven flow. Once the channel is filled, the flow can be stopped and the particles allowed to come to rest. At this point, the electrode array can be energized to $1V_{peak}$ and 1 MHz. When the array is energized, particles will experience a negative dielectrophoresis force levitating them above the array. In addition, the spatial variation in phase will tend to drive the particles perpendicular to the array (in either direction, depending on the sign of $\Im[f_{CM}]$). Terminal particle velocity can also be measured, and will be directly related to the magnitude of $\Im[f_{CM}]$.

6.4.2.4   ROT spectra

Electrorotation techniques provide information about electrical phenotypes on a particle-by-particle basis. Because it traps particles in one position and can measure frequency-dependent properties in situ, electrorotation is an ideal technique for observing the electrical response of a single particle to various stimuli.

Experiment ID #7: electrorotation

Discussion: In order to measure the rotation spectra of a sample, individual particles must be trapped and examined under a wide range of frequencies. This is typically accomplished using a quadrupolar set of electrodes.

Chan et al. utilized a quadrupolar electrode array (Section 6.4.1.2) to conduct electrorotation studies on uni- and multilamellar liposomes [24]. Liposome membranes were constructed from the phospholipid 1,2-dioleoylsn-glycero-3-phosphocholine (DOPC) and cholesterol. Liposomes encapsulated a solution of Percoll, which increased their density and made handling easier. The use of constructed liposomes was designed to validate multilayer, effective permittivity models[12].

Based on their data, Chan et al. determined that a single-shell effective permittivity model can be used for uni- and multilamellar liposomes. However, if the liposome contains distinct compartments, the single-shell model fails, and multishell modeling must be employed. In addition, it was confirmed that the electrorotation response is a function of internal as well as membrane properties.

Methods: Electrorotation experiments involve trapping a particle in a single location, subjecting it to a rotating external electric field, and observing the rate of particle rotation as a function of the electric field frequency. Consider the experimental conditions summarized in Tables 6.6 and 6.8. A polynomial, quadrupolar electrode array can be fabricated, after which particles can be introduced to the chamber surrounding the electrode array at a low concentration (10^6 liposomes/mL). Once introduced, particles can be allowed to settle. After settling, the trap can be energized, liposomes can be trapped within the array, and the trapped liposomes can be observed to measure their rate of rotation. As the frequency changes, particles will go through regimes where the rotation rate matches the electric field as well as regimes where particle rotation lags electric field rotation or does not rotate at all. When conducting these measurements, only the liposomes that are many diameters from one another and near the center of the array should be recorded. These precautions ensure that particle-particle interactions are minimized and that the electric field near the rotating particles is spatially uniform.

6.4.3 Insulative dielectrophoresis

Insulative techniques, which rely on compressions or expansions in channel geometry to generate electric field nonuniformities, have been used for a number of applications. Constrictions create locally high electric field regions that can induce deflection [82, 83], trapping [54, 84–86], or sorting via negative dielectrophoresis [36, 87].

In order to understand iDEP systems, it can be beneficial to examine a simple constriction in the electric current path (similar to the theoretical development presented in

---

[12] The effective permittivity technique replaces ε_p for a homogeneous particle with $\varepsilon_{p,effective}$. The derivation involves several iterations of the Laplace equation solution technique. Ultimately, the dipole moment becomes proportional to $\varepsilon_{p,effective}$. A more detailed discussion can be found in [12].

Section 6.3.3.2), and from there, build up our understanding of more complicated systems such as an angled constriction, and finally our experimental example: a curved constriction.

Perpendicular constriction: In a system with a simple constriction (Figure 6.5), there are two modes that can be explored: trapping particles against fluid drag via negative DEP [36] or deflection of particles from a known, controlled position to another output position based on positive- or negative-dielectrophoretic response [82].

Constriction configurations such as these are useful for trapping or concentrating particles against fluid drag forces using negative dielectrophoresis, but trapped particles must be released before forming a bolus that may clog or alter fluid flow patterns. Constrictions such as these have also been used to alter the flow patterns of particles passing through the constriction [82]. A variable constriction was constructed by introducing an immiscible fluid in a side channel. As the immiscible (insulating) fluid was introduced in the side of the main flow channel, surface tension formed a circular constriction. The amount of constriction could be varied by changing the amount of immiscible fluid in the side channel, thereby changing the size of the "bubble" constricting the main channel [88].

Angled constriction: Angled constrictions are an extension of the rectangular constrictions considered earlier. A constriction in channel width can become a constriction in channel depth with a simple change of variables. However, practical fabrication considerations lead to an additional degree of freedom that was not possible with constrictions in width: the angle of incidence, θ. The DEP forces generated by the insulating constriction act primarily perpendicular to the constriction, while fluid flow (in the low Reynolds number limit) retains its direction. A summation of forces at the constriction yields an expression that depends on the magnitude of the dielectrophoretic force and the angle of the constriction. If the DEP force is large enough, for example:

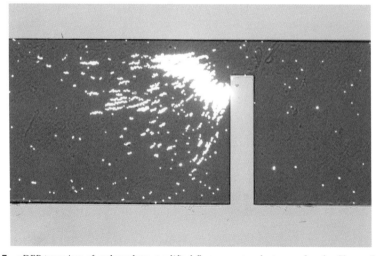

Figure 6.5 nDEP trapping of carboxylate-modified fluorescent polystyrene beads. Channel width is 1 mm, 200 μm in the constricted region, and channel depth is 100 μm. (From the authors' lab; unpublished work.)

$$\left|\vec{F}_{DEP}\right|\cos\theta > \left|\vec{F}_{Drag}\right| \tag{6.33}$$

particles will be stopped at the constriction and "deflected" parallel to the constriction. Angled constrictions have been used to preferentially separate particles and concentrate particles of interest in a single step [87].

Experimental ID #8: postarray

Discussion: Trapping experiments using post-array type devices offer the most common example of insulative dielectrophoretic techniques. Post-array configurations have been used for microbe isolation and detection [54, 84, 85].

The mechanism of action in a post array is largely similar to that of a perpendicular constriction: a constriction in the current path causes a concomitant increase in the local electric field magnitude, leading to an electric field gradient that drives DEP (Section 6.3.3.2). Post-array trapping techniques involve flow of a suspension of analyte past a large array of insulating posts. Fluid flow can be either pressure-driven from a syringe pump or electrokinetically driven via an external electric field. The advantage of electrokinetically driven flow is that a single DC voltage can be applied to induce fluid flow as well as dielectrophoretic trapping. This approach was used successfully by Lapizco-Encinas et al. to separate live and dead *Escherichia coli*.

Like electrode-based devices, iDEP devices have a wide range of variability in the geometry used to shape electric fields. In the case of post-array type devices, the geometric variables that affect the electric field are the post shape (circular, elliptical, square, etc.), gap between posts, and the array angle to the applied electric field. The post shape affects the rate of change of the local electric field between the posts; the gap itself determines the magnitude of the electric field between the posts; and the angle will have a relatively minor impact on the variation of the electric field magnitude[13]. Lapizco-Encinas et al. report that a square array of circular posts yielded the best results (See Tables 6.6 and 6.8).

Methods: Trapping and separation of live and dead *E. coli* can be observed in an insulating post-array, given device and solution characteristics summarized in Tables 6.6 and 6.8, respectively. Once the device is fabricated and filled with particle solution, care must be taken—*as must always be done in such experiments*—to eliminate pressure-driven flow due to reservoir height differences and/or surface tension phenomena. Once the reservoirs are loaded and pressure differences eliminated, flow can be initiated by applying an electric field via electrodes in the external reservoirs. As the applied potential increases, fluid flow rates ($\propto \vec{E}$) and dielectrophoretic forces ($\propto \nabla\left(\vec{E}\cdot\vec{E}\right)$) will

---

[13] Since dielectrophoresis depends on the gradient of the squared electric field magnitude, changes in either the peak value of the electric field—between the posts—or the bulk value—"away" from the posts—will alter the dielectrophoretic force. In this case, the angling of the array will tend to shift insulating posts into what would normally be the "bulk" region for subsequent posts in the array. This will lead to an increase in the "bulk" value of the electric field, and a decrease in the gradient, leading to a minor decrease in the dielectrophoretic force. Lapizco-Encinas et al. do not report significant variation in performance as a function of the angle of the electrode array. In nontrapping regimes, however, Cummings and coworkers have found that the angle of the array will play a significant role in the "streaming" dielectrophoresis effects observed under nontrapping conditions [38–89].

increase. As the applied potential reaches 160V ($|\vec{E}| = 16\,\text{V/mm}$), nDEP forces will begin to dominate fluid drag forces, and live bacteria will trap between posts in the array. Increasing the potential to 400V ($|\vec{E}| = 40\,\text{V/mm}$) will lead to trapping of both live and dead cells. Finally, at a potential of 600V ($|\vec{E}| = 60\,\text{V/mm}$), discrete banding of samples based on differences in dielectrophoretic mobility should appear.

Recall that the dielectrophoretic force is proportional to the permittivity and conductivity of both particle and media, as well as the frequency of the applied electric field. Closer examination of (6.1) yields that as $\omega \to 0$, the complex permittivity becomes i/σ, and f_{CM} becomes:

Table 6.7 Table 6.6 continued

ID	V0-peak	Channel (length x width x depth)	Characteristic Length Scale
1	1V	2 cm x 100 μm x 50 μm	Interelectrode spacing, 10 μm
2	1V	Not Applicable	Interelectrode Spacing, 25 μm
3	2V	Not Applicable	Electrode width, 10 μm; Ring diameter, 50 μm
4	0–20 V_{peak}	Not Applicable	Electrode width, 50–250 μm; Electrode spacing, 50–250 μm
5	5.3V	($\sim$ 1) cm x 300-800 μm x 40-140 μm	Alignment electrode width, 15 μm; Curved electrode width, 7 μm
6	1 V_{peak}	(($\sim$ 1) cm x 300 μm x 70 μm	Electrode width, 10 μm; Electrode spacing, 10 μm
7	Not Available*	Not Applicable	Diagonal electrode spacing, 400 μm
8	160-600 V_{DC}	10.2 mm x Not Applicable x 10 μm	Post diameter, 200 μm; Center-to center distance, 250 μm
9	50 V_{DC}, 750 V_{AC}	1 cm x 2500 μm x 100 μm	Constriction, 10 μm; Constriction ratio, r = 10 : 1

* It is the author's opinion that voltages on the order of 1V should be sufficient.

Table 6.8 Experimental Solution Characteristics for DEP Applications Described in Section 6.4.*

ID	Particle	Size	Concentration
1	Carboxylate-modified polystyrene	1 μm diameter	1.9 $\times$ 10^9 particles/mL
2	Carboxylate-modified polystyrene	2 μm diameter	2.4 $\times$ 10^7 particles/mL
3	Silver-coated polystyrene	20 μm diameter	Not Applicable
4	Jurkat and AC3 stromal cells	($\sim$ 10) μm	5 $\times$ 10^5 cells/mL
5	Unmodified latex	6.4, 10, 15, and 20 μm diameter	Not Available
6	Unmodified polystyrene	10 mm diameter	Not Available
7	Uni-, multilamellar liposomes	Variable, 2.5-12 μm†	Not Applicable, 10^6 liposomes somes/mL
8	Viable and heat-treated *Escherichia coli* cells	1 μm	6 $\times$ 10^6 cells/mL
9	Carboxylate-modified polystyrene	1.75 μm and 2 μm diameter	2 $\to$ 2.4 $\times$ 10^7 particles/mL

* All solution percentages are w/v. IDs correspond to data listed in Table 6.6.
† The expression for electric torque contains a^3 dependence, but this balances against the viscous torque (drag), leading to a rotation rate that is independent of size.

$$f_{CM} = \frac{\sigma_p - \sigma_m}{\sigma_p + 2\sigma_m} \tag{6.34}$$

It has been reported that, in deionized water, the membrane conductivity of live cells is significantly lower than that of dead cells, leading to a significant difference in f_{CM} for the two populations and an observable separation (Figure 6.6). The key advantages of post-array techniques stem from their simplicity of operation. After fabrication, the only necessary component is a power supply that will drive flow and preferential trapping.

Safety considerations: In order to obtain the electric fields required for separation (on the order of 200 V/mm) a high voltage (on the order of 2 kV) power supply must be used. High voltages are hazardous; experimenters should be trained in the proper usage of all equipment and safety measures.

Experimental ID #9: curved constriction

Discussion: Further refinement of the angled constriction in channel depth leads to a continuously varying angle of constriction throughout the channel. The resulting curved constriction in channel depth (Figure 6.7) operates on the same basic principle. Dielectrophoretic forces are still perpendicular to the constriction, but now this force is directed along the radius of constriction curvature. Particles are either trapped and "deflected" at the constriction or allowed to pass. Again, summing forces on a deflected particle at the constriction yields a component parallel to the constriction. As a deflected particle traverses the constriction, the angle between the direction of bulk flow and the DEP force (normal to constriction curvature) changes, until the inequality (6.33) is violated or until the particle reaches the channel wall [36].

Hawkins et al. [36] demonstrated particle trapping in a system with a simple constriction using a relatively low-voltage DC signal to drive electrokinetic flow and a high-voltage AC signal to achieve particle trapping via negative dielectrophoresis. They also demonstrated a separation utilizing a curved constriction in channel depth to separate polystyrene particles on the basis of size.

Methods: Sample experimental parameters used to separate particles using a curved constriction in channel depth are similar to those described earlier for trapping in an

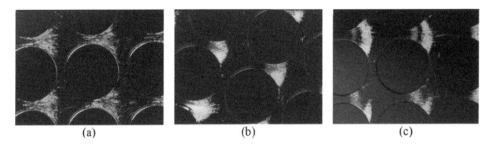

(a)   (b)   (c)

Figure 6.6 Trapping and subsequent separation of live (green) and dead (red) *Escherichia coli* in an insulating post array. (a) 16 V/mm, live cells only are trapped, (b) 40 V/mm, trapping both live and dead cells with observable banding, and (c) 60 V/mm, trapping of both live and deadcells with separation of populations. (From [84].) Copyright American Chemcial Society. Used with permission.

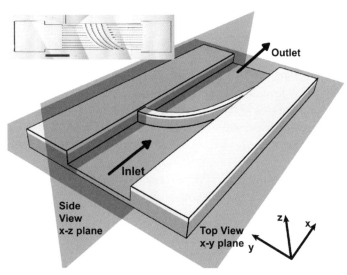

Figure 6.7 A schematic of a constriction in channel depth. Input and output reservoirs, channel top not shown. Inset: a top view of device fabricated in Zeonor 1020R polymer substrate. (From [36].)

insulating post array and are listed in Tables 6.6 and 6.8. External reservoirs can be utilized for sample introduction and application of electric potential via platinum electrodes. As with the post-array experiment (Section 6.4.2), pressure-driven flow due to reservoir height mismatching or surface tension effects should be eliminated by careful observation and the use of large external reservoirs.

Once the sample is loaded, a low, DC potential of 25V can be applied to induce electrophoretic and electroosmotic flow. After flow is established, an AC signal at 1 kHz can be applied in addition to the DC potential. The magnitude of the AC signal is typically 5 to 10 times larger than the DC signal ($125 - 250\ V_{peak}$). At low AC potentials ($< 100\ V_{peak}$), little change will be observed. As the AC potential is increased, deflection of particles at the perpendicular section of the curved constriction will be observed. Once particle deflection is observed, the field amplitude can be increased further to observe additional deflection of particles. In an experiment where two particle populations (e.g., 1.75- and 2-μm diameter) are initially uniformly mixed and distributed throughout the channel, deflection will lead to three distinct regions in the output section of the device. Low θ values will lead to downstream regions containing no particles, as they have all been deflected by the constriction. The next region, with higher values of θ will contain smaller particles only. The lower dielectrophoretic force associated with smaller diameters will lead to less deflection. The last region, downstream of higher θ values, will contain both particle types, with a higher population of larger particles due to deflection.

6.4.4 Summary of experimental parameters

Tables 6.6, 6.7, 6.8, and 6.9 summarize sample experimental parameters for devices discussed in this section.

6.5 Troubleshooting

Tables 6.10, 6.11, and 6.12 summarize a few of the common problems associated with microfluidic dielectrophoresis experiments and some potential solutions.

6.6 Application notes

Dielectrophoresis applications range from fractionating particles based upon their "electrical phenotype" [69] to precise manipulation of single particles for property interrogation to new strategies for the creation of engineered tissues and organs. With the emergence of multistep lab-on-a-chip devices and the use of miniaturized bioanalytical techniques in a microdevice, DEP has become an attractive option for particle and cell manipulation. Owing to the favorable scaling of dielectrophoretic effects at shrinking length scales, the utility of electrokinetically driven flows at these length scales, and the flexibility of microfabrication techniques, integration of these effects is common.

We categorize applications in this section based upon the goals of the user. The first category includes applications that control the relationship between the particle location and the fluid (e.g., holding or trapping particles relative to a fluid volume). The second category includes applications which control the relationship between particle types, usually dynamically, as the particles progress through a device or fluid stream (e.g., particle sorting and fractionation). The third category includes applications in which control the position of particles relative to each other. While many examples overlap two or more of these categories, these categories nonetheless serve as a framework for classifying relevant applications based upon user goals. This section will serve only as an overview for readers, highlighting specific examples rather than providing a comprehensive review of the applications of DEP. These examples provide context for the detailed discussions throughout the chapter on the methods used to implement DEP.

6.6.1 Particle trapping

The most common use of DEP is to fractionate a particle suspension by statically trapping particles within a fluidic channel. Variations in electric field are induced in a fluid

Table 6.9 Comparison of Conclusions from Table 6.8

ID	Fluid	Flow Rate
1	Deionized water	0.05 mL/hr
2	Deionized water	0 mL/hr
3	50% sucsrose/DI water 1% BSA, 1% Triton X-100	0 mL/hr for trapping; 0.6–3.0 mL/hr for clearing; 0.06 mL/hr for release
4	450 mM D-sorbitol	Not Applicable (Slow)
5	Deionized water*	800 μm/sb[†]
6	Deionized water[‡]	0 mL/hr
7	5% mannitol	Not Applicable
8	Deionized water	Variable, electrokinetically driven
9	Deionized water	Variable, electrokinetically driven

* Phosphate buffered saline solution was added to stabilize the conductivity of the solution at 1.7 mS/m.
† Volumetric flow rates were not available for this experiment.
‡ An electrolyte solution was added to bring the solution conductivity to 1 mS/m.

Table 6.10 Tips and Suggestions for Common Problems Encountered in Dielectrophoresis Experiments

Problem	Potential Causes	Solutions
Inconsistent flow rate	In pressure-driven flow: Check pump for failure Check pump flow rate Check device for bubbles at inlet, outlet, and electrodes	In pressure-driven flow: Fix pump Use pump rated for low flow rates Fill device with compatible, low surface tension fluid (e.g. methanol), then thoroughly flush with working fluid For bubbles forming at electrodes: Reduce fluid conductivity Choose different electrode material (see Section 6.4.0.1)
	In electrokinetic flow: check power supply check electrodes for contact check for bubbles in device check for bubbles at electrode	In electrokinetic flow: Make sure load (channel) resistance is high for high-voltage applications use a lower conductivity working fluid reduce channel cross section increase channel length remove bubbles from current path (see above) if electrode contact area is small, bubbles here can also break the current path, take measures to reduce electrolysis reduce fluid conductivity choose different electrode material (see Section 6.4.0.1)

Table 6.11 More Troubleshooting

Problem	Potential Causes	Solutions
Particle adhesion	Electric field too high Hydrophobic interactions	Lower applied potential See Section 6.4.0.1
No or low particle DEP response	$f_{CM} \approx 0$ Check fluid conductivity Check particle conductivity Check electric field frequency	Change σ_m (usually decrease) Change ω
	In trapping experiments: if $f_{CM} > 0$, channel depth too large if $f_{CM} < 0$, flow rate too high	Decrease channel depth Decrease flow rate
	Particles not close enough to electrodes (or constrictions in iDEP)	Flow focusing (Section 6.4.0.1)
Particle aggregation/ clogging	Particle concentration too high DEP chaining	Dilute particles Lower $\left| \vec{E} \right|$
Bubble generation during experiment	Electrolysis or boiling Leaking channel	See Section 6.4.0.1 Seal channel completely

channel and exert a positive or negative DEP force, causing the particle of interest to be statically confined at or near a peak or valley in the electric field distribution. Particle confinement often occurs as a colloidal suspension propagates through the channel. While positive DEP (pDEP) is most frequently used, owing to the simplicity of system design, negative DEP (nDEP) has been used as well, owing to the lower fields it applies to cells. In order to hold or trap specific populations of particles from a fluid suspension, (1) the DEP forces on the particles of interest must be greater than the hydrodynamic and

Table 6.12 More Troubleshooting

Problem	Potential Causes	Solutions
Particle response/aggregation via DEP too high or unexpected*	Nonuniformities or defects in channel or electrode geometry leading to locally large ∇E^2	Refine fabrication process, use flat surfaces for embossing and remove sharp corners in design see Section 6.4.0.1
Recirculating flow near electrodes or constrictions	$\left|\vec{E}\right|$ too high, leading to thermal variation in σ_m	Lower $\left|\vec{E}\right|$
	electrodes or constrictions	Increase feature spacing
	iDEP constrictions too close together	Decrease $\left|\vec{E}\right|$
	($\left|\vec{E}\right|$ not decaying fully to bulk value between constrictions)	Increase feature spacing
	AC electroosmosis	
	Closely spaced angled constrictions in pressure-driven flow	

*It is difficult to address all experimental challenges. Here we attempt to point the experimenter toward a few potential sources of problems.

gravity forces on the particles, and (2) differences in dielectrophoretic mobility must exist to discriminate populations of particles.

Becker et al. created an "electroaffinity column" based upon differences in dielectrophoretic mobility for separating leukaemia cells from blood [90]. They created a wide channel with offset, castellated electrodes (see Section 6.4.1.2) along the bottom and injected diluted human blood that was spiked with HL-60 leukemia cells. Erythrocytes (red blood cells) and leukemia cells were captured from the blood with the application of a 5 V_{peak}, 200-kHz AC electrical signal to the electrodes. An elution volume was flowed through the device and the frequency of the electrical signal was decreased to 80 kHz, causing the release of blood cells but retention of leukemia cells. At 80 kHz, Becker et al. calculated the polarizabilities of the leukemia cells to yield a strongly positive DEP force and the erythrocytes to have a repulsive force. As a result, blood cells were eluted and the retained leukemia cell population was of high purity. The prototype electroaffinity column created by Becker et al. was able to sort approximately 1000 cells/sec. This technology is scalable, enabling higher throughput of cell sorting by increasing the size of the channel and electrode area and thus allowing a greater volumetric flow rate of sample through the device.

Another application of DEP was demonstrated by Yasukawa et al. [80], who used an nDEP "cage" to confine particles within a desired volume and subsequently perform an enzyme-linked immunosorbent assay (ELISA). An ELISA is a quantitative assay that involves the conjugation of fluorophores or chromogens to target surface molecules by antibody coupling for the measurement of selected molecules. The authors patterned electrodes on the top and bottom surface of a channel to create a "caged" area. The downstream electrodes were activated, and polystyrene latex microspheres were concentrated into a stream using preliminary focusing electrodes (see Section 6.4.1.3). Once the desired quantity of microspheres was collected, the upstream DEP electrode was activated to close the "cage" and prevent additional particles from entering. The authors then performed an immunoreaction to attach antibodies and subsequently fluorphores to the microspheres by adding reagents to the fluid stream. Fluorescence intensity was measured with conventional microscopic techniques. At the conclusion of the assay, both electrodes were deactivated and the labeled particles were released down the channel in the fluid stream.

In a variation on the typical pDEP capture of particles, Urdaneta and Smela [91] used electrodes at multiple frequencies to capture two separate populations of yeast cells (live and dead) from a heterogenous cell suspension. Urdaneta and Smela took advantage of being able to tune the effective Clausius-Mossotti factor (f_{CM}) or frequency response and thus create regions of p- and nDEP that were opposite for each cell type. They created three planar electrodes, one grounded and two electrodes at varying frequencies and amplitudes, in two different geometries within a fluidic compartment. An equal part mixture of live and dead yeast cells was injected into the device and the two cell populations were separated with applied signals at 5 kHz and 5 MHz, respectively. Multiple-frequency DEP (MFDEP) has been used for electro-rotational spectra [92], particle levitation [93], and traveling wave DEP (twDEP) [94] applications.

Separation and trapping of cells by their electrical phenotype using DEP enables the efficient capture of a high purity population of cells. As mentioned earlier, Becker et al., as well as others, used the "electroaffinity" column to demonstrate the ability to separate different cell types from each other, including the clinically applicable separation of tumor cells from nonpathologic cells [90]. In addition to separating a specific cell type from a heterogenous population, DEP has also been utilized to separate same cell populations to discern physiologic differences such as activation of mitosis [95, 96], exposure to drugs [97–100], induced cell differentiation [3, 101, 102], and cell death [91, 103–105]. Using DEP-based devices to potentially separate the same type of cells based upon physiologic parameters is a powerful tool to identify responses of cells to various soluble agents. Besides the direct application of DEP trapping to biological problems, capturing cells also enables a practical method of manipulating cells and particles within a multistep microfluidic lab-on-a-chip device. Investigators have used DEP "gate" electrodes to enable the passage of selected particles based upon dielectrophoretic mobility [106]. This idea can be extended to incorporate multiple gate electrodes to control spatial and temporal particle motion through a microfluidic chip. Likewise, the ability to localize cells within a specified area, either on a 2-D surface or within a 3-D cage, potentially creates a platform to investigate responses of a population of phenotypically similar cells to different chemical environments.

6.6.2 Particle sorting and fractionation

Many continuous flow separation techniques use dielectrophoresis to sort, fractionate, or enrich particle populations dynamically within a fluid stream. These methods use either positive or negative DEP (or both) in microfluidic devices to deflect particles transversely such that an enriched population of the desired particles passes into one of several downstream exits. The heterogenous electric fields are created either through variations in fluidic channel geometry (insulative DEP, iDEP) or electrodes (electrode-based DEP, eDEP) along the channel in various configurations. Application of a voltage across a channel with varying geometry, or across electrodes along the length of the fluidic channel, creates locally nonuniform fields and a resulting DEP response from the particles or cells in the fluid stream (see Sections 6.2 and 6.3).

Many continuous-flow dielectrophoresis systems utilize insulative configurations. Cummings and Singh [89] explored iDEP using insulating postarrays and dc electric fields to trap particles or align them in streams. The authors used posts of various shapes and adjusted the orientation of the postarrays within the electric field to induce a p- or nDEP response and thus concentrate or deplete streams of microspheres or reversibly immobilize the microspheres on the insulating posts. In another application, Hawkins et al. [36] used a curved ridge to create an insulative constriction region within a microchannel to separate 1.75-μm from 2-μm microspheres in a heterogenous population (Figure 6.8). Electrokinetically driven flow within the device was established via a dc-offset, ac electric field and particles were deflected along the curved constriction to change the particle's position within the channel (see Section 6.4.2). The particle deflection depends on the orientation of the constriction and thus Hawkins et al. created a tunable system to separate particles based upon dielectrophoretic mobility.

Continuous flow separations have also been demonstrated in electrode-based systems. Hu et al. [77] devised a device that uses DEP to sort cells in a manner analogous to a fluorescence-activated cell sorting (FACS) device (Figure 6.8). In a manner analogous to fluorescent staining for FACS, the authors conjugated beads to some bacterial cells via antibody coupling and mixed the bead conjugated bacteria with nonconjugated bacterial cells. The newly made bead-bacterial cell conjugates are sensitive to a DEP force due to the increased size of the complex. Hu et al. injected the bacterial cell suspension into their device and utilized angled electrodes (see Section 6.4.1.3) to induce a nDEP force on the labeled bacterial cells to deflect them into a separate collection channel. The device collected 95% of the labeled cells with a throughput of 2–3 $\times$ 10^7 cells/hr, a rate comparable to conventional cell sorters. Braschler et al. [107] introduced fine control of a particle's deflection along the length of a channel using multiple frequency eDEP (Figure 6.9). Two electrode arrays are patterned on either side of a fluidic channel with each array excited at different frequencies. The difference in frequencies and the ability to tune the system allows the user to balance the DEP forces on the particle and control a particles' position within the channel as it travels down the length of the electrode array.

Continuous flow techniques have also been used to create temporal separation of particles within a flow. Taking advantage of differing dielectrophoretic mobilities, several groups have used DEP for field flow fractionation (FFF-DEP) [108, 109]. As a particle

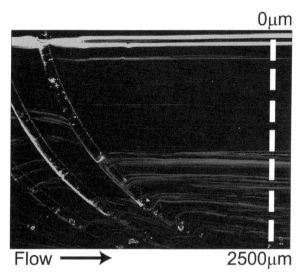

Figure 6.8   Time-lapse (60 seconds) fluorescence microscopy image. 1.75-μm diameter polystyrene spheres (green) are sorted to the center of the channel, and 2-μm diameter polystyrene spheres (red) are sorted to the bottom of the channel. Copyright American Chemical Society. Used with permission.

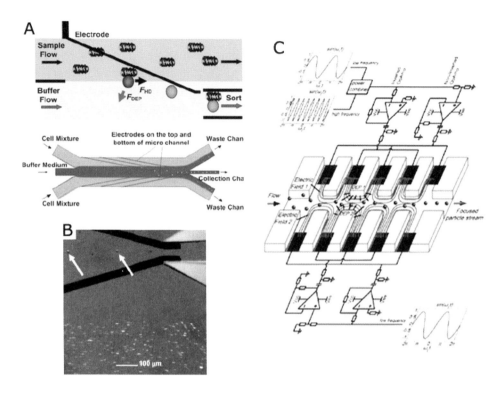

Figure 6.9 DEP was used to continuously separate bacteria conjugated with beads from a heterogenous sample (a) by Hu et al. [77]. Angled electrodes deflect bacteria conjugated beads into a buffer stream that exits the device through a collection channel (b). Braschler et al. [107] used MFDEP to focus a stream of particles within a channel (c). The differential application of two frequencies to electrodes along each side of a channel enables tunable control of a particle's position. (a) copyright National Academy of Sciences. (b, c) copyright Royal Society of Chemistry. Used with permission.

stream progresses through a microfluidic channel, DEP forces oppose sedimentation (gravity) forces and particles with a weaker DEP response will settle to a lower-velocity streamline within the Poiseuille flow profile. Differences in fluid velocity across the cross-section of the channel cause elution of the DEP-responsive particles at different times. Thus, in a manner similar to a separation column, particles with high DEP response will be eluted from the device subsequent to those with weaker DEP response.

An advantage of DEP is the ability to separate particles or cells without the need for secondary affinity coupling steps to achieve separation. Traditional techniques such as FACS, flow cytometry, ELISA and affinity chromatography require a pretreatment with fluorescent molecules (FACS, flow cytometry), or speciflc antigens conjugated to the particles (ELISA, flow chromatography) prior to separation. This bypass of prelabeling eliminates use of additional reagents and saves processing time. Like many of the commonly cited benefits of microfluidic systems—low Reynolds number, laminar flow, low reagent volumes—dielectrophoresis benefits from the same favorable scaling at small scales. The DEP force is proportional to the gradient of the electric field squared, or the inverse of the characteristic length cubed, and thus the DEP force increases dramatically in these microfluidic environments.

The use of DEP as a continuous-flow separation technique has been demonstrated for several types of particles including polymeric spheres [36, 82], bacteria [110], yeast [111], and mammalian cells [112]. Similar to trapping applications, investigators have used DEP in continuous flow applications to separate populations of the same cell with physiologic differences. Examples include separation of stem cells from their differentiated progeny [113] and mammalian cells based upon their cell-cycle phase [112]. These continuous-flow, DEP fractionation devices enable rapid sorting of cells and particles which can be scaled up to fractionate larger populations. Separation of large populations of phenotypically similar cells is of great interest to biologists and tissue engineers who investigate drug interactions with phenotypically varied cells and create in vitro tissue models from pure cell populations.

6.6.3   Single-particle trapping

In addition to bulk processing of large numbers of particles, the appropriate electrode design enables manipulation of single particles or small groups thereof. These methodologies have enabled significant advances in microscale bioanalytical techniques. From a biological perspective, the utility of DEP to manipulate individual particles for the discrimination of properties is promising and opens up new tools to investigate single cells or cell-cell interactions. In general, strategies to address the fine manipulation of particles involve either (1) the use of electrode arrays that are individually addressable, or (2) the use of light on photosensitive conducting materials to dynamically alter the electrode configuration.

An electrode post-array was used by Voldman et al. [68] to create a microfabricated cytometer to trap and release individual cells with temporal control (Figures 6.10 and 6.2(b), Section 6.4.1.2). They used an array of posts to create several quadrupole DEP traps. The nDEP traps captured calcein labeled HL-60 cells and maintained the position of the cell for more than 40 minutes. While the cells were trapped, the investigators were able to perform fluorescence measurements in the device and selectively release individual cells at different time points. Similarly, Taff and Voldman [70] improved on the original cell sorting array by creating addressable traps using a circle-dot pDEP electrode design that is highly scalable (see Section 6.4.1.2). Shih et al. [114] developed a device for sorting particles with trapping, detection, and sorting regions. The cells were captured in quadrupolar traps, analyzed in real time in a second region of the chip, and then passed into a wide channel with several electrodes patterned on the lower surface (Figure 6.10). The electrodes are addressed appropriately to create nDEP in all areas except a path to the correct outlet channel. The authors were able to successfully sort microspheres with a protein coating from uncoated microspheres into one of five output channels.

While electrode arrays are usually implemented in a 2-D configuration, Albrecht et al. [115] used pDEP and photopolymerizable hydrogels to precisely position chondrocytes within a 3-D environment (Figure 6.10). They injected cells within a prepolymer solution between two plates patterned with electrodes. The electrodes were energized and DEP caused clusters of cells to form based upon the electrode geometry. Once the cells were patterned, the prepolymer was polymerized using UV light, locking cells into position within a hydrogel. This methodology of using DEP to spatially control cell-cell spacing is a significant step forward in the field of tissue engineering, because conventional techniques are not able to control intercellular spacing. These 3-D, in vitro

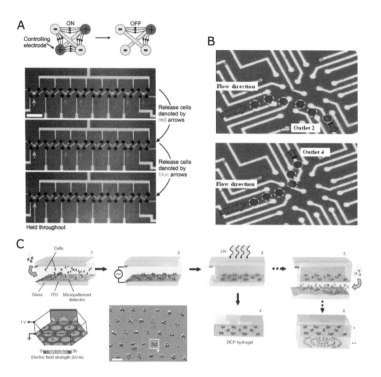

Figure 6.10 A dynamic cytometer was created by Voldman and colleagues [68] with an array of quadrupole traps (a). Individual cells were captured within the respective traps, fluorescence measurements made, and then selectively released. Shih et al. [114] used nDEP to dynamically control which outlet particles are sorted into (b). DEP has also been used to pattern cells with controlled spacing in a 3-D environment (c) for the investigation of the role of cell signaling on cell growth and tissue formation. Albrecht and colleagues [115] created aggregates of cells with a patterned electrode with evenly spaced trapping regions. Prepolymer was flowed into the channel and photopolymerized to create a hydrogel encasing the cell aggregates. Serial patterning and polymerization can then be used to create engineered tissue analogs with controlled cell localization within a 3-D environment. (a) copyright American Chemical Society, (b) copyright IEEE, and (c) copyright Nature Publishing Group. Used with permission.

models allow the investigation of local micro-scale cell signaling and the resulting impact on cell growth and tissue formation.

As evidenced by these examples, DEP's strength is in manipulation at the single-particle/single-organism level. While techniques such as optical tweezers and ultrasonic particle manipulation already exist to trap and isolate single particles, DEP systems, in addition to being simpler, cheaper, and potentially more parallelizable, have the advantage of inducing either positive or negative forces. Under the appropriate experimental conditions, this ability leads to selective trapping or controlled aggregation of particles within novel trap designs or 3-D cages. In addition to these direct applications, DEP has also proven useful to control cells and particles in multiple-process lab-on-a-chip devices using gates, traps, and particle aligners to regulate spatial and temporal trafficking through microdevices [116, 117].

References

[1] Pohl, H. A., "The motion and precipitation of suspensions in divergent electric fields," *Journal of Applied Physics,* Vol. 22, no. 7 1951, pp. 869–871.

[2] Pohl, H. A., *Dielectrophoresis: The behavior of neutral matter in nonuniform electric fields,* New York, NY: Cambridge University Press 1978.

[3] Wang, X. B., et al., "A unified theory of dielectrophoresis and traveling-wave dielectrophoresis," *Journal of Physics D-Applied Physics,* Vol. 27, no. 7 1994, pp. 1571–1574.

[4] Wang, X. J., Wang, X. B., and Gascoyne, P. R. C., "General expressions for dielectrophoretic force and electrorotational torque derived using the maxwell stress tensor method," *Journal of Electrostatics,* Vol. 39, no. 4 1997, pp. 277–295.

[5] Kang, K. H. and Li, D. Q., "Force acting on a dielectric particle in a concentration gradient by ionic concentration polarization under an externally applied dc electric field," *Journal of Colloid and Interface Science,* Vol. 286, no. 2 2005, pp. 792–806.

[6] Liu, H. and Bau, H. H., "The dielectrophoresis of cylindrical and spherical particles submerged in shells and in semi-infinite media," *Physics of Fluids,* Vol. 16, no. 5 2004, pp. 1217–1228.

[7] Rosales, C. and Lim, K. M., "Numerical comparison between maxwell stress method and equivalent multipole approach for calculation of the dielectrophoretic force in single-cell traps," *Electrophoresis,* Vol. 26, no. 11 2005, pp. 2057–2065.

[8] Al-Jarro, A., et al., "Direct calculation of maxwell stress tensor for accurate trajectory prediction during dep for 2d and 3d structures," *Journal of Physics D-Applied Physics,* Vol. 40, No. 1 2007, pp. 71–77.

[9] Jones, T. B., Wang, K. L., and Yao, D. J., "Frequency-dependent electromechanics of aqueous liquids: Electrowetting and dielectrophoresis," *Langmuir,* Vol. 20, No. 7 2004, pp. 2813–2818.

[10] Liu, Y., et al., "Immersed electrokinetic finite element method," *International Journal for Numerical Methods in Engineering,* Vol. 71, No. 4 2007, pp. 379–405.

[11] Singh, P. and Aubry, N., "Trapping force on a finite-sized particle in a dielectrophoretic cage," *Physical Review E,* Vol. 72, No. 1 2005.

[12] Jones, T. B., *Electromechanics of Particles,* New York, NY: Cambridge University Press 1995.

[13] Castellarnau, M., et al., "Dielectrophoresis as a tool to characterize and differentiate isogenic mutants of escherichia coli," *Biophysical Journal,* Vol. 91, No. 10 2006, pp. 3937–3945.

[14] Ehe, A. Z., et al., "Bioimpedance spectra of small particles in liquid solutions: Mathematical modeling of erythrocyte rouleaux in human blood," *Cross-Disciplinary Applied Research in Materials Science and Technology,* Vol. 480-481 2005, pp. 251–255.

[15] Gimsa, J., "A comprehensive approach to electro-orientation, electrodeformation, dielectrophoresis, and electrorotation of ellipsoidal particles and biological cells," *Bioelectrochemistry,* Vol. 54, No. 1 2001, pp. 23–31.

[16] Gimsa, J., et al., "Dielectric-spectroscopy of human erythrocytes -investigations under the influence of nystatin," *Biophysical Journal,* Vol. 66, No. 4 1994, pp. 1244–1253.

[17] Green, N. G. and Jones, T. B., "Numerical determination of the effective moments of non-spherical particles," Journal of Physics D-Applied Physics, Vol. 40, No. 1 2007, pp. 78–85.

[18] Maswiwat, K., et al., "Simplified equations for the transmembrane potential induced in ellipsoidal cells of rotational symmetry," *Journal of Physics D-Applied Physics,* Vol. 40, No. 3 2007, pp. 914–923.

[19] Rivette, N. J. and Baygents, J. C., "A note on the electrostatic force and torque acting on an isolated body in an electric field," *Chemical Engineering Science,* Vol. 51, No. 23 1996, pp. 5205–5211.

[20] Simeonova, M. and Gimsa, J., "Dielectric anisotropy, volume potential anomalies and the persistent maxwellian equivalent body," *Journal of Physics-Condensed Matter,* Vol. 17, No. 50 2005, pp. 7817–7831.

[21] Archer, S., Morgan, H., and Rixon, F. J., "Electrorotation studies of baby hamster kidney fibroblasts infected with herpes simplex virus type 1," *Biophysical Journal,* Vol. 76, No. 5 1999, pp. 2833–2842.

[22] Bakirov, T. S., et al., "Analysis of cell depolarization mechanism at the initial stages of virus-cell interaction," *Doklady Akademii Nauk,* Vol. 363, No. 2 1998, pp. 258–259.

[23] Becker, F. F., et al., "Separation of human breast-cancer cells from blood by differential dielectric affinity," *Proceedings of the National Academy of Sciences of the United States of America,* Vol. 92, No. 3 1995, pp. 860–864.

[24] Chan, K. L., et al., "Electrorotation of liposomes: verification of dielectric multi-shell model for cells," *Biochimica Et Biophysica Acta-Lipids and Lipid Metabolism,* Vol. 1349, No. 2 1997, pp. 182–196.

[25] Egger, M. and Donath, E., "Electrorotation measurements of diamide-induced platelet activation changes," *Biophysical Journal,* Vol. 68, No. 1 1995, pp. 364–372.

[26] Falokun, C. D. and Markx, G. H., "Electrorotation of beads of immobilized cells," *Journal of Electrostatics,* Vol. 65, No. 7 2007, pp. 475–482.

[27] Falokun, C. D., Mavituna, F., and Markx, G. H., "Ac electrokinetic characterisation and separation of cells with high and low embryogenic potential in suspension cultures of carrot (daucus carota)," *Plant Cell Tissue and Organ Culture,* Vol. 75, No. 3 2003, pp. 261–272.

[28] Gascoyne, P., et al., "Microsample preparation by dielectrophoresis: isolation of malaria," *Lab on a Chip,* Vol. 2, No. 2 2002, pp. 70–75.

[29] Gimsa, J., et al., "Dielectrophoresis and electrorotation of neurospora slime and murine myeloma cells," *Biophysical Journal,* Vol. 60, No. 4 1991, pp. 749–760.

[30] Huang, Y., et al., "Membrane changes associated with the temperature-sensitive p85(gag-mos)-dependent transformation of rat kidney cells as determined by dielectrophoresis and electrorotation," *Biochimica Et Biophysica Acta-Biomembranes,* Vol. 1282, No. 1 1996, pp. 76–84.

[31] Huang, Y., et al., "Electrorotational studies of the cytoplasmic dielectric-properties of friend murine erythroleukemia-cells," *Physics in Medicine and Biology,* Vol. 40, No. 11 1995, pp. 1789–1806.

[32] Chwang, A. and Wu, T., "Hydromechanics of low-reynolds-number flow. 4. translation of spheroids," *Journal of Fluid Mechanics,* Vol. 75 1976, pp. 677–689.

[33] Brenner, H., "The slow motion of a sphere through a viscous fluid towards a plane surface," *Chemical Engineering Science,* Vol. 16 1961, pp. 242–251.

[34] Batchelor, G., "Brownian diffusion of particles with hydrodynamic interaction," *Journal of Fluid Mechanics,* Vol. 74 1976, pp. 1–29.

[35] Goldman, A., Cox, R., and Brenner, H., "The slow motion of two identical arbitrarily oriented spheres through a viscous fluid," *Chemical Engineering Science,* Vol. 21 1966, pp. 1151–1170.

[36] Hawkins, B. G., et al., "Continuous-flow particle separation by 3d insulative dielectrophoresis using coherently shaped, dc-biased, ac electric flelds," *Analytical Chemistry,* Vol. 79, No. 19 2007, pp. 7291–7300.

[37] Chang, D. E., Loire, S., and Mezic, I., "Separation of bioparticles using the traveling wave dielectrophoresis with multiple frequencies," *Proceedings of the 42nd IEEE Conference on Decision and Control,* Vol. 6, Maui, Hawaii, USA 2003, pp. 6448–6453.

[38] Cummings, E. B. and Singh, A. K., "Dielectrophoresis in microchips containing arrays of insulating posts: Theoretical and experimental results," *Analytical Chemistry,* Vol. 75, No. 18 2003, pp. 4724–4731.

[39] Schnelle, T., Muller, T., and Fuhr, G., "Trapping in ac octode field cages," *Journal of Electrostatics,* Vol. 50, No. 1 2000, pp. 17–29.

[40] Schnelle, T., et al., "Combined dielectrophoretic field cages and laser tweezers for electrorotation," *Applied Physics B-Lasers and Optics,* Vol. 70, No. 2 2000, pp. 267–274.

[41] Castellanos, A., et al., "Electrohydrodynamics and dielectrophoresis in microsystems: scaling laws," *Journal of Physics D-Applied Physics,* Vol. 36, No. 20 2003, pp. 2584–2597.

[42] Voldman, J., et al., "Design and analysis of extruded quadrupolar dielectrophoretic traps," *Journal of Electrostatics,* Vol. 57, No. 1 2003, pp. 69–90.

[43] Armani, M. D., et al., "Using feedback control of microflows to independently steer multiple particles," *Journal of Microelectromechanical Systems,* Vol. 15, No. 4 2006, pp. 945–956.

[44] Choi, S. and Park, J. K., "Microfluidic system for dielectrophoretic separation based on a trapezoidal electrode array," *Lab on a Chip,* Vol. 5, No. 10 2005, pp. 1161–1167.

[45] Dalton, C. and Kaler, K. V. I. S., "A cost effective, re-configurable electrokinetic microfluidic chip platform," *Sensors and Actuators B-Chemical,* Vol. 123, No. 1 2007, pp. 628–635.

[46] Lagally, E. T., Lee, S. H., and Soh, H. T., "Integrated microsystem for dielectrophoretic cell concentration and genetic detection," *Lab on a Chip,* Vol. 5, No. 10 2005, pp. 1053–1058.

[47] Li, S., et al., "Aligned single-walled carbon nanotube patterns with nanoscale width, micron-scale length and controllable pitch," *Nanotechnology,* Vol. 18, No. 45 2007.

[48] Liu, Y. S., et al., "Electrical detection of germination of viable model bacillus anthracis spores in microfluidic biochips," *Lab on a Chip,* Vol. 7, No. 5 2007, pp. 603–610.

[49] Rajaraman, S., et al., "Rapid, low cost microfabrication technologies toward realization of devices for dielectrophoretic manipulation of particles and nanowires," *Sensors and Actuators B-Chemical,* Vol. 114, No. 1 2006, pp. 392–401.

[50] Varshney, M., et al., "A label-free, microfluidics and interdigitated array microelectrode-based impedance biosensor in combination with nanoparticles immunoseparation for detection of escherichia coli O157 : H7 in food samples," *Sensors and Actuators B-Chemical,* Vol. 128, No. 1 2007, pp. 99–107.

[51] Wang, L., Flanagan, L., and Lee, A., "Side-wall vertical electrodes for lateral field microfluidic applications," *Journal of Microelectromechanical Systems,* Vol. 16, No. 2 2007, pp. 454–461.

[52] Yang, L. J., et al., "A multifunctional micro-fluidic system for dielectrophoretic concentration coupled with immuno-capture of low numbers of listeria monocytogenes," *Lab on a Chip,* Vol. 6, No. 7 2006, pp. 896–905.

[53] Cummings, E. B., et al., "Fast and selective concentration of pathogens by insulator-based dielectrophoresis," *Abstracts of Papers of the American Chemical Society,* Vol. 230 2005, pp. U404–U405.

[54] Lapizco-Encinas, B. H., et al., "An insulator-based (electrodeless) dielectrophoretic concentrator for microbes in water," *Journal of Microbiological Methods,* Vol. 62, No. 3 2005, pp. 317–326.

[55] White, F. M., *Fluid Mechanics,* New York, NY: McGraw Hill, 5 edn. 2003.

[56] Xuan, X., "Joule heating in electrokinetic flow," *Electrophoresis,* Vol. 298 2008, pp. 33–43.

[57] Green, N. G., et al., "Electrothermally induced fluid flow on microelectrodes," *Journal of Electrostatics,* Vol. 53, No. 2 2001, pp. 71–87.

[58] Ding, W., "Micro/nano-particle manipulation and adhesion studies," *Journal of Adhesion Science and Technology,* Vol. 22, No. 5-6 2008, pp. 457–480.

[59] Kim, S., Lee, S., and Suh, K., "Cell research with physically modified microfluidic channels: A review," *Lab on a Chip,* Vol. 8, No. 7 2008, pp. 1015–1023.

[60] Radisic, M., Iyer, R., and Murthy, S., "Micro-and nanotechnology in cell separation," *International Journal of Nanomedicine,* Vol. 1, No. 1 2006, pp. 3–14.

[61] Taylor, J., Stubley, G., and Ren, C., "Experimental determination of sample stream focusing with fluorescent dye," *Electrophoresis,* Vol. 29, No. 14 2008, pp. 2953–2959.

[62] Esch, M. B., et al., "Influence of master fabrication techniques on the characteristics of embossed microfluidic channels," *Lab on a Chip,* Vol. 3, No. 2 2003, pp. 121–127.

[63] Sun, T., Morgan, H., and Green, N. G., "Analytical solutions of ac electrokinetics in interdigitated electrode arrays: Electric field, dielectrophoretic and traveling-wave dielectrophoretic forces," *Physical Review E,* Vol. 76, No. 4 2007.

[64] Sanchis, A., et al., "Dielectric characterization of bacterial cells using dielectrophoresis," *Bioelectromagnetics,* Vol. 28, No. 5 2007, pp. 393–401.

[65] Abe, M., et al., "Three-dimensional arrangements of polystyrene latex particles with a hyperbolic quadruple electrode system," *Langmuir,* Vol. 20, No. 12 2004, pp. 5046–5051.

[66] Fuhr, G., et al., "High-frequency electric field trapping of individual human spermatozoa," *Human Reproduction,* Vol. 13, No. 1 1998, pp. 136–141.

[67] Ikeda, I., et al., "Fabrication of planar multipole microelectrodes for dielectrophoresis by laser ablation," *Bunseki Kagaku,* Vol. 51, No. 9 2002, pp. 767–772.

[68] Voldman, J., et al., "A microfabrication-based dynamic array cytometer," *Analytical Chemistry,* Vol. 74, No. 16 2002, pp. 3984–3990.

[69] Voldman, J., "Electrical forces for microscale cell manipulation," *Annual Review of Biomedical Engineering,* Vol. 8 2006, pp. 425–454.

[70] Taff, B. M. and Voldman, J., "A scalable addressable positive-dielectrophoretic cell-sorting array," *Analytical Chemistry,* Vol. 77, No. 24 2005, pp. 7976–7983.

[71] Mietchen, D., et al., "Automated dielectric single cell spectroscopy -temperature dependence of electrorotation," *Journal of Physics D-Applied Physics,* Vol. 35, No. 11 2002, pp. 1258–1270.

[72] Morgan, H., et al., "Single cell dielectric spectroscopy," *Journal of Physics D-Applied Physics,* Vol. 40, No. 1 2007, pp. 61–70.

[73] Sebastian, A., Buckle, A. M., and Markx, G. H., "Formation of multilayer aggregates of mammalian cells by dielectrophoresis," *Journal of Micromechanics and Microengineering,* Vol. 16, No. 9 2006, pp. 1769–1777.

[74] Schnelle, T., et al., "The influence of higher moments on particle behaviour in dielectrophoretic field cages," *Journal of Electrostatics,* Vol. 46, No. 1 1999, pp. 13–28.

[75] Wang, X. B., et al., "Dielectrophoretic manipulation of cells with spiral electrodes," *Biophysical Journal,* Vol. 72, No. 4 1997, pp. 1887–1899.

[76] Vahey, M. D. and Voldman, J., "An equilibrium method for continuous-flow cell sorting using dielectrophoresis," *Analytical Chemistry,* Vol. 80, No. 9 2008, pp. 3135–3143.

[77] Hu, X. Y., et al., "Marker-specific sorting of rare cells using dielectrophoresis," *Proceedings of the National Academy of Sciences of the United States of America,* Vol. 102, No. 44 2005, pp. 15757–15761.

[78] Holmes, D. and Morgan, H., "Cell sorting and separation using dielectrophoresis," *Electrostatics 2003,* No. 178 2004, pp. 107–112.

[79] Holmes, D., Morgan, H., and Green, N. G., "High throughput particle analysis: Combining dielectrophoretic particle focussing with confocal optical detection," *Biosensors & Bioelectronics,* Vol. 21, No. 8 2006, pp. 1621–1630.

[80] Yasukawa, T., et al., "Flow sandwich-type immunoassay in microfluidic devices based on negative dielectrophoresis," *Biosensors & Bioelectronics,* Vol. 22, No. 11 2007, pp. 2730–2736.

[81] Cui, L. and Morgan, H., "Design and fabrication of travelling wave dielectrophoresis structures," *Journal of Micromechanics and Microengineering,* Vol. 10, No. 1 2000, pp. 72–79.

[82] Kang, K. H., et al., "Continuous separation of microparticies by size with direct current-dielectrophoresis," *Electrophoresis,* Vol. 27, No. 3 2006, pp. 694–702.

[83]   Kang, K. H., et al., "Effects of dc-dielectrophoretic force on particle trajectories in microchannels," *Journal of Applied Physics,* Vol. 99, No. 6 2006.

[84]   Lapizco-Encinas, B. H., et al., "Dielectrophoretic concentration and separation of live and dead bacteria in an array of insulators," *Analytical Chemistry,* Vol. 76, No. 6 2004, pp. 1571–1579.

[85]   Lapizco-Encinas, B. H., et al., "Insulator-based dielectrophoresis for the selective concentration and separation of live bacteria in water," *Electrophoresis,* Vol. 25, No. 10-11 2004, pp. 1695–1704.

[86]   Mela, P., et al., "The zeta potential of cyclo-olefin polymer microchannels and its effects on insulative (electrodeless) dielectrophoresis particle trapping devices," *Electrophoresis,* Vol. 26, No. 9 2005, pp. 1792–1799.

[87]   Barrett, L. M., et al., "Dielectrophoretic manipulation of particles and cells using insulating ridges in faceted prism microchannels," *Analytical Chemistry,* Vol. 77, No. 21 2005, pp. 6798–6804.

[88]   Barbulovic-Nad, I., et al., "Dc-dielectrophoretic separation of microparticles using an oil droplet obstacle," *Lab on a Chip,* Vol. 6 2006, pp. 274–279.

[89]   Cummings, E. B., "Streaming dielectrophoresis for continuous-flow microfluidic devices," *IEEE Engineering in Medicine and Biology Magazine,* Vol. 22, No. 6 2003, pp. 75–84.

[90]   Becker, F. F., et al., "The removal of human leukemia-cells from blood using interdigitated microelectrodes," *Journal of Physics D-Applied Physics,* Vol. 27, No. 12 1994, pp. 2659–2662.

[91]   Urdaneta, M. and Smela, E., "Multiple frequency dielectrophoresis," *Electrophoresis,* Vol. 28, No. 18 2007, pp. 3145–3155.

[92]   Arnold, W., Schwan, H., and Zimmermann, U., "Surface conductance and other properties of latex-particles measured by electrorotation," *Journal of Physical Chemistry,* Vol. 91, No. 19 1987, pp. 5093–5098.

[93]   Kaler, K. V. I. S., et al., "Dual-frequency dielectrophoretic levitation of canola protoplasts," *Biophysical Journal,* Vol. 63, No. 1 1992, pp. 58–69.

[94]   Pethig, R., Talary, M. S., and Lee, R. S., "Enhancing traveling-wave dielectrophoresis with signal superposition," *IEEE Engineering in Medicine and Biology Magazine,* Vol. 22, No. 6 2003, pp. 43–50.

[95]   Huang, Y., et al., "Membrane dielectric responses of human t-lymphocytes following mitogenic stimulation," *Biochimica Et Biophysica Acta-Biomembranes,* Vol. 1417, No. 1 1999, pp. 51–62.

[96]   Hu, X., Arnold, W., and Aimmermann, U., "Alterations in the electrical-properties of lymphocyte-t and lymphocyte-b membranes induced by mitogenic stimulation -activation monitored by electro-rotation of single cells," *Biochimica et biophysica acta,* Vol. 1021, No. 2 1990, pp. 191–200.

[97]   Labeed, F. H., et al., "Assessment of multidrug resistance reversal using dielectrophoresis and flow cytometry," *Biophysical Journal,* Vol. 85, No. 3 2003, pp. 2028–2034.

[98]   Hoettges, K., et al., "Dielectrophoresis-activated multiwell plate for label-free high-throughput drug assessment," *Analytical Chemistry,* Vol. 80, No. 6 2008, pp. 2063–2068.

[99]   Duncan, L., et al., "Dielectrophoretic analysis of changes in cytoplasmic ion levels due to ion channel blocker action reveals underlying differences between drug-sensitive and multidrug-resistant leukaemic cells," *Physics in Medicine and Biology,* Vol. 53, No. 2 2008, pp. N1–N7.

[100]   Hughes, M. P. and Hoettges, K. F., "Dielectrophoresis for drug discovery and cell analysis: novel electrodes for high-throughput screening," *Biophysical Journal,* Vol. 88, No. 1 2005, pp. 172A–172A.

[101]   Gascoyne, P. R. C., et al., "Membrane-changes accompanying the induced-differentiation of friend murine erythroleukemia-cells studied by dielectrophoresis," *Biochimica Et Biophysica Acta,* Vol. 1149, No. 1 1993, pp. 119–126.

[102]   Chin, S., et al., "Rapid assessment of early biophysical changes in k562 cells during apoptosis determined using dielectrophoresis," *International Journal of Nanomedicine,* Vol. 1, No. 3 2006, pp. 333–337.

[103]   Docoslis, A., et al., "A novel dielectrophoresis-based device for the selective retention of viable cells in cell culture media," *Biotechnology and Bioengineering,* Vol. 54, No. 3 1997, pp. 239–250.

[104]   Docoslis, A., Kalogerakis, N., and Behie, L. A., "Dielectrophoretic forces can be safely used to retain viable cells in perfusion cultures of animal cells," *Cytotechnology,* Vol. 30, No. 1-3 1999, pp. 133–142.

[105]   Labeed, F. H., Coley, H. M., and Hughes, M. P., "Differences in the biophysical properties of membrane and cytoplasm of apoptotic cells revealed using dielectrophoresis," *Biochimica Et Biophysica Acta-General Subjects,* Vol. 1760, No. 6 2006, pp. 922–929.

[106]   James, C. D., et al., "Surface micromachined dielectrophoretic gates for the front-end device of a biodetection system," *Journal of Fluids Engineering-Transactions of the Asme,* Vol. 128, No. 1 2006, pp. 14–19.

[107]   Braschler, T., et al., "Continuous separation of cells by balanced dielectrophoretic forces at multiple frequencies," *Lab on a Chip,* Vol. 8, No. 2 2008, pp. 280–286.

[108]   Markx, G. H., Rousselet, J., and Pethig, R., "Dep-fff: Field-flow fractation using non-uniform electric fields," *Journal of Liquid Chromatography & Related Technologies,* Vol. 20, No. 16-17 1997, pp. 2857–2872.

[109]   Huang, Y., et al., "Introducing dielectrophoresis as a new force field for field-flow fractionation," *Biophysical Journal,* Vol. 73, No. 2 1997, pp. 1118–1129.

[110]   Markx, G. H. and Pethig, R., "Dielectrophoretic separation of cells-continuous separation," *Biotechnology and Bioengineering,* Vol. 45, No. 4 1995, pp. 337–343.

[111]   Li, J. Q., et al., "Fabrication of carbon nanotube field-effect transistors by fluidic alignment technique," *IEEE Transactions on Nanotechnology,* Vol. 6, No. 4 2007, pp. 481–484.

[112]   Kim, Y., et al., "Novel platform for minimizing cell loss on separation process: Droplet-based magnetically activated cell separator," *Review of Scientific Instruments,* Vol. 78, No. 7 2007.

[113]   Flanagan, L., et al., "Unique dielectric properties distinguish stem cells and their differentiated progeny," *Stem Cells,* Vol. 26, No. 3 2008, pp. 656–665.

[114]   Shih, T. C., Chu, K. H., and Liu, C. H., "A programmable biochip for the applications of trapping and adaptive multisorting using dielectrophoresis array," *Journal of Microelectromechanical Systems,* Vol. 16, No. 4 2007, pp. 816–825.

[115]   Albrecht, D. R., et al., "Probing the role of multicellular organization in three-dimensional microenvironments," *Nature Methods,* Vol. 3, No. 5 2006, pp. 369–375.

[116]   Fiedler, S., et al., "Dielectrophoretic sorting of particles and cells in a microsystem," *Analytical Chemistry,* Vol. 70, No. 9 1998, pp. 1909–1915.

[117]   Muller, T., et al., "A 3-D microelectrode system for handling and caging single cells and particles," *Biosensors & Bioelectronics,* Vol. 14, No. 3 1999, pp. 247–256.

[118]   Chiou, P. Y., Ohta, A. T., and Wu, M. C., "Massively parallel manipulation of single cells and microparticles using optical images," *Nature,* Vol. 436, No. 7049 2005, pp. 370–372.

[119]   Ohta, A. T., et al., "Dynamic cell and microparticle control via optoelectronic tweezers," *Journal of Microelectromechanical Systems,* Vol. 16, No. 3 2007, pp. 491–499.

Optical Microfluidics for Molecular Diagnostics

Luke P. Lee and Frank B. Myers

Abstract

Recently there have been significant advances in integrated optical microfluidic devices for molecular diagnostic applications. This chapter reviews the developments of low-cost integrated optical microfluidic devices for high sensitivity, selectivity, and portable diagnostic system design, and shows how integrated optical microfluidics are applied for the detection and manipulation of biological samples in microfluidic devices. Advanced optical microfluidic diagnostic systems with sample preparation, analyte enrichment, and optical detection are highlighted. We will discuss the current status of label-free nanoplasmonic optical technology and its associated optical microfluidics along with surface enhanced Raman scattering (SERS)-based detection paradigms in microfluidics.

Key terms optical detection
molecular diagnostics
microfluidics
fluorescence
plasmonics
surface-enhanced raman spectroscopy
point-of-care

7.1 Introduction

The potential role of microfluidics in point-of-care (POC) diagnostics is widely acknowledged, and many reviews have explored its potential applications in clinical diagnostics [1], personalized medicine, [2] global healthcare, [3, 4] and forensics [5]. Despite this, relatively few successful commercial implementations have been demonstrated [6]. To realize the commercialization of microfluidic POC diagnostics challenges in integrating low-cost, sensitive, and portable optical detection systems must be addressed. Furthermore, to demonstrate the practicality of new techniques, an effort should be made to integrate sample preparation from raw clinical samples and compare detection sensitivity to conventional methods. Here, we review optical microfluidic systems, which address these goals and introduce novel techniques for realizing practical POC diagnostics.

Lateral-flow assays (LFAs) and electrochemical sensors dominate the POC diagnostics market today. Immunochromatographic LFAs, commonly called dipstick tests, rely on capillary flow and qualitative visual readout. LFAs are commercially available for a variety of diagnostic tests (pregnancy, cardiac markers, infectious diseases, etc.). These devices are successful because they are inexpensive to manufacture, robust, and easy to use. Although sensitivity is relatively poor compared to conventional immunological laboratory assays (such as the enzyme-linked immunosorbent assay, ELISA), LFAs are ideal for applications where analyte abundance is relatively high, complex sample preparation is not needed, and a simple yes/no diagnostic is sufficient. In many cases, however, more sophisticated assays are required that call for multistep protocols and complex fluid handling. Nucleic acid amplification and analysis is perhaps the most pertinent example, as more genotypic assays are emerging for pathogen detection and therapy prescription. For example, efforts are underway to develop genotypic POC assays for HIV, [7] tuberculosis [8], and enteric diseases [4], all of which are desperately needed in the developing world.

Electrochemical assays, on the other hand, have been very successful for quantitative analysis of certain small-molecule analytes, blood chemistry, and urinalysis. Conventional glucose meters are based on electrochemical detection methods, as is the handheld i-STAT used in hospital ICUs, which is capable of rapidly analyzing 25 different blood parameters. However, electroactive enzyme labeling is generally required for proteins and nucleic acids, and background interference with nonspecific redox species in the sample is a concern. Electrochemical techniques are heavily influenced by temperature variations at the electrode, chemical factors like pH and ionic concentrations, redox by-product accumulation near the electrodes, and electrode surface conditions—a factor that may limit the shelf life and require more stringent storage conditions for microfluidic disposables.

Because of the limitations of other techniques and the ubiquity of optical instrumentation in the laboratory, optical detection remains the preferred technique for quantitative proteomic or genomic diagnostics. Optical detection may have a per-test cost advantage over electrochemical detection because electrodes do not have to be integrated onto the disposable. Also, optical detection may be easier to multiplex since commercial CCD or CMOS image sensors could detect hundreds or thousands of simultaneous reactions at once, where each pixel corresponded to a different location on the detection array. Multiplexing on this scale is difficult to achieve with electrodes. While optical detection is quite straightforward in a laboratory environment where microscopes, lasers, spectrophotometers, lenses, and filters can be precisely arranged and aligned, these systems are difficult to miniaturize into a low-cost, portable, and robust system. Currently, most microfluidic devices are demonstrated using conventional optical systems. However, thanks to the rapid reduction in the cost of quality optoelectronic components like CCDs and laser diodes, as well as recent innovations in microfluidic integration and nanoscale materials for label-free biosensing, optical detection is becoming more practical for POC diagnostics.

This review examines current research and product development in POC diagnostic devices, which utilize optical detection within microfluidic platforms. We emphasize those efforts, which demonstrate low-cost integration and quantitative results on real-world samples. The first half of the review focuses on the miniaturization of conventional optical measurements, both on-chip and off-chip, while the second half focuses on emerging detection paradigms which rely on nanoparticles and nanoengineered materials. We focus primarily on developments from 2005 to 2008.

7.2   Integrated optical systems

Conventional optical detection methods, including absorbance, fluorescence, chemiluminescence, interferometry, and surface plasmon resonance, have all been applied in microfluidic biosensors. However, optical detection generally requires expensive hardware that is difficult to miniaturize, and it suffers at lower length scales. The shorter optical path lengths through the sample reduce sensitivity and higher surface-to-volume ratios lead to increased noise from nonspecific adsorption to chamber walls. To address these issues, many integrated optical systems are being explored in which waveguides, filters, and even optoelectronic elements are integrated onto the microfluidic device to improve sensitivity while reducing cost. In conjunction with these on-chip integrated components, many groups are incorporating low-cost optics, laser diodes, LEDs, CCD cameras, and photodiodes into portable diagnostic platforms. It is also worth noting that optical systems can not only be used for detection but also for actuation through various optical forces. Furthermore, through microscale manipulation of fluids one can achieve tunable and reconfigurable on-chip optical systems. These fascinating techniques have given rise to the new field of optofluidics, which has been reviewed elsewhere [9, 10]. Here, we focus exclusively on detection methods.

7.2.1   Absorbance detection

UV/visible absorption spectroscopy is a well-established technique in macroscale analytical chemistry and laboratory diagnostics. In this technique, the attenuation of incident light as a function of wavelength is measured using a spectrophotometer. The resulting spectrum reveals peaks in absorption, which can help identify the composition and concentration of the sample. In most cases, changes in optical density or color are sufficient for diagnosis, so instrumentation for absorbance measurements tends to be much simpler than for other methods. However, the major limitation with absorbance measurements in microfluidics is that as sample volumes decrease, the optical path length through the sample decreases, and this directly impacts sensitivity as described by the Beer-Lambert law.

One interesting way to address this problem is by incorporating "air mirrors" that take advantage of the refractive index difference between PDMS and air. In one example, air cavities were fabricated adjacent to the flow cell, which reflected light back into the fluid (Figure 7.1) [11]. Biconvex microlenses for light collimation were also integrated, and the system showed an impressive limit of detection (LOD) of 41 nM for absorption measurements of fluorescein. Collimation lenses are necessary to achieve reasonable optical path lengths because light diverges from the source. Furthermore, stray light can reach the detector and reduce sensitivity. Ro et al. introduced planocovenx collimation lenses along with rectangular apertures to block stray light at both the input and output fiber channels [12]. They demonstrated a tenfold increase in sensitivity using this technique, but the device required a three-layer PDMS fabrication process to realize the lenses, slits, and flow cell.

Steigert et al. developed a completely integrated centrifugal CD-based microfluidic system for alcohol detection from whole blood using absorbance measurements [13]. A cocktail of reagents were introduced, which lead to the production of a colorimetric dye in the presence of ethanol. A laser was focused perpendicular to the cyclic olefin copolymer (COC) substrate where it was reflected 90 degrees into the detection chamber by a micropatterned V-groove adjacent to the chamber. Another V-groove on the opposite

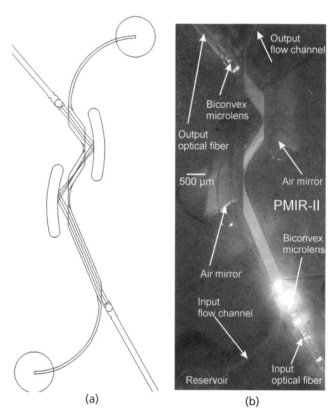

(a) (b)

Figure 7.1 Air mirrors and air lenses, which take advantage of the difference in refractive indices of PDMS and air, are positioned around an absorbance flow cell for increased optical path length. (a) Ray-tracing diagram of two mirrors reflecting light through the flow cell. (b) Photograph showing fluorescein illumination in the channel. (*From* [11] with permission from the Royal Society of Chemistry.)

side reflected the light back to a spectrophotometer. This allowed the laser to interrogate the entire cross sectional length of the chamber (10 mm), enabling an LOD of 0.04% for ethanol, which is sufficient for the target applications in emergency medicine. The device leveraged a number of interesting microfluidic techniques unique to centrifugal systems, namely: capillary hold and burst valves for precise metering of sample and reagents, sample separation by centrifugation, and rapid mixing via oscillatory rotation. Furthermore, one could conceivably multiplex this assay by having multiple chambers around a single disk and synchronizing the laser/detector with the disk as it spins. This would allow a single fixed laser and photodiode to interrogate an entire array of chambers. Although CD-based microfluidics have their disadvantages, particularly when more complex valving is required, for many applications they are an attractive option, especially given the ease with which centrifugal separations can be integrated. Given their ubiquity and low cost, there is substantial interest in using conventional CD/DVD drives to perform molecular diagnostics. By functionalizing probe molecules onto the surface of CDs and observing the data error levels as a CD-ROM reads the disc, groups have shown that it is possible to colorimetrically detect ligand binding [14, 15]. However, optical detection with a CD-ROM has yet to be shown with a microfluidic system integrated on the disc.

Since nucleic acids and proteins generally absorb strongly in the UV, absorption spectroscopy is commonly employed in liquid chromatography and electrophoresis to monitor the progression of biomolecules down a separation column in a label-free manner. Miniaturization provides a number of benefits for these techniques including higher separation efficiencies, reduced nondiffusional band broadening, reduced separation times, and reduced electric potentials (in the case of capillary electrophoresis). Gustoffson et al. integrated optical waveguides and a fluidic separation channel into a silicon substrate using a single etching step followed by thermal oxidation to form UV-transparent SiO_2 waveguides [16]. The separation channel consisted of micromachined pillars and was functionalized with octylsilane to facilitate electrochromatographic separation of electrically neutral species. A glass lid containing fluidic access holes was fusion bonded on top of the oxidized silicon, forming closed channels. Two 90-degree bends in the channel allowed the waveguides to couple into a 1-mm straight segment at the distal end of the separation column. The waveguide structures featured micromachined fiber couplers at the end to ensure optimal alignment with external optical fibers. A major limitation of this fabrication procedure was that thermal oxidation led to rounding of the pillars, which contributed to band broadening due to the nonuniform cross section along the height of the separation channel. Because of this, separation bands were an order of magnitude wider than previous microchip electrochromatography systems. Still, the device introduced an interesting design for integrated optics in microchip chromatography systems and a revision of the fabrication procedure could most likely improve device performance.

Despite the relatively poor sensitivity of absorbance measurements in microfluidics compared to fluorescence, its instrumentation simplicity gives it an advantage in certain applications. Indeed, several absorbance-based microfluidic POC products are available. The Nanogen i-Lynx, for example, performs quantitative absorbance measurements on LFAs for monitoring cardiac markers [17]. Cholestech's GDX System uses a microfluidic disposable to measure glycosylated hemoglobin from whole blood [18]. The system performs lysis of red blood cells, affinity chromatography, and absorbance measurements to determine the ratio of glycosylated to nonglycosylated hemoglobin, an important diagnostic for diabetes patients.

7.2.2 Fluorescence detection

Fluorescence is the most common optical method for molecular sensing in microfluidic systems due to the well-established, highly sensitive, and highly selective fluorescent labeling techniques from conventional genomic and proteomic analysis. However, autofluorescence is a problem with many polymer materials as well as nonspecific biomolecules in the sample, so material selection and sample purity are very important. Fluorescence is typically laser-induced, because lasers have a low divergence and can easily be focused into a small detection region. Furthermore, laser diodes are inexpensive and easily integrated into a portable device [19–22]. As an even lower cost alternative, LEDs can be used for fluorescence excitation. LEDs are less ideal because of their high divergence and relatively broad emission spectra, but many groups have shown good results with LED-induced fluorescence by incorporating integrated lenses, waveguides and filters into their microfluidic designs. Seo demonstrated a self-aligned 2-D compound microlens that dramatically reduced spherical aberration compared to single-lens designs [23, 24]. They showed that a 1mm alignment offset of the LED corresponded to only an 8.4% decrease in fluorescence (Figure 7.2). Furthermore, the fabrication was simple, requiring only a single layer of PDMS.

Liquid-core waveguide arrangements have been shown, where a capillary serves as both the fluid channel (often for capillary electrophoresis) and a waveguide for collecting fluorescence emission [25]. To further increase signal-to-noise ratio (SNR) in this configuration, Zhang et al. used a synchronized dual-wavelength LED modulation

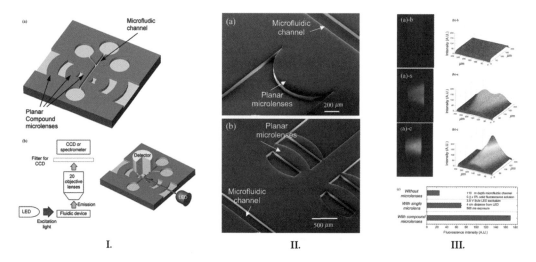

Figure 7.2 Disposable integrated optical microfluidic device with self-aligned planar compound microlenses. (I) conceptual design of device showing (a) schematic of lenses and microfluidic channels and (b) experimental detection setup with LED. (II) SEM images of fabricated (a) single lens and (b) compound lens devices. (III) CCD images of fluorescent emission through single and compound microlenses. (a) Fluorescence images from microfluidic channel with LED excitation (b, s, and c denote without microlens, with single microlens and with compound microlens, respectively). (b) Intensity profile of focused beam. (c) Relative fluorescent amplification showing best results is achieved with compound lens. (*From* [24] with permission from Elsevier.)

approach to subtract background noise from stray excitation light [26]. They demonstrated a 100-fold improvement in SNR over previous work, with an LOD of 10 nM for FITC-labeled arginine. LED and optical fibers have also been encapsulated directly into a PDMS device for capillary electrophoresis [27]. In this example, the epoxy housing of the LED was removed, and the LED was placed in close proximity to the flow cell with a short-pass filter in between. The authors demonstrated a 600-nM LOD for fluorescein.

Other groups have explored the possibility of implementing long-pass filters [28] and waveguides [29] directly in PDMS. In the case of filters, the PDMS prepolymer is doped with an organic dye, and in the case of waveguides, a prepolymer with a higher refractive index than the device material is chosen. Schmidt et al. demonstrated an innovative microfluidic fluorescence spectrometer for flow cytometry consisting of a linear variable optical band-pass filter sandwiched between a flow cell and a CMOS image sensor [30]. In this configuration, each pixel on the CMOS sensor corresponded to a different color, so as particles flowed down the channel, a complete spectrum could be observed.

Other groups have attempted to completely integrate a fluorescent detection system on-chip. Vezenov et al. demonstrated a tunable optically pumped dye laser implemented in a microfluidic cavity [31]. Balslev et al. developed a monolithic optoelectronic/microfluidic hybrid platform incorporating an optically pumped dye laser, waveguides, fluid channels with passive mixers, and silicon photodiodes [32]. Chediak et al. demonstrated a novel microassembly process for integrating (In, Ga)N blue LEDs, CdS thin-film filters, Si PIN photodetectors, and a PDMS microfluidic device (Figure 7.3) [33]. Monolithic integration of the entire optoelectronic system may not be the most cost-effective option when the device must be disposable. But as new manufacturing processes mature for organic optoelectronics, which utilize low-cost solution deposition techniques, monolithic integration may become more attractive. Towards this vision, Pais et al. recently demonstrated a microfluidic device sandwiched between a planar organic LED and an organic photodiode. They characterized this device with rhodamine 6G (LOD of 100 nM) and fluorescein (LOD of 10 μM) [34]. Although this detection limit is significantly higher than that achieved with conventional LEDs/photodiodes, organic optoelectronics are improving steadily and may soon represent a cost-effective alternative for detection in POC microfluidic devices.

In general, diagnostic sensitivity is directly related to the degree to which an analyte of interest can be concentrated within the detection region. In capillary electrophoresis (CE), for example, the more highly concentrated the "plug" of nucleic acids or proteins before electrophoretic separation, the more sharply defined the resulting separation bands will be. Herr et al. developed a preconcentration method for CE using in-situ photopolymerized gel membranes [35]. Fluorescently-labeled antibodies and analyte proteins were first electrophoretically driven against a region of gel with a relatively high degree of cross-linking (and therefore a smaller effective pore size), which acted as a size-exclusion membrane. The electric field was then switched, and the immunocomplexes and unbound antibodies flowed down a separation channel containing a gel with larger pores. This caused the antigen-antibody immunocomplexes and unbound antibodies to separate, and their relative fluorescent intensity was measured with a laser diode and PMT at a certain point on the channel. The ratio of bound to unbound antibodies correlated with protein concentration. This was incorporated into a portable saliva-based diagnostic platform for periodontal disease, which included all optical ele-

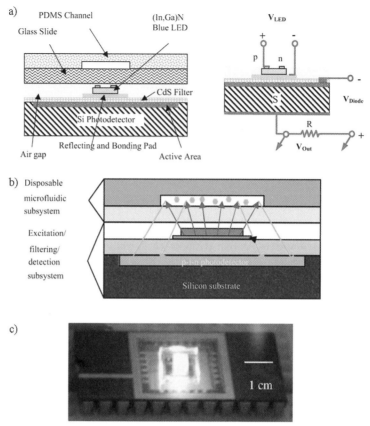

Figure 7.3 (a) Schematic of the heterogeneous integration of a CdS filter with an (In, Ga)N LED, a Si PIN photodetector, and a microfluidic device. (b) highlighting light paths for the excitation and Stokes-shifted emission signals. (c) image of the prototype microsystem with the LED on and exciting the microfluidic device. (*From* [33] with permission from Elsevier.)

ments [22]. In another example of how analyte preconcentration can greatly enhance sensitivity, Christodoulides et al. demonstrated a hundredfold LOD improvement over conventional ELISA diagnostics for the detection of C-reactive protein (CRP) in saliva [36]. Their device consisted of microwells etched in silicon, which trapped agarose beads that were functionalized with CRP antibodies. A sandwich assay was performed on these beads. The microwells extended completely through the silicon substrate, and the sample and secondary labels were flowed perpendicular to the substrate via transparent PMMA manifolds. Because of this configuration, CRP was forced into very close proximity with the beads, yielding excellent sensitivity.

Nucleic acid amplification and hybridization detection represent a growing field in microfluidic diagnostics. Conventionally, double-stranded DNA is most often detected using a fluorescent intercalating dye, so it is no surprise that this technique is also dominant in microfluidic systems. By far, the most common amplification technique employed is the polymerase chain reaction (PCR). Miniaturized PCR presents a number of advantages to conventional PCR: high thermal cycling rates (and therefore lower analysis times), low sample volumes, and integration with upstream sample preparation and downstream analysis components within a monolithic system to improve sensitivity and avoid sample contamination. On-chip PCR with capillary electrophoresis (CE) and fluorescent detection was first demonstrated by Woolley et al. in 1996 [37], and several recent examples have continued to improve on PCR with CE, [20, 38, 39] as well as hybridization detection [40, 41].

Liu et al. demonstrated a system for forensic analysis consisting of a glass micro-fluidic disposable for quadruplex Y-chromosome single tandem repeat (STR) typing that incorporated resistive heaters and pneumatic valves for performing on-chip PCR and CE detection [20]. A portable instrument was developed that contained power supplies for electrophoresis, a diode laser, an objective lens, and four dichroic filters and photomultiplier tubes (PMTs) for four-color fluorescent detection. They demonstrated the system with oral swab and human bone extracts, although sample preparation was performed off-chip. The system provided excellent detection sensitivity (down to 20 copies of DNA) with a 1.5 hour run time. Another impressive system was demonstrated by Easley et al., which not only incorporated PCR and CE detection but also sample preparation and DNA/RNA purification. The device detected *Bacillus anthracis* from whole blood and *Bordetella pertussis* from nasal aspirate [39]. For fluorescent detection, the device employed external optics, a high-powered laser, and PMTs. Both of these systems are quite useful for certain applications (such as forensic labs), but in order to make them viable solutions for POC applications, more work is necessary to miniaturize, automate, and reduce the cost of the optical detection systems.

A complete handheld nucleic acid detection system has been developed that incorporates amplification not with PCR but with a lesser-known isothermal amplification technique called NASBA (nucleic acid sequence-based amplification). The system quantitatively detected the microorganism *Karenia brevis* using low-cost optical components including LEDs, photodiodes, and optical filters. Even temperature was maintained optically using an infrared heater and thermometer [42, 43]. The system was designed to interface with a PDA for real-time monitoring of fluorescence. Unfortunately, the LOD was not characterized, but nevertheless this is an impressive demonstration of practical, low-cost optical nucleic acid detection. NASBA, in fact, may be preferable to PCR for POC viral diagnostic applications because it does not require a reverse transcriptase step for RNA amplification.

In general, fluorescence remains the most widely used method of optical detection. Fluorescent detection has a number of advantages over other techniques, namely high sensitivity and a wealth of available fluorophores and labeling chemistries. However, fluorescent dyes remain rather costly, have a limited shelf life, and are often influenced by chemical factors such as pH, which may vary from sample to sample. Furthermore, the labeling step itself requires complex fluid handling, mixing, and washing, which are difficult to automate in a rapid assay. For this reason, label-free analysis techniques, which approach the sensitivity and specificity of fluorescent labeling, are highly sought-after.

7.2.3 Chemiluminescence detection

Chemiluminescence is another attractive option for detection in which analyte binding causes photochemical emission, either directly or with the help of an enzyme label. The advantage of this technique for POC applications is that excitation instrumentation is not required, and therefore background interference is virtually eliminated. However, very sensitive detectors are typically required. Bhattacharyya and Klapperich developed a disposable microfluidic chemiluminescent immunoassay for detecting CRP in serum [44]. Their device demonstrated excellent correlation with conventional ELISA measurements throughout the clinically relevant range, with an LOD of 0.1 mg/L when using an external photomultiplier tube (PMT) luminescent reader. Their assay time was 25 minutes (compared to >12 hours for ELISA). They also demonstrated proof-of-concept detection with an integrated photosensitive film. Yacoub-George et al. developed a multiplexed chemiluminescent capillary enzyme immunoassay sensor capable of simultaneously detecting toxins, bacteria, and viruses [45]. The assay was completed in 29 minutes. Using PMTs, they demonstrated LODs of 0.1 ng/mL for staphylococcal enterotoxin B, 10^4 colony-forming units per mL for E. Coli O157:H7, and 5×10^5 plaque-forming units per mL for bacteriophage M13.

Marchand et al. demonstrated a disposable microfluidic card with an integrated commercial CMOS active pixel sensor (APS) functionalized with DNA oligonucleotide probes for chemiluminescent hybridization detection [46]. Since the probes are directly attached to the sensor surface, this offers excellent photonic capture efficiency. The form factor of the device was equivalent to "smart card" credit cards, leveraging existing pick-and-place manufacturing technology for the integration of the CMOS chip. However, although the cost of silicon CMOS image sensors continues to fall, it is difficult to say if they will become inexpensive enough to be incorporated on a disposable for POC applications. But if such a platform could be regenerated and reused, this might be a very attractive option for genomic analysis.

7.2.4 Interferometric detection

Interferometric sensing potentially offers label-free detection with very high sensitivity. In general, interferometric biosensor techniques rely on the splitting of a single coherent light source into two paths. Both paths are adjacent to the sample media, but only one is functionalized to be sensitive to the analyte of interest. Analyte binding causes a refractive index change along this optical path, resulting in a phase shift with respect to the nonfunctionalized reference path. This phase shift can have a dramatic effect on the interference pattern resulting from the projection of these two beams, so by imaging this interference pattern, one can directly observe binding events. The reference path is what gives interferometry its sensitivity because it accounts for common-mode interference such as nonspecific adsorption, temperature fluctuations, and intensity fluctuations. Ymeti et al. demonstrated an immunosensor using a Young interferometer configuration in which a microfluidic chip with integrated waveguides projected an interference pattern onto a CCD image sensor (Figure 7.4) [47]. In this device, three waveguides were functionalized with different antibodies and one waveguide was nonfunctionalized and served as a reference. When antigens bound, they interacted with the evanescent wave at the surface of the waveguide, changing the refractive index and leading to a shift in the interference pattern. The sensor was evaluated with herpes simplex virus. The sensor demonstrated a response time of just a few minutes and an LOD in the femtomolar range, approaching a single virion. Another report describes a similar Young interferometer biosensor assay to detect avian influenza (H5N1) [48]. One potential drawback of interferometers of this type is that the micromachined geometries and material properties of the waveguides must be precisely matched.

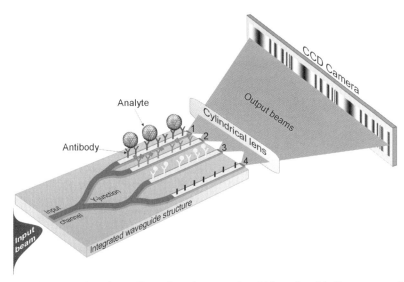

Figure 7.4   A biosensor based on a Young Interferometer in which analyte binding causes a change in the interference pattern projected onto a CCD camera. (*From* [47] with permission from the American Chemical Society.)

Blanco et al. developed a biosensor based on a Mach-Zehnder interferometer, which was implemented in a CMOS-compatible silicon process and was integrated with SU-8 microfluidic channels [49]. The advantage of the Mach-Zehnder configuration is that the interfering light paths are recombined on the device so that only an intensity variation is measured at the output. However, because this configuration requires monomode propagation, a 4-nm silicon nitride core was used in the waveguide, which led to substantial insertion loss (9 dB) when coupled with external fibers.

Reflectometric interferometry is another method, which can very precisely detect analyte binding at a surface. In this method, a laser is projected towards a functionalized surface and the reflected interference pattern is measured. Pröll et al. used this method to measure oligonucleotide hybridization. Because interferometry is less sensitive to temperature variations that other methods, they were able to perform melting curve studies and identify single nucleotide polymorphisms (SNPs) with single base-pair resolution [50]. The same group recently demonstrated this method for immunological detection [51].

Backscattering interferometry was demonstrated in a rectangular PDMS microfluidic channel, in which the walls of the channel caused light to interfere and produce fringe patterns whose position depended on the refractive index at the channel floor. An inexpensive, low-power, unfocused light source was used. By functionalizing the channel floor, the authors achieved femtomolar detection limits for both streptavidin-biotin and protein A-IgG immunodetection [52]. More recently, the same group demonstrated that the technique could also be applied to detect molecular binding in free solution rather than at a functionalized surface (Figure 7.5) [53]. Because surface functionalization is laborious, expensive, and limits device shelf life, free solution techniques are appealing. The authors were able to detect interaction of calmodulin with the protein calcineurin as well as Ca^{2+} ions, a small molecule inhibitor, and the M13 peptide. Furthermore, they detected immunocomplex formation between interleukin-2 and its antibody as well as

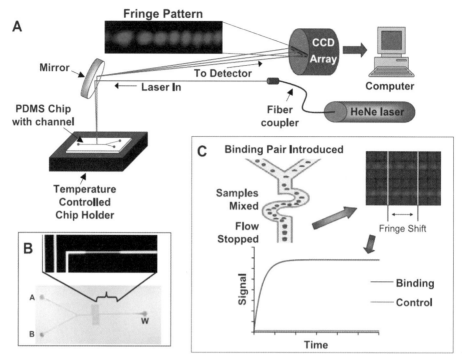

Figure 7.5 Back-scattering interferometry through a rectangular PDMS channel produces fringe patters that correspond to antigen-antibody binding within the channel volume. No surface functionalization is necessary. (a) Optical elements of the system. (b) Photograph of microfluidic device with closeup of constriction region used for mixing. (c) Diagram showing device operation and observed fringe shift resulting from binding. (*From* [53] with permission from the American Association for the Advancement of Science.)

protein A with IgG. The technique showed remarkable sensitivity; for interleukin-2, the authors estimate a detection limit of just 12,600 molecules. The major drawback of this technique is that because it is a volumetric measurement it may not be suitable for complex samples (such as human serum), which have other constituents that might affect bulk refractive index.

7.2.5 Surface plasmon resonance detection

Surface plasmon resonance (SPR) detection relies on the measurement of a refractive index change at a metal surface, which has been functionalized with probe molecules. When light is incident on a thin metal film at a specific angle through a prism (Kretschmann configuration), it excites a propagating surface plasmon at the surface of the metal. At this angle, the reflectance intensity decreases sharply. This SPR angle is highly sensitive to the dielectric environment on the opposite side of the metal, so when ligands bind to the surface of the metal, they cause this SPR angle to change. The angle is measured either by rotating a narrowly focused laser beam and looking for a reflectance minimum or by using a slightly divergent laser beam and imaging the reflectance angle spectrum exiting the prism.

Several laboratory-scale SPR detection systems for immunosensing and DNA hybridization detection exist, most notably the Biacore. Recently, efforts have focused on reducing the size and complexity of SPR sensors using integrated microfluidic systems. The Sensata Spreeta SPR sensor is a commercially available, low-cost SPR sensor, which can be adapted to a variety of applications [54]. The device provides an impressive LOD of 80 pM when tested with IgG. Waswa et al. demonstrated the use of the Spreeta for the immunological detection of *Escherichia coli* O157:H7 in milk, apple juice, and ground beef extract [55]. They showed an LOD of 10^2 colony-forming units per mL and demonstrated that the sensor was nonresponsive to other organisms (*E. coli K12 and Shigella). E. coli* is an enteric pathogen, which poses a serious health risk and must be monitored closely in the food supply. Conventional methods rely on culturing that takes 1 to 3 days, a high degree of expertise, and specialized facilities. The Spreeta sensor was able to reliably quantify *E. coli* in 30 minutes. A similar assay was carried out by Wei et al. using the Spreeta to detect *Campylobacter jejuni* in poultry meat [56].

Chinowsky et al. developed a portable briefcase-style SPR immunosensor platform [57]. The system uses a folded light path incorporating a range of optical elements (mirrors, prisms, lenses) and an LED light source, which is translated mechanically in order to generate the reflection spectrum on a CCD. A unique microfluidic card made of a combination of laser-cut laminated Mylar sheets, PDMS, and glass incorporates a diffusion-based H-filter, a herringbone micromixer, and gold-coated SPR detection regions. The materials were carefully chosen with manufacturability and cost in mind [4]. The sensor was demonstrated for phenytoin detection from saliva [58]. The sensitivity of the system is comparable to that obtained with the commercial Biacore system. Another briefcase-style sensor was developed by the same group, which was capable of simultaneously detecting small molecules, proteins, viruses, bacteria, and spores [59].

Recently, Feltis et al. demonstrated a low-cost handheld SPR-based immunosensor for the toxin ricin [60]. They demonstrated an LOD of 200 ng/mL. Although this is substantially higher than the LOD obtained with the Biacore sensor (10 ng/mL), it is quite sufficient for this particular application (200 ng/mL is 2,500 times lower than the minimum lethal dose), and the result is available in 10 minutes.

Luo et al. developed a multilayer PDMS array consisting of multiple gold spots for the real-time observation of immunocomplex formation using SPR [61]. By using gold nanoparticles as secondary labels in a sandwich assay format, they demonstrated an LOD of ~38 pM with biotin-BSA/antibiotin as the antigen/antibody pair. There was a trade-off, however, between assay time and sensitivity. Without the nanoparticles, the LOD was 0.21 nM, but the assay took only 10 minutes to complete, compared to 60 minutes for the case with nanoparticles.

SPR is a well-established technique with excellent sensitivity. However, the instrumentation required is a bit complex, and a layer of metal (typically gold) must be deposited on the disposable device, thus raising cost. Also, the technique suffers from strong temperature dependence.

7.3   Nanoengineered optical probes

Nanoengineered materials are emerging as powerful aids to optical detection. New labels such as quantum dots and up-converting phosphors have been developed which offer tunabilities, intensities, and longevities better than conventional organic dyes. Other approaches leverage the localized surface plasmon resonance (LSPR) effect to create highly sensitive sensors that depend not on sample volume but on near-field surface interactions, which make them excellent candidates for miniaturized systems.

7.3.1 Quantum dots

Quantum dots (QDs) are fluorescent semiconductor particles with finely tunable narrow-line emission spectra. This makes them ideal for multiplexed detection, potentially eliminating the need to immobilize capture probes in a spatial array. Not only would this reduce fabrication cost, but also since diffusion times severely limit the throughput of conventional microarrays, this would provide a significant advantage for rapid diagnostics. Also, compared to organic dyes, quantum dots are brighter, do not photobleach, and are more resistant to chemical degradation [62]. Quantum dot "bar-coding" is a popular technique employed in recent years in which combinations of quantum dots with different emission peaks are encapsulated in polymer microbeads such that as many as 10^6 unique biosensor probes, each functionalized with a different capture antibody or oligonucleotide, can be realized. Importantly, all of these bar codes can be excited with one wavelength. So, when used with a secondary fluorescent label, analyte binding of multiple barcodes can be measured simultaneously, without a priori knowledge of spatial arrangement. This technique was recently applied in a multiplexed microfluidic biosensor for the simultaneous detection of HIV, HBV, and HCV (Figure 7.6) [63]. A similar technique was applied for gene expression analysis in which 100 different genes were detected simultaneously with QD bar codes with detection

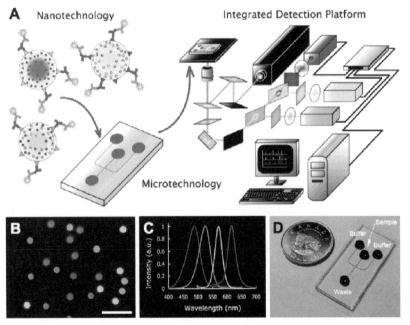

Figure 7.6 (a) Quantum dot barcodes are used for multiplexed immunosensing in a microfluidic device with external optical detection. (b) Fluorescent image of quantum dots on chip. (c) Quantum dot emission spectra. (d) Fabricated microfluidic chip. (*From* [63] with permission from the American Chemical Society.)

limit, dynamic range, and analysis time all an order of magnitude better than conventional microarray methods [64]. Quantum dots do have their problems, however. They are relatively difficult to synthesize and functionalize and are consequently much more expensive than fluorophores. Also, they suffer from a limited shelf life. There are only a few examples of quantum dots being used as biosensors in microfluidics [63, 65–67], and no examples integrating them into portable POC diagnostic devices.

7.3.2 Up-converting phosphors

A new labeling technique currently being commercialized by OraSure Technologies is upconverting phosphor technology (UPT) [68–71]. This technique involves using ceramic nanospheres containing rare earth metals, which absorb multiple photons of infrared light for each emitted photon in the visible spectrum (anti-Stokes shift). Since this phenomenon bears no natural analog in most materials or biological media, it provides excellent SNR as compared with fluorescence, which is often limited by autofluorescence. Furthermore, UPT particles are available in different colors, are excited in the near IR range (where absorption from biological media is minimal), and do not fade or photobleach. UPT has been demonstrated for bacterial and viral identification using both genetic [69, 71] and immunological [68, 70] methods (Figure 7.7). These examples use nitrocellulose LFA strips and a portable reader, which detects the UPT particles with an IR laser diode and a PMT.

7.3.3 Silver-enhanced nanoparticle labeling

A novel method for DNA microarray analysis was introduced in 2000 [72], whereby gold nanoparticles are used as secondary labels and then subsequently enlarged via catalytic silver deposition to the point that they can be imaged using a conventional camera or slide scanner. It was further shown that the same technique could be applied to protein detection [73]. This technique has huge implications for POC diagnostics. For DNA analysis, it is so sensitive that it does not require nucleic acid amplification, greatly simplifying the instrumentation required. It also allows for simultaneous detection of DNA and

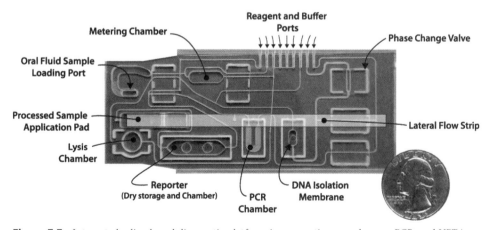

Figure 7.7 Integrated saliva-based diagnostic platform incorporating sample prep, PCR, and UPT in a lateral flow detection strip. (From [71] with permission from Blackwell Publishing.)

proteins in the same array using the same processing steps. Recently, the technique has been commercialized by Nanosphere. The company received FDA approval in 2007 for its Verigene platform to be used in two genetic tests, one to identify markers for blood coagulation disorders and the other to assess a patient's candidacy for associated medication. Whole blood is dropped in a disposable cartridge, which contains all required reagents for sample preparation, nanoparticle labeling, and silver deposition. The processing is completely automated and results are read using a slide scanner. The complete test takes only 90 minutes.

7.3.4 Localized surface plasmon resonance

Biosensors based on the localized surface plasmon resonance (LSPR) phenomenon represent one of the most active research areas in optical microfluidic biosensors today, and several reviews on this topic have been presented recently [74–76]. LSPR refers to the collective resonant oscillation of conduction electrons at the surface of a metal nanoparticle under the perturbation of incident light. Whereas conventional SPR sensing requires a prism or grating coupler to excite propagating plasmons on the surface of a metal, LSPR sensing requires no special coupling instrumentation and is typically performed with a white light source. A recent review compares biosensing with localized versus propagating surface plasmons [77]. The specific resonant frequency of a particle depends on the nanoparticle's size, shape, composition, and, most importantly for biosensing, the molecules in close proximity with the particle. This resonance leads to highly enhanced scattering at a specific wavelength. When molecules bind to the surface of the nanoparticle, they cause marked shifts in this resonance. LSPR is therefore label-free: it directly transduces a binding event to a spectral shift. Because LSPR sensing is a near-field phenomenon (<20 nm), it is less sensitive to background interference (e.g., from nonspecific adsorbates on chamber walls) than absorbance or fluorescence. LSPR scattering is also very intense. For example, as pointed out by Anker et al., the LSPR scattering cross-section of a gold nanoparticle is a million-fold greater than that of a single fluorescein molecule, and a thousand-fold greater than an equivalent volume of fluorescein [75]. LSPR measurements are generally based on spectroscopic or refractive index changes rather than intensity changes, and are therefore less sensitive to background noise and normalization error from instrumental, chemical, or environmental variations as compared with conventional optical detection [78]. LSPR biosensing can be performed on ensembles of nanoparticles as well as individual particles. Many structures for LSPR have been explored, but few practical POC implementations have been shown. This is mainly due to the difficulty in producing nanoscale structures and particles, which are sufficiently uniform and robust. Research has focused on developing new low-cost bulk fabrication methods for LSPR substrates and developing new surface treatment chemistries, particularly thiol-anchored self-assembled monolayers (SAMs), to improve nanoparticle stability [79].

Pioneered by Van Duyne's group, nanosphere lithography (NSL) has emerged as one of the most promising techniques for bulk, high-uniformity manufacturing of LSPR structures on a flat substrate, particularly for surface-enhanced Raman scattering [79]. In this technique, silica or polystyrene nanospheres self-assemble into close-packed monolayers on a substrate and act as a shadow mask for metal deposition. Since the initial demonstration of NSL, several innovative structures have been demonstrated using this technique including nanocrescents [80], nanodisks [81], films-over-nanospheres (FONs) [82, 83] and triangular prism arrays [79, 84]. Despite the performance and simplicity of the NSL technique, it still does not achieve wafer-level uniformity because self-assembly of nanoscale colloids only provide defect-free areas of 10 to 100 μm^2. Wafer-scale fabrication of plasmonic substrates is currently a major area of research. Other groups have explored chemically synthesized colloidal particles including nanospheres [85, 86], nanorods [78, 87], nanoshells [88], and nanorice [89]. These are attractive for POC diagnostics because they can be introduced into the device after it has been manufactured, allowing a single device to be mass produced for a variety of different applications.

LSPR biosensors have been demonstrated for detecting DNA hybridization (Figure 7.8) [82, 90], antigen/antibody binding [78, 82, 83], and small molecules [85, 86]. In one excellent example of LSPR integration in a biosensor, Endo et al. created a multiplexed microarray biochip based on the FON technique for measuring immuno-globulins, CRP, and fibrinogen [82]. They used a nanospotter to deposit 300 different antibody spots on the array, and achieved a detection limit of 100 pg/mL. The same group previously demonstrated a sensor for detecting casein from whole milk [83]. Gold nanorods immobilized on a substrate were also used for the detection of a model protein (streptavidin) in blood serum [87]. Schofield et al. showed that ligand-induced colloidal aggregation of GNPs upon binding of cholera toxin provided a dramatic shift in LSPR wavelength, such that the resulting color change could easily be distinguished with the naked eye [86]. Wu et al. is developing an integrated LSPR biosensor array in microfluidics for cellular analysis systems. Recently, Hiep et al. demonstrated an LSPR biosensor on a microfluidic device [91], but portable, fully integrated diagnostic systems incorporating sample preparation and spectroscopic detection have yet to be realized.

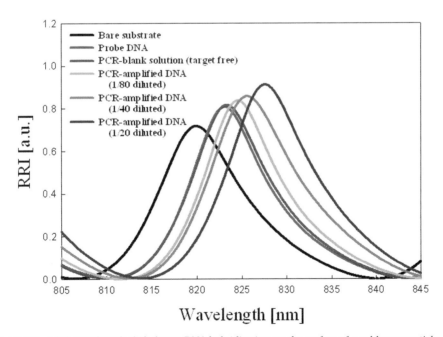

Figure 7.8 LSPR wavelength shift due to DNA hybridization on the surface of a gold nanoparticle, showing various dilutions of PCR product.(*From* [90] with permission from the American Chemical Society.)

Probably the main reason for this is that very sensitive spectrophotometers are required to resolve the relatively small spectral shifts. For example, for that particular device there is only a ~5-nm shift in LSPR wavelength between the hybridized and unhybridized cases. This kind of distinction would be difficult to resolve with a low-cost, portable spectrophotometer.

7.3.5 SPR with nanohole gratings

As with LSPR sensors, SPR sensors based on propagating plasmons can also benefit from nanofabrication and nanoscale optics. Nanohole gratings have been studied as possible substitutes to smooth metal surfaces in SPR sensing. The main advantage of these nanostructures over smooth metal films for POC diagnostics is that they can be interrogated with white light at normal incidence, greatly simplifying optical instrumentation [92]. Rather than measuring the SPR angle, one instead measures the transmission spectrum through the nanohole array. High numerical aperture optics can be used, so a very large multiplexed array of biosensor probes can be simultaneously imaged. However, the detection limit of the nanohole array technique is poor compared with conventional SPR. In recent microfluidic implementations characterized with glutathione S-transferase (GST) [93] and bovine serum albumin (BSA) [94], the LODs were 13 and 26 nM, respectively. While more work needs to be done to improve the sensitivity of these devices and demonstrate their utility in real biological samples, this is a promising technology for future optical microfluidic POC diagnostic systems.

7.3.6 Surface-enhanced Raman spectroscopy

Raman scattering is an extremely weak phenomenon in which a small fraction of photons impinging on a molecule lose some of their energy to one or more quantum vibrational modes in the molecule and thereby scatter inelastically, leading to a spectrum of lower energy (Stokes-shifted) peaks that can help uniquely identify the molecule. Raman spectroscopy is commonly used in analytical chemistry with large cuvet volumes, powerful lasers, and precision optics, but certain LSPR substrate geometries, which include nanoscale gaps or sharp features, exhibit extremely high electric field intensities in their adjacent volumes, giving rise to a dramatic increase in the intensity of Raman scattering for molecules located within these volumes. This technique, called surface-enhanced Raman scattering (SERS), was first discovered by Fleischmann et al. in 1974 [95] with the use of a roughened silver substrate. More recently, nanoengineered LSPR substrates have been applied to achieve even greater Raman enhancement [96, 97]. Liu and Lee demonstrated a batch-fabrication process for patterned SERS substrates in microfluidic devices with enhancement factors 10^7 times greater than unpatterned (i.e., smooth metal) substrates (Figure 7.9) [98]. Most research incorporating SERS substrates in microfluidic devices focuses on either characterizing new substrates (typically using model Raman analytes like Rhodamine 6G) or on detecting small organic molecule analytes for environmental, defense, and industrial applications [99]. A few examples of DNA hybridization [100, 101] and bacteria detection [102] exist. But as with LSPR-based sensors, portable diagnostic systems have been slow to emerge.

Many SERS biosensors employ reporter molecules with strong, distinct Raman spectra to label molecules of interest. This means that, as with QDs, spectral multiplexing may be achievable while using the same excitation wavelength. When used in a sand-

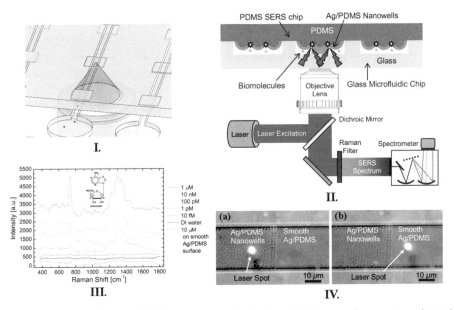

Figure 7.9 Integrated nanowell SERS array demonstrating batch fabrication of nanopatterned metallic substrates and microfluidic devices. (a) Conceptual design. (b) Schematic showing optical instrumentation. (c) Resulting Raman spectra from the device with different concentrations of rhodamine 6G. (d) The device incorporated both patterned and unpatterned Ag substrates to clearly demonstrate the enhancement provided by the patterning method (measured to be 107).(*From* [98] with permission from the American Institute of Physics.)

wich assay configuration where capture probes are attached to a SERS substrate, these SERS labels will only be visible when they are brought in close proximity to the substrate (i.e., by antigen/antibody conjugation or DNA hybridization). This may provide an advantage over fluorophores or other labels for POC systems, because it virtually eliminates background signal from unbound labels, thus boosting sensitivity. Also, as with UPTs, SERS excitation is typically done in the near IR range, where interference from biological media is minimal. These two advantages may obviate the need for a washing step, decreasing procedural complexity. Furthermore, SERS nanoparticle tags have been demonstrated with sensitivities one order of magnitude greater than those of fluorophores [103].

A variation of SERS, surface-enhanced resonance Raman scattering (SERRS), involves the resonant excitation of a molecule to an excited electronic state, and adds an additional factor of 100 to 1,000 to scattering enhancement. Oxonica (formerly Nanoplex Technologies), has commercialized the first SERRS-based POC immunoassay. The device consists of a disposable LFA containing glass-coated nanoparticles that feature a gold core coated with a SERRS reporter molecule [104]. Thanks to the glass coating, these particles are extremely robust and show a readily identifiable spectral signature no matter what media they are in or what molecules are bound to their surface. Oxonica produces a variety of SERRS tags with unique spectra, enabling simple multiplexing on the same LFA strip. They demonstrate a >10× sensitivity increase over conventional lateral flow immunoassays for the subtyping of influenza. The company is also pursuing bench-top ELISA devices which use SERRS labels rather than fluorophores for a one-step, no-wash assay.

It is also possible to achieve label-free, probe-free detection of certain pathogens using SERS. Shanmukth et al. demonstrated that SERS could be used to uniquely identify different respiratory viruses, and even different strains of the same virus, without using antibodies or labels [105]. Measurements were performed in filtered cell lysate, and it was shown that distinct Raman peaks could be observed that corresponded to different viruses and viral strains due to their unique surface proteins. Wu et al. are developing integrated optical microfluidics with SERS array. Antinanocrescent SERS probe array is integrated on the bottom surface of microfluidic cell culture array for the detection of secreted molecules from cell culture array (Figure 7.10). Although reliable diagnostics will most likely always require capture probe molecules, these results indicate that SERS can be used as an additional, orthogonal layer of detection which can help discriminate between specifically- and nonspecifically-bound ligands.

Many groups have strived to understand the physical basis for SERS in order to develop better substrates, and several reviews are available on this subject [97, 99]. SERS shows promise in POC diagnostics because of its ability to provide information-rich spectra which can lead to highly specific molecular identification. As with LSPR sensing, the challenge lies in producing sufficiently uniform nanostructures and low-cost portable spectrometers, which provide sufficient spectral resolution.

a)

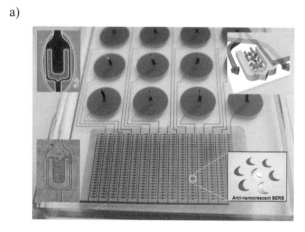

b)

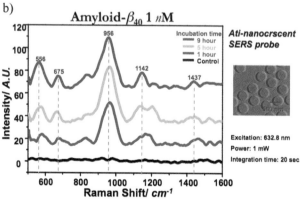

Figure 7.10 Integrated optical microfluidics with SERS array. (a) Antinanocrescent SERS probe array is integrated on the bottom surface of a microfluidic cell culture array for the detection of secreted biomolecules. (b) Representative SERS spectra from amyloid beta secretions from the cell culture array. (Figures courtesy of the Berkeley BioPOETS group.)

7.4 Conclusions

Optical detection remains the most powerful technique for genomic and proteomic analysis in microfluidics, and integrating sensitive optical systems that are robust and inexpensive remains an ongoing challenge. Many groups have addressed this by building portable versions of conventional instruments (such as fluorimeters and spectrometers), while others have instead attempted to integrate some or all of the detection hardware on the microfluidic device itself. Still others are exploring nanophotonic probes and plasmonics to enable entirely new detection paradigms. A variety of label-free detection techniques have been explored which may reduce the number of reagents and fluid handling steps required to perform an assay and hold great promise for POC diagnostics.

A variety of the techniques illustrated here measure analyte binding at a surface and have the advantage that they are less sensitive to interference from nonspecific molecules in the bulk fluid. However, these techniques require surface functionalization that can increase cost, reduce shelf life, and introduce manufacturing variability to the device. Volumetric measurement techniques, on the other hand, are much simpler to implement, but they tend to suffer from shorter optical path lengths and are more susceptible to nonspecific interference. As device integration improves, there is likely to be a push for greater orthogonality in testing, potentially incorporating both proteomic and genomic detection, along with multiple detection mechanisms. In this regard, spectroscopic detection methods like SERS are very attractive because they can add another layer of confirmation over conventional chromatography or labeling techniques.

The optical detection methods in microfluidics reviewed here have significant potential for integration into the next generation of POC devices, which will undoubtedly have a substantial impact on global health, clinical diagnostics, and emergency medicine. Future efforts will likely involve the integration of optical detection systems within automated microfluidic platforms, which incorporate sample preparation from raw biological samples (e.g., whole blood or saliva). As these technologies move closer to commercial viability, there will be an emphasis on using conventional mass fabrication processes and materials (such as injection molding of thermoplastics). Although PDMS is ideal for quick prototyping and exhibits excellent optical properties (low autofluorescence and high transmissivity), it may not be a good candidate for the batch fabrication of optical microfluidic devices for biomedical applications. As the optoelectronics industry continues to drive down the cost and improve the quality of components like diode lasers and CCD image sensors, leveraging these off-the-shelf technologies will be key to realizing practical POC diagnostic devices. Also, it is very important to explore label-free nanoplasmonic-based optical detection and the effective integration of optical antennas in microfluidics.

7.5 Summary

- Point-of-care molecular diagnostic instruments based on optical detection strategies represent a promising area of research for global health applications.
- Optical instrumentation is difficult to miniaturize, and many challenges must be overcome before optical POC devices are practical. While many groups have

focused on integrating conventional detection into portable devices, others have explored entirely new optical detection paradigms that leverage microscale and nanoscale photonic phenomena.

- Label-free spectroscopic detection strategies represent a promising alternative to more traditional fluorescent and colorimetric assays because they do not require complicated labeling procedures with molecules that have a limited shelf life.

Acknowledgments

The authors would like to thank Rick Henrikson, Tanner Nevill, David Breslauer, Ben Ross, and Adrian Sprenger for helpful discussions. FBM gratefully acknowledges the support of an NDSEG fellowship.

References

[1]   Dupuy, A., S. Lehmann, and J. Cristol, "Protein biochip systems for the clinical laboratory," *Clinical Chemistry and Laboratory Medicine*, Vol. 43, Dec. 2005, pp. 1291–1302.

[2]   L. Bissonnette and M. Bergeron, "Next revolution in the molecular theranostics of infectious diseases: microfabricated systems for personalized medicine," *Expert Review of Medical Diagnostics*, Vol. 6, May 2006, pp. 433–450.

[3]   C.D. Chin, V. Linder, and S.K. Sia, "Lab-on-a-chip devices for global health: Past studies and future opportunities," *Lab on a Chip*, Vol. 7, 2007, pp. 41–57.

[4]   P. Yager, T. Edwards, E. Fu, K. Helton, K. Nelson, M.R. Tam, and B.H. Weigl, "Microfluidic diagnostic technologies for global public health," *Nature*, Vol. 442, Jul. 2006, pp. 412–418.

[5]   K.M. Horsman, J.M. Bienvenue, K.R. Blasier, and J.P. Landers, "Forensic Dna Analysis on Microfluidic Devices: a Review," *Journal of Forensic Sciences*, Vol. 52, Jul. 2007, pp. 784–799.

[6]   G.M. Whitesides, "The origins and the future of microfluidics," *Nature*, Vol. 442, Jul. 2006, pp. 368–373.

[7]   F. Rouet, D.K. Ekouevi, M. Chaix, M. Burgard, A. Inwoley, T.D. Tony, C. Danel, X. Anglaret, V. Leroy, P. Msellati, F. Dabis, and C. Rouzioux, "Transfer and Evaluation of an Automated, Low-Cost Real-Time Reverse Transcription-PCR Test for Diagnosis and Monitoring of Human Immunodeficiency Virus Type 1 Infection in a West African Resource-Limited Setting," *J. Clin. Microbiol.*, Vol. 43, Jun. 2005, pp. 2709–2717.

[8]   H. Park, H. Jang, E. Song, C.L. Chang, M. Lee, S. Jeong, J. Park, B. Kang, and C. Kim, "Detection and Genotyping of Mycobacterium Species from Clinical Isolates and Specimens by Oligonucleotide Array," *J. Clin. Microbiol.*, Vol. 43, Apr. 2005, pp. 1782–1788.

[9]   D. Psaltis, S.R. Quake, and C. Yang, "Developing optofluidic technology through the fusion of microfluidics and optics," *Nature*, Vol. 442, Jul. 2006, pp. 381–386.

[10]   Hunt and Wilkinson, "Optofluidic integration for microanalysis," *Microfluidics and Nanofluidics*, Vol. 4, Jan. 2008, pp. 53–79.

[11]   A. Llobera, S. Demming, R. Wilke, and S. Buttgenbach, "Multiple internal reflection poly(dimethylsiloxane) systems for optical sensing," *Lab on a Chip*, Vol. 7, 2007, pp. 1560–1566.

[12]   K. Ro, K. Lim, B. Shim, and J. Hahn, "Integrated Light Collimating System for Extended Optical-Path-Length Absorbance Detection in Microchip-Based Capillary Electrophoresis," *Analytical Chemistry*, Vol. 77, Aug. 2005, pp. 5160–5166.

[13]   J. Steigert, M. Grumann, T. Brenner, L. Riegger, J. Harter, R. Zengerle, and J. Ducree, "Fully integrated whole blood testing by real-time absorption measurement on a centrifugal platform," *Lab on a Chip*, Vol. 6, 2006, pp. 1040–1044.

[14]   Y. Li, L.M.L. Ou, and H. Yu, "Digitized Molecular Diagnostics: Reading Disk-Based Bioassays with Standard Computer Drives," *Analytical Chemistry*, Sep. 2008.

[15]   J.J. La Clair and M.D. Burkart, "Molecular screening on a compact disc," *Organic & Biomolecular Chemistry*, Vol. 1, Sep. 2003, pp. 3244–9.

[16]   O. Gustafsson, K.B. Mogensen, P.D. Ohlsson, Y. Liu, S.C. Jacobson, and J.P. Kutter, "An electrochromatography chip with integrated waveguides for UV absorbance detection," *Journal of Micromechanics and Microengineering*, Vol. 18, 2008, p. 055021.

[17]   "Nanogen i-Lynx™ Reader."

[18]   "Cholestech GDX Technical Note."

[19]   F. Xu, P. Datta, H. Wang, S. Gurung, M. Hashimoto, S. Wei, J. Goettert, R. McCarley, and S. Soper, "Polymer Microfluidic Chips with Integrated Waveguides for Reading Microarrays," *Analytical Chemistry*, Vol. 79, Dec. 2007, pp. 9007–9013.

[20]   P. Liu, T. Seo, N. Beyor, K. Shin, J. Scherer, and R. Mathies, "Integrated Portable Polymerase Chain Reaction-Capillary Electrophoresis Microsystem for Rapid Forensic Short Tandem Repeat Typing," *Analytical Chemistry*, Vol. 79, Mar. 2007, pp. 1881–1889.

[21]   D. Liu, X. Zhou, R. Zhong, N. Ye, G. Chang, W. Xiong, X. Mei, and B. Lin, "Analysis of multiplex PCR fragments with PMMA microchip," *Talanta*, Vol. 68, Jan. 2006, pp. 616–622.

[22]   A.E. Herr, A.V. Hatch, W.V. Giannobile, D.J. Throckmorton, H.M. Tran, J.S. Brennan, and A.K. Singh, "Integrated Microfluidic Platform for Oral Diagnostics," *Ann NY Acad Sci*, Vol. 1098, Mar. 2007, pp. 362–374.

[23]   J. Seo and L.P. Lee, "Fluorescence amplification by self-aligned integrated microfluidic optical systems," *TRANSDUCERS, Solid-State Sensors, Actuators and Microsystems, 12th International Conference on*, 2003, 2003, pp. 1136–1139 vol.2.

[24]   J. Seo and L.P. Lee, "Disposable integrated microfluidics with self-aligned planar microlenses," *Sensors and Actuators B: Chemical*, Vol. 99, May. 2004, pp. 615–622.

[25]   W. Du, Q. Fang, Q. He, and Z. Fang, "High-Throughput Nanoliter Sample Introduction Microfluidic Chip-Based Flow Injection Analysis System with Gravity-Driven Flows," *Analytical Chemistry*, Vol. 77, Mar. 2005, pp. 1330–1337.

[26]   T. Zhang, Q. Fang, S. Wang, L. Qin, P. Wang, Z. Wu, and Z. Fang, "Enhancement of signal-to-noise level by synchronized dual wavelength modulation for light emitting diode fluorimetry in a liquid-core-waveguide microfluidic capillary electrophoresis system," *Talanta*, Vol. 68, Nov. 2005, pp. 19–24.

[27]   K. Miyaki, Y. Guo, T. Shimosaka, T. Nakagama, H. Nakajima, and K. Uchiyama, "Fabrication of an integrated PDMS microchip incorporating an LED-induced fluorescence device," *Analytical and Bioanalytical Chemistry*, Vol. 382, Jun. 2005, pp. 810–816.

[28]   O. Hofmann, X. Wang, A. Cornwell, S. Beecher, A. Raja, D.D.C. Bradley, A.J. deMello, and J.C. deMello, "Monolithically integrated dye-doped PDMS long-pass filters for disposable on-chip fluorescence detection," *Lab on a Chip*, Vol. 6, 2006, pp. 981–987.

[29]   C.L. Bliss, J.N. McMullin, and C.J. Backhouse, "Rapid fabrication of a microfluidic device with integrated optical waveguides for DNA fragment analysis," *Lab on a Chip*, Vol. 7, 2007, pp. 1280–1287.

[30]   O. Schmidt, M. Bassler, P. Kiesel, C. Knollenberg, and N. Johnson, "Fluorescence spectrometer-on-a-fluidic-chip," *Lab on a Chip*, Vol. 7, 2007, pp. 626–629.

[31]   D. Vezenov, B. Mayers, R. Conroy, G. Whitesides, P. Snee, Y. Chan, D. Nocera, and M. Bawendi, "A Low-Threshold, High-Efficiency Microfluidic Waveguide Laser," *Journal of the American Chemical Society*, Vol. 127, Jun. 2005, pp. 8952–8953.

[32]   S. Balslev, A.M. Jorgensen, B. Bilenberg, K.B. Mogensen, D. Snakenborg, O. Geschke, J.P. Kutter, and A. Kristensen, "Lab-on-a-chip with integrated optical transducers," *Lab on a Chip*, Vol. 6, 2006, pp. 213–217.

[33]   J.A. Chediak, Z. Luo, J. Seo, N. Cheung, L.P. Lee, and T.D. Sands, "Heterogeneous integration of CdS filters with GaN LEDs for fluorescence detection microsystems," *Sensors and Actuators A: Physical*, Vol. 111, Mar. 2004, pp. 1–7.

[34]   A. Pais, A. Banerjee, D. Klotzkin, and I. Papautsky, "High-sensitivity, disposable lab-on-a-chip with thin-film organic electronics for fluorescence detection," *Lab on a Chip*, Vol. 8, 2008, pp. 794–800.

[35]   A.E. Herr, A.V. Hatch, D.J. Throckmorton, H.M. Tran, J.S. Brennan, W.V. Giannobile, and A.K. Singh, "Microfluidic immunoassays as rapid saliva-based clinical diagnostics," *Proceedings of the National Academy of Sciences*, Vol. 104, Mar. 2007, pp. 5268–5273.

[36]   N. Christodoulides, S. Mohanty, C.S. Miller, M.C. Langub, P.N. Floriano, P. Dharshan, M.F. Ali, B. Bernard, D. Romanovicz, E. Anslyn, P.C. Fox, and J.T. McDevitt, "Application of microchip assay system for the measurement of C-reactive protein in human saliva," *Lab on a Chip*, Vol. 5, 2005, pp. 261–269.

[37]   A. Woolley, D. Hadley, P. Landre, A. deMello, R. Mathies, and M. Northrup, "Functional Integration of PCR Amplification and Capillary Electrophoresis in a Microfabricated DNA Analysis Device," *Analytical Chemistry*, Vol. 68, Dec. 1996, pp. 4081–4086.

[38]   F. Huang, C. Liao, and G. Lee, "An integrated microfluidic chip for DNA/RNA amplification, electrophoresis separation and on-line optical detection," *Electrophoresis*, Vol. 27, 2006, pp. 3297–3305.

[39]   C.J. Easley, J.M. Karlinsey, J.M. Bienvenue, L.A. Legendre, M.G. Roper, S.H. Feldman, M.A. Hughes, E.L. Hewlett, T.J. Merkel, J.P. Ferrance, and J.P. Landers, "A fully integrated microfluidic genetic analysis system with sample-in-answer-out capability," *Proceedings of the National Academy of Sciences*, Vol. 103, Dec. 2006, pp. 19272–19277.

[40] M. Hashimoto, F. Barany, and S.A. Soper, "Polymerase chain reaction/ligase detection reaction/hybridization assays using flow-through microfluidic devices for the detection of low-abundant DNA point mutations," *Biosensors and Bioelectronics*, Vol. 21, Apr. 2006, pp. 1915–1923.

[41] Z. Wang, A. Sekulovic, J.P. Kutter, D.D. Bang, and A. Wolff, "Towards a portable microchip system with integrated thermal control and polymer waveguides for real-time PCR," *Electrophoresis*, Vol. 27, 2006, pp. 5051–5058.

[42] M.C. Smith, G. Steimle, S. Ivanov, M. Holly, and D.P. Fries, "An integrated portable hand-held analyser for real-time isothermal nucleic acid amplification," *Analytica Chimica Acta*, Vol. 598, Aug. 2007, pp. 286–294.

[43] S. Kedia, S.A. Samson, A. Farmer, M.C. Smith, D. Fries, and S. Bhansali, "Handheld interface for miniature sensors," 2005, pp. 241–252.

[44] A. Bhattacharyya and C. Klapperich, "Design and testing of a disposable microfluidic chemiluminescent immunoassay for disease biomarkers in human serum samples," *Biomedical Microdevices*, Vol. 9, Apr. 2007, pp. 245–251.

[45] E. Yacoub-George, W. Hell, L. Meixner, F. Wenninger, K. Bock, P. Lindner, H. Wolf, T. Kloth, and K.A. Feller, "Automated 10-channel capillary chip immunodetector for biological agents detection," *Biosensors and Bioelectronics*, Vol. 22, Feb. 2007, pp. 1368–1375.

[46] G. Marchand, P. Broyer, V. Lanet, C. Delattre, F. Foucault, L. Menou, B. Calvas, D. Roller, F. Ginot, R. Campagnolo, and F. Mallard, "Opto-electronic DNA chip-based integrated card for clinical diagnostics," *Biomedical Microdevices*, Vol. 10, Feb. 2008, pp. 35–45.

[47] A. Ymeti, J. Greve, P. Lambeck, T. Wink, S. vanHovell, T. Beumer, R. Wijn, R. Heideman, V. Subramaniam, and J. Kanger, "Fast, Ultrasensitive Virus Detection Using a Young Interferometer Sensor," *Nano Letters*, Vol. 7, Feb. 2007, pp. 394–397.

[48] J. Xu, D. Suarez, and D. Gottfried, "Detection of avian influenza virus using an interferometric biosensor," *Analytical and Bioanalytical Chemistry*, Vol. 389, Oct. 2007, pp. 1193–1199.

[49] F.J. Blanco, M. Agirregabiria, J. Berganzo, K. Mayora, J. Elizalde, A. Calle, C. Dominguez, and L.M. Lechuga, "Microfluidic-optical integrated CMOS compatible devices for label-free biochemical sensing," *Journal of Micromechanics and Microengineering*, Vol. 16, 2006, pp. 1006–1016.

[50] F. Pröll, B. Möhrle, M. Kumpf, and G. Gauglitz, "Label-free characterisation of oligonucleotide hybridisation using reflectometric interference spectroscopy," *Analytical and Bioanalytical Chemistry*, Vol. 382, 2005, pp. 1889–1894.

[51] C. Albrecht, N. Kaeppel, and G. Gauglitz, "Two immunoassay formats for fully automated CRP detection in human serum," *Analytical and Bioanalytical Chemistry*, Vol. 391, Jul. 2008, pp. 1845–1852.

[52] D. Markov, K. Swinney, and D. Bornhop, "Label-Free Molecular Interaction Determinations with Nanoscale Interferometry," *Journal of the American Chemical Society*, Vol. 126, Dec. 2004, pp. 16659–16664.

[53] D.J. Bornhop, J.C. Latham, A. Kussrow, D.A. Markov, R.D. Jones, and H.S. Sorensen, "Free-Solution, Label-Free Molecular Interactions Studied by Back-Scattering Interferometry," *Science*, Vol. 317, Sep. 2007, pp. 1732–1736.

[54] T.M. Chinowsky, J.G. Quinn, D.U. Bartholomew, R. Kaiser, and J.L. Elkind, "Performance of the Spreeta 2000 integrated surface plasmon resonance affinity sensor," *Sensors and Actuators B: Chemical*, Vol. 91, Jun. 2003, pp. 266–274.

[55] J. Waswa, J. Irudayaraj, and C. DebRoy, "Direct detection of E. Coli O157:H7 in selected food systems by a surface plasmon resonance biosensor," *LWT—Food Science and Technology*, Vol. 40, Mar. 2007, pp. 187–192.

[56] D. Wei, O.A. Oyarzabal, T. Huang, S. Balasubramanian, S. Sista, and A.L. Simonian, "Development of a surface plasmon resonance biosensor for the identification of Campylobacter jejuni," *Journal of Microbiological Methods*, Vol. 69, Apr. 2007, pp. 78–85.

[57] T.M. Chinowsky, M.S. Grow, K.S. Johnston, K. Nelson, T. Edwards, E. Fu, and P. Yager, "Compact, high performance surface plasmon resonance imaging system," *Biosensors and Bioelectronics*, Vol. 22, Apr. 2007, pp. 2208–2215.

[58] E. Fu, T. Chinowsky, K. Nelson, K. Johnston, T. Edwards, K. Helton, M. Grow, J. Miller, and P. Yager, "SPR imaging-based salivary diagnostics system for the detection of small molecule analytes," *Oral-Based Diagnostics*, Vol. 1098, 2007, pp. 335–344.

[59] T.M. Chinowsky, S.D. Soelberg, P. Baker, N.R. Swanson, P. Kauffman, A. Mactutis, M.S. Grow, R. Atmar, S.S. Yee, and C.E. Furlong, "Portable 24-analyte surface plasmon resonance instruments for rapid, versatile biodetection," *Biosensors and Bioelectronics*, Vol. 22, Apr. 2007, pp. 2268–2275.

[60] B. Feltis, B. Sexton, F. Glenn, M. Best, M. Wilkins, and T. Davis, "A hand-held surface plasmon resonance biosensor for the detection of ricin and other biological agents," *Biosensors and Bioelectronics*, Vol. 23, Feb. 2008, pp. 1131–1136.

[61] Y. Luo, F. Yu, and R.N. Zare, "Microfluidic device for immunoassays based on surface plasmon resonance imaging," *Lab on a Chip*, Vol. 8, 2008, pp. 694–700.

[62]   J. Klostranec and W. Chan, "Quantum Dots in Biological and Biomedical Research: Recent Progress and Present Challenges," *Advanced Materials*, Vol. 18, 2006, pp. 1953–1964.

[63]   J. Klostranec, Q. Xiang, G. Farcas, J. Lee, A. Rhee, E. Lafferty, S. Perrault, K. Kain, and W. Chan, "Convergence of Quantum Dot Barcodes with Microfluidics and Signal Processing for Multiplexed High-Throughput Infectious Disease Diagnostics," *Nano Letters*, Vol. 7, Sep. 2007, pp. 2812–2818.

[64]   P. Eastman, W. Ruan, M. Doctolero, R. Nuttall, G. deFeo, J. Park, J. Chu, P. Cooke, J. Gray, S. Li, and F. Chen, "Qdot Nanobarcodes for Multiplexed Gene Expression Analysis," *Nano Letters*, Vol. 6, May. 2006, pp. 1059–1064.

[65]   K. Yun, D. Lee, H. Kim, and E. Yoon, "A microfluidic chip for measurement of biomolecules using a microbead-based quantum dot fluorescence assay," *Measurement Science and Technology*, Vol. 17, 2006, pp. 3178–3183.

[66]   W. Liu, L. Zhu, Q. Qin, Q. Zhang, H. Feng, and S. Ang, "Microfluidic device as a new platform for immunofluorescent detection of viruses," *Lab on a Chip*, Vol. 5, 2005, pp. 1327–1330.

[67]   L.J. Lucas, J.N. Chesler, and J. Yoon, "Lab-on-a-chip immunoassay for multiple antibodies using microsphere light scattering and quantum dot emission," *Biosensors and Bioelectronics*, Vol. 23, Dec. 2007, pp. 675–681.

[68]   Z. Yan, L. Zhou, Y. Zhao, J. Wang, L. Huang, K. Hu, H. Liu, H. Wang, Z. Guo, Y. Song, H. Huang, and R. Yang, "Rapid quantitative detection of Yersinia pestis by lateral-flow immunoassay and up-converting phosphor technology-based biosensor," *Sensors and Actuators B: Chemical*, Vol. 119, Dec. 2006, pp. 656–663.

[69]   J. Wang, Z. Chen, P.L.A.M. Corstjens, M.G. Mauk, and H.H. Bau, "A disposable microfluidic cassette for DNA amplification and detection," *Lab on a Chip*, Vol. 6, 2006, pp. 46–53.

[70]   P.L.A.M. Corstjens, Z. Chen, M. Zuiderwijk, H.H. Bau, W.R. Abrams, D. Malamud, R.S. Niedbala, and H.J. Tanke, "Rapid Assay Format for Multiplex Detection of Humoral Immune Responses to Infectious Disease Pathogens (HIV, HCV, and TB)," *Annals of the New York Academy of Sciences*, Vol. 1098, Mar. 2007, pp. 437–445.

[71]   W. Abrams, C. Barber, K. McCann, G. Tong, Z. Chen, M. Mauk, J. Wang, A. Volkov, P. Bourdelle, P. Corstjensd, M. Zuiderwijk, K. Kardos, S. Li, H. Tanke, R. Niedbala, D. Malamud, and H. Bau, "Development of a microfluidic device for detection of pathogens in oral samples using upconverting phosphor technology (UPT)," *Oral-Based Diagnostics*, Vol. 1098, 2007, pp. 375–388.

[72]   T.A. Taton, C.A. Mirkin, and R.L. Letsinger, "Scanometric DNA Array Detection with Nanoparticle Probes," *Science*, Vol. 289, Sep. 2000, pp. 1757–1760.

[73]   J. Nam, C.S. Thaxton, and C.A. Mirkin, "Nanoparticle-Based Bio-Bar Codes for the Ultrasensitive Detection of Proteins," *Science*, Vol. 301, Sep. 2003, pp. 1884–1886.

[74]   P. Jain, X. Huang, I. El-Sayed, and M. El-Sayed, "Review of Some Interesting Surface Plasmon Resonance-enhanced Properties of Noble Metal Nanoparticles and Their Applications to Biosystems," *Plasmonics*, Vol. 2, 2007, pp. 107–118.

[75]   J.N. Anker, W.P. Hall, O. Lyandres, N.C. Shah, J. Zhao, and R.P. Van Duyne, "Biosensing with plasmonic nanosensors," *Nat Mater*, Vol. 7, Jun. 2008, pp. 442–453.

[76]   M. Stewart, C. Anderton, L. Thompson, J. Maria, S. Gray, J. Rogers, and R. Nuzzo, "Nanostructured Plasmonic Sensors," *Chemical Reviews*, Vol. 108, Feb. 2008, pp. 494–521.

[77]   C. Yonzon, E. Jeoung, S. Zou, G. Schatz, M. Mrksich, and R. VanDuyne, "A Comparative Analysis of Localized and Propagating Surface Plasmon Resonance Sensors: The Binding of Concanavalin A to a Monosaccharide Functionalized Self-Assembled Monolayer," *Journal of the American Chemical Society*, Vol. 126, Oct. 2004, pp. 12669–12676.

[78]   K.M. Mayer, S. Lee, H. Liao, B.C. Rostro, A. Fuentes, P.T. Scully, C.L. Nehl, and J.H. Hafner, "A Label-Free Immunoassay Based Upon Localized Surface Plasmon Resonance of Gold Nanorods," *ACS Nano*, Vol. 2, Apr. 2008, pp. 687–692.

[79]   C. Haynes and R. Van Duyne, "Nanosphere Lithography: A Versatile Nanofabrication Tool for Studies of Size-Dependent Nanoparticle Optics," *Journal of Physical Chemistry B*, Vol. 105, Jun. 2001, pp. 5599–5611.

[80]   G. Liu, Y. Lu, J. Kim, J. Doll, and L. Lee, "Magnetic Nanocrescents as Controllable Surface-Enhanced Raman Scattering Nanoprobes for Biomolecular Imaging," *Advanced Materials*, Vol. 17, 2005, pp. 2683–2688.

[81]   Y.B. Zheng, B.K. Juluri, X. Mao, T.R. Walker, and T.J. Huang, "Systematic investigation of localized surface plasmon resonance of long-range ordered Au nanodisk arrays," *Journal of Applied Physics*, Vol. 103, Jan. 2008, pp. 014308–9.

[82]   T. Endo, K. Kerman, N. Nagatani, H. Hiepa, D. Kim, Y. Yonezawa, K. Nakano, and E. Tamiya, "Multiple Label-Free Detection of Antigen-Antibody Reaction Using Localized Surface Plasmon Resonance-Based Core-Shell Structured Nanoparticle Layer Nanochip," *Analytical Chemistry*, Vol. 78, Sep. 2006, pp. 6465–6475.

[83] H. Minh Hiep, T. Endo, K. Kerman, M. Chikae, D. Kim, S. Yamamura, Y. Takamura, and E. Tamiya, "A localized surface plasmon resonance based immunosensor for the detection of casein in milk," *Science and Technology of Advanced Materials*, Vol. 8, May 2007, pp. 331–338.

[84] L. Sherry, R. Jin, C. Mirkin, G. Schatz, and R. VanDuyne, "Localized Surface Plasmon Resonance Spectroscopy of Single Silver Triangular Nanoprisms," *Nano Letters*, Vol. 6, Sep. 2006, pp. 2060–2065.

[85] T. Lin, K. Huang, and C. Liu, "Determination of organophosphorous pesticides by a novel biosensor based on localized surface plasmon resonance," *Biosensors and Bioelectronics*, Vol. 22, Oct. 2006, pp. 513–518.

[86] C. Schofield, R. Field, and D. Russell, "Glyconanoparticles for the Colorimetric Detection of Cholera Toxin," *Analytical Chemistry*, Vol. 79, Feb. 2007, pp. 1356–1361.

[87] S. Marinakos, S. Chen, and A. Chilkoti, "Plasmonic Detection of a Model Analyte in Serum by a Gold Nanorod Sensor," *Analytical Chemistry*, Vol. 79, Jul. 2007, pp. 5278–5283.

[88] S. Bishnoi, C. Rozell, C. Levin, M. Gheith, B. Johnson, D. Johnson, and N. Halas, "All-Optical Nanoscale pH Meter," *Nano Letters*, Vol. 6, Aug. 2006, pp. 1687–1692.

[89] H. Wang, D. Brandl, F. Le, P. Nordlander, and N. Halas, "Nanorice: A Hybrid Plasmonic Nanostructure," *Nano Letters*, Vol. 6, Apr. 2006, pp. 827–832.

[90] D. Kim, K. Kerman, M. Saito, R. Sathuluri, T. Endo, S. Yamamura, Y. Kwon, and E. Tamiya, "Label-Free DNA Biosensor Based on Localized Surface Plasmon Resonance Coupled with Interferometry," *Analytical Chemistry*, Vol. 79, Mar. 2007, pp. 1855–1864.

[91] H.M. Hiep, T. Nakayama, M. Saito, S. Yamamura, Y. Takamura, and E. Tamiya, "A Microfluidic Chip Based on Localized Surface Plasmon Resonance for Real-Time Monitoring of Antigen–Antibody Reactions," *Japanese Journal of Applied Physics*, Vol. 47, 2008, pp. 1337–1341.

[92] A. DeLeebeeck, L. Kumar, V. deLange, D. Sinton, R. Gordon, and A. Brolo, "On-Chip Surface-Based Detection with Nanohole Arrays," *Analytical Chemistry*, Vol. 79, Jun. 2007, pp. 4094–4100.

[93] J. Ji, J. O'Connell, D. Carter, and D. Larson, "High-Throughput Nanohole Array Based System To Monitor Multiple Binding Events in Real Time," *Analytical Chemistry*, Vol. 80, Apr. 2008, pp. 2491–2498.

[94] L. Pang, G.M. Hwang, B. Slutsky, and Y. Fainman, "Spectral sensitivity of two-dimensional nanohole array surface plasmon polariton resonance sensor," *Applied Physics Letters*, Vol. 91, 2007, pp. 123112–3.

[95] M. Fleischmann, P.J. Hendra, and A.J. McQuillan, "Raman spectra of pyridine adsorbed at a silver electrode," *Chemical Physics Letters*, Vol. 26, May 1974, pp. 163–166.

[96] K.A. Willets and R.P. Van Duyne, "Localized Surface Plasmon Resonance Spectroscopy and Sensing," *Annual Review of Physical Chemistry*, Vol. 58, Apr. 2007, pp. 267–297.

[97] M.J. Banholzer, J.E. Millstone, L. Qin, and C.A. Mirkin, "Rationally designed nanostructures for surface-enhanced Raman spectroscopy," *Chemical Society Reviews*, Vol. 37, 2008, pp. 885–897.

[98] G.L. Liu and L.P. Lee, "Nanowell surface enhanced Raman scattering arrays fabricated by soft-lithography for label-free biomolecular detections in integrated microfluidics," *Applied Physics Letters*, Vol. 87, 2005, pp. 074101–3.

[99] L. Chen and J. Choo, "Recent advances in surface-enhanced Raman scattering detection technology for microfluidic chips," *Electrophoresis*, Vol. 9999, 2008.

[100] P. Monaghan, K. McCarney, A. Ricketts, R. Littleford, F. Docherty, W. Smith, D. Graham, and J. Cooper, "Bead-Based DNA Diagnostic Assay for Chlamydia Using Nanoparticle-Mediated Surface-Enhanced Resonance Raman Scattering Detection within a Lab-on-a-Chip Format," *Analytical Chemistry*, Vol. 79, Apr. 2007, pp. 2844–2849.

[101] M. Sha, S. Penn, G. Freeman, and W. Doering, "Detection of Human Viral RNA via a Combined Fluorescence and SERS Molecular Beacon Assay," N*anoBioTechnology*, Vol. 3, Feb. 2007, pp. 23–30.

[102] D. Hou, S. Maheshwari, and Hsueh-Chia Chang, "Rapid bioparticle concentration and detection by combining a discharge driven vortex with surface enhanced Raman scattering.," *Biomicrofluidics*, Vol. 1, Jan. 2007.

[103] G. Sabatte, R. Keir, M. Lawlor, M. Black, D. Graham, and W. Smith, "Comparison of Surface-Enhanced Resonance Raman Scattering and Fluorescence for Detection of a Labeled Antibody," *Analytical Chemistry*, Vol. 80, Apr. 2008, pp. 2351–2356.

[104] S. Mulvaney, M. Musick, C. Keating, and M. Natan, "Glass-Coated, Analyte-Tagged Nanoparticles: A New Tagging System Based on Detection with Surface-Enhanced Raman Scattering," *Langmuir*, Vol. 19, May. 2003, pp. 4784–4790.

[105] S. Shanmukh, L. Jones, J. Driskell, Y. Zhao, R. Dluhy, and R. Tripp, "Rapid and Sensitive Detection of Respiratory Virus Molecular Signatures Using a Silver Nanorod Array SERS Substrate," *Nano Letters*, Vol. 6, Nov. 2006, pp. 2630–2636.

Neutrophil Chemotaxis Assay from Whole Blood Samples

Nitin Agrawal, Mehmet Toner, and Daniel Irimia*

BioMEMS Resource Center, Center for Engineering in Medicine and Surgical Services, Massachusetts General Hospital, Shriners Hospital for Children, and Harvard Medical School, Boston, MA 02114, E-mail: *dirimia@hms.harvard.edu

Abstract

Neutrophils constitute an integral part of the immune system and possess abilities to sense chemical gradients and migrate towards sites of infection and inflammation in the body. These processes, also known as chemotaxis, are critical for early immune responses and for preserving the sterility of the body's internal environment. To better understand the complexities involved in chemotaxis process in vivo, usually neutrophils are isolated and chemotaxis assays are performed in vitro under a higher degree of microenvironment control. However, most in vitro chemotaxis studies often require large volumes of blood to isolate neutrophils and the isolation protocols could alter the original neutrophils activity levels. In this chapter, a microfluidic device is described that captures neutrophils from just one drop of whole blood and quantitatively measures neutrophil chemotaxis towards stable gradients of chemokines. To better replicate in vivo conditions, a systematic approach to creating various temporal sequences of linear and non-linear spatial gradients in microfluidic devices has been developed and discussed in this chapter.

Key terms	chemotaxis
	neutrophil
	chemokine
	gradient
	microfluidic
	P-selectin

8.1   Introduction

Chemical gradients coordinate many activities of the cellular components of the immune system, from the release of various cells out of the bone marrow to the migration of neutrophils, lymphocytes, and monocytes into the tissue and lymph nodes [1]. Of particular importance for the protection of the body against infections is the rapid increase of chemokine concentrations within the tissue following microbial infections or sterile tissue injury [2]. At the same time, chemokines and other molecules are exposed to the endothelium facilitating the rolling and adhesion of leukocytes necessary for the migration of white blood cells from circulating blood to the sites of infection and inflammation in the tissue [3]. For example, neutrophils adhere in larger numbers to the endothelium expressing different selectin molecules, and infiltrate the inflamed tissues following gradients of chemokines [4]. Although techniques for studying the responses of leukocytes in vivo have been developed, the complexity of the interactions between different cell types and local chemical gradients makes the interpretation of the observations and results difficult [5] and better in vitro systems capable to replicate complex environments in controlled and precise manner are always needed.

Among the new technologies with an increasingly broader impact in biology, microfluidic devices proved extremely useful for neutrophil migration studies [6]. Following the first demonstration of neutrophil chemotaxis in linear IL8 gradients [7], several studies reported observations of neutrophil migration in different chemokines [8], gradient profiles [9], or combinations of different chemokine gradients [10]. Compared to other in vitro chemotaxis assays, (e.g., transwell [11], under agarose [12, 13], or Zigmond chamber [14]) microfluidic systems provide better control and reproducibility of chemokine gradients [7]. Moreover, the same microfluidic technology can be used to isolate cells directly from whole blood [15, 16], circumventing the need for lengthy, work-intensive leukocyte isolation procedures. Here we describe in detail a protocol for isolating human neutrophils from whole blood for chemotaxis assays that is capable of both neutrophil isolation and chemotaxis assay in one integrated microfluidic device. Neutrophils are captured on selectin coated surfaces and the binding between neutrophils and substrate is such that neutrophils are able to move directionally in response to a chemoattractant gradient [17]. Moreover, the microfluidic device also allows one to quickly expose resting neutrophils to gradients or to switch between gradients of different chemokines and study the transitory responses [18].

8.2   Device design

The microfluidic device has three main components working together in an integrated system (Figure 8.1). There are two gradient generator networks that produce two distinct chemical gradients that can be successively employed for chemotaxis studies. There is one chemotaxis channel (cell chamber) in which neutrophils are isolated and then observed during migration in the presence of chemical gradients. Finally, there is a system of microstructured membrane valves that control the communication between the two gradient generator networks, the chemotaxis channel, and the accessory channel for handling the whole blood.

The method described here for generating chemical gradients in microfluidic systems makes use of a series of dividers in the longitudinal direction of a channel

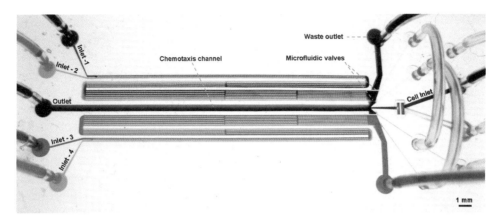

Figure 8.1 Neutrophil isolation and chemotaxis device. The device is loaded with food dyes of different colors to demonstrate the gradient formation. The device combines two gradient generator networks and a chemotaxis channel through the use of a system of microstructured valves. For consistently establishing the chemotactic gradient from top to bottom across the chamber, inlets 2 and 4 contains the two chemokine solutions while inlets 1 and 3 contains buffer solutions. Inlet configuration can be altered appropriately to obtain the desired gradient directionality.

(Figure 8.2). The dividers restrict the diffusion between parallel streams of distinct concentrations that are flowing through the channel in the longitudinal direction [19]. Considering the small (less than 1 mm) dimensions of the channel and slow flow rates (less than 1 mm/s) the flow can be considered laminar (low Reynolds number flow). Under such conditions and in the absence of the dividers, unrestricted diffusion transport between the streams would result in a characteristic concentration profile in the direction transversal to the channel, that progressively evolve downstream towards uni-

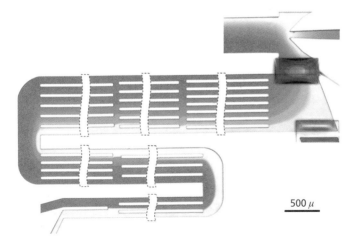

Figure 8.2 Micrographs showing different sections of the gradient generator. Two streams with red and yellow food dyes are used to demonstrate the generation of a linear gradient. A system of dividers controls the mixing of the two streams as they flow along the gradient generator network. In conditions of laminar flow, streams of different concentrations are split and recombined, and adjacent streams allowed to mix by diffusion, resulting in an increasing number of streams of intermediate concentrations. At the end of the gradient generator network, eight different streams of increasing concentrations are merged together to form a linear gradient. The output of the gradient generator network is directed to the chemotaxis channel through a system of valves.

form concentration profile. The presence of dividers restricts the diffusive mixing of the two initial solutions of concentrations C_0 and C_1 in such a way that different concentration profiles can be formed.

To better understand the formation of a chemical gradient in the microfluidic devices with dividers, consider the first pair of two dividers that splits the flow stream into three. While the two streams between the dividers and the side walls have concentrations identical with the original streams, the stream between the dividers will constitute a mixture of the two streams and have a concentration in between the two, depending on the mixing ratio. By properly positioning the dividers, precise fractions from the two original streams are separated and allowed to mix between the first pair of dividers. At the end of the first set of dividers, three streams of the initial two concentrations and the resulting intermediary concentration are present. A group of three dividers is now used to split the three streams into four streams. The two streams close to the walls have again concentrations that are identical to the concentrations C_0 and C_1 of the two initial streams, while the two streams in the middle will have intermediary concentration between C_0 and C_1. The four resulting streams at the end of the three divider set will be arranged in order of their concentration from C_0 to C_1. Through the iteration of the divider stages, an increasing number of streams, with concentrations values between the two initial streams are produced. The number of dividers is increased by one with each additional stage. Merging all the streams into the main channel will result in a concentration profile across the channel that evolves from two streams of distinct concentrations, C_0 to C_1, to a concentration profile of choice. Indeed, one can demonstrate mathematically that for any chosen concentration profile at the output, at least one set of dividers exists to generate any particular concentration profile defined by a function that is continuous and monotonic [19].

The practical approach for numerically calculating the precise position inside the channel of all the dividers and for replicating chemical gradients having target concentration profiles across the width of the channel is described below and its implementation using MATLAB code that is presented in the Appendix 8A. The steps for calculating the position of the dividers in the channel in order to reproduce a desired output concentration gradient are as follows:

1. First, estimate the number of the dividers at the last level for target accuracy of reproducing the desired output concentration gradient. Then choose the position and spacing for each of the dividers across the channel, with narrower spacing usually considered in the regions of higher gradient steepness. The MATLAB code offers the choice of either equally spaced dividers or dividers spaced for uniform concentration jump between adjacent channels.

2. Start from the last row of dividers and then calculate the positions of preceding dividers based on the positioning equations that are repeatedly applied. Several choices for parameter α representing the mixing ratios are possible when calculating the positions of all the dividers down to level 0 where one divider separates the two initial solutions. Consequently, for one final concentration profile, at least in theory, an infinite number of possible arrangements of the dividers is possible.

3. Additional aspects of the practical implementation are the thickness and length of dividers. At any level n, with $n+1$ dividers, the total thickness of the dividers has to be subtracted from the width of the channel before calculating the physical spacing between dividers.

4. Finally, calculate the length of the dividers in the longitudinal direction parallel to the flow of the dividers in order to match the velocity of the streams and the time required for 99% complete mixing by diffusion of the merged streams at each level.

8.3   Materials

A list of all materials used in these experiments including their suppliers and part numbers is given in Table 8.1.

8.4   Methods

8.4.1   Device fabrication

The microfluidic devices are manufactured in Polydimethylsiloxane (PDMS) following standard photolithographic procedures (described in earlier chapters of the book). Here, only few aspects that are specific to the device for neutrophil isolation and chemotaxis are presented in detail. The devices consist of two different layers of PDMS, a thinner layer (approximately $150\,\mu$m) with the microfluidic network (network-layer) and a thick layer (approximately 4 mm) with control channels for actuating the valves (control-layer). To manufacture the devices, two silicon masters are prepared using standard SU8 photolithography. Once the wafers are ready, device fabrication involves the following steps:

1. Carefully mix about 30 grams of PDMS with 3 grams of curing agent and pour about 25 grams of the mixture on the top of the first, control-layer wafer, fixed with tape in a Petri dish.

2. Mount the second, network-layer wafer on a spinner and pour the remaining 3 to 5 grams of PDMS mix at the center of wafer. Spin the wafer at 1,000 rpm for 30 seconds to form a PDMS membrane of about 150 to $200\,\mu$m thickness.

Table 8.1   List of Materials for Neutrophil Chemotaxis Assay from Whole Blood

Item	Vendor	Model/Part #
Syringe pump	Harvard Apparatus	PHD 2000
Microscope	Olympus	CKX41
CCD camera	Pixelink	PL-A742
PDMS	Ellsworth Adhesives	SYLGARD 184
1-mL BD syringes	Fisher Scientific	309602
Tubing	Small Parts Inc.	TGY-010-C
Needles	Small Parts Inc.	NE-301PL-C
Lancets, 1.25x28G BD Purple	Fisher Scientific	366579
strightBD vacutainer with sodium heparin, 2 ml	Fisher Scientific	026888
PBS	Fisher Scientific	21-040-CV
HBSS	Invitrogen	14025-134
Recombinant human IL8	R&D Systems Inc.	208-IL-010
fMLP	Sigma Aldrich	F3506
Recombinant human P-selectin	R&D Systems Inc.	ADP3-050

3. Bake both wafers overnight in an oven at 65°C.

4. From the control-layer wafer, cut and peel the thick PDMS layer and cut the individual devices.

5. Take a sharpened needle and punch holes for tubing connections to the valve control channels.

6. Expose the channel-side of the punched devices and the network-layer wafer to oxygen plasma and then bond the top layer pieces onto the membrane of bottom layer. In order to align the control channels with the valve features, the bonding should be carefully performed using an aligner or manually under a stereo microscope. If air bubbles are trapped between the two layers of PDMS during bonding, they should be eliminated by gently tapping the top layer with tweezers.

7. Bake the wafer with control-layer devices on a hot plate at 75°C for 10 minutes for irreversible bonding between the two layers.

8. After bonding, carefully peel of the composite device with both layers of PDMS bonded together and cut the membrane around top layer devices using a scalpel.

9. Punch the remaining holes for tubing connections to the fluidic network.

10. Take two short pieces of the tubing (2-cm long) and use these to interconnect valve #1 with #3 and valve #2 with #4. (Figure 8.3) This way the two pairs of interconnected valves can be actuated simultaneously using only two actuating syringes.

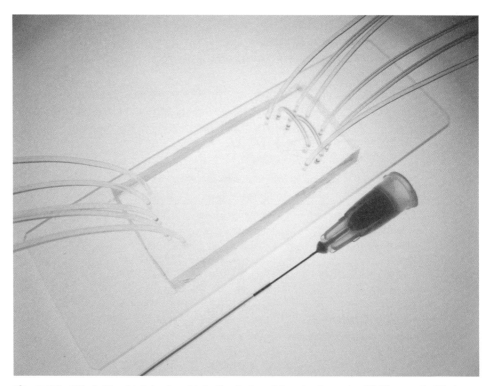

Figure 8.3 Whole blood is introduced into the device while actuating valve #5. The needle filled with blood is attached to the cell inlet on the right and blood loaded in the chemotaxis channel. Blood is gradually loaded by applying slight pressure on the top of the needle with index finger. Excess blood passing through the chemotaxis channel is directed to waste outlet.

11. Before exposing the devices to oxygen plasma for bonding with glass slide, connect three pieces of approximately 10" tubing to the three valve inlets. Connect needles to the opposite ends, to be used for actuation of the valves using syringes. The microfluidic valves in the bottom layer membrane are placed in the same plane as the fluidic network, meaning that the valves are in closed position at rest. Therefore, to prevent them from irreversibly bonding to the glass substrate, valves are actuated during bonding to slightly lift them up above the glass substrate. For this, vacuum is applied to the valves using syringe and tubing connections to maintain them in open position.

12. After the plasma treatment, carefully connect actuation syringes to the devices and apply suction to keep the valves open and block the syringe plunger with a metal screw or a similar object.

13. Bond the devices to glass slides while the valves are actuated and bake the device on a hot plate at 75°C for 10 minutes.

14. Under a microscope, individually pull and push the syringe plunger several times for each valve. This procedure will prevent further bonding of the valve to the substrate and maintain fully functional valves.

8.4.2   Surface treatment

Cell chamber treatment is performed prior to loading the whole blood in the device to facilitate capture of neutrophils. P-selectin has been known to predominantly capture neutrophils among the entire leukocyte subpopulation. In vivo, the P-selectin is expressed on the endothelium during chemotactic exposure and plays significant role in neutrophil trafficking. We have previously determined that neutrophil capture yield is maximum at around 25 μg/mL of P-selectin concentration [17].

1. Prepare fresh P-selectin solution of 25-μg/mL concentration in PBS and preserve at 4°C.

2. Pull the plunger half way and connect three 10-mL actuating syringes to the device through tubing connections approximately 1 foot long.

3. Operate two of the syringes to close valves #1 through #4 while valve #5 is actuated to its open position. Place a paper tape around the plunger to block it at the pushed position while place a metal screw between the plunger and the syringe to keep it pulled.

4. Aspirate 2- to 3-μL P-selectin solution using a 10-μL pipette and gradually load the solution by inserting the pipette tip into the cell chamber inlet or outlet port.

5. Incubate the device at room temperature for one hour before aspirating the excess P-selectin solution.

6. Open all the valves and perfuse the entire microfluidic network with previously prepared 2% HSA (human serum albumin) in Hank's buffer saline solution (HBSS) to block non-specific binding sites and incubate for at least 15 minutes. Waste channel ports should be individually used for this purpose.

7. A small volume of 2% HSA solution (~500 μL) should be previously prepared by adding HSA in HBSS at the ratio of 2 grams/100 mL and can be preserved at 4°C for about 7 days.

8.4.3 Chemotaxis assay

A simplified chemotaxis assay is described here for direct on-chip isolation and migration of neutrophils from a droplet of whole blood. The switching gradient device is best suited for performing combinatorial chemotaxis assays where neutrophil exposure to rapidly changing chemotactic conditions is desired (Figure 8.4). The cell chamber is initially treated with an appropriate concentration of cell adhesion molecule (P-selectin) to capture primary neutrophils from whole blood in the microfluidic channel, as described above.

8.4.3.1 Media preparation

1. *HBSS buffer:* Take 2 mL HBSS and add 0.004 grams HSA to make 0.2% HSA solution (0.2 grams/100 mL) in HBSS.
2. *Chemotaxis solutions:* Take two vials each containing 500 μL of the above prepared HBSS buffer and add appropriate volume of the IL8 and fMLP stock solutions to achieve 10- and 20-nM final respective concentrations.

8.4.3.2 Device priming

While initially loading the working solutions and buffers in a PDMS microfluidic device, several small air pockets are usually trapped inside the channels that must be removed

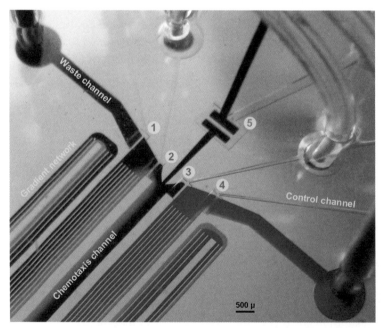

Figure 8.4 A system of five microfluidic valves control the flow in the blood inlet, gradient generator networks, and the chemotaxis channel. In this image, valves #1 and #3 are closed while #2 and #4 are open such that the output of the top gradient generator network is directed towards the chemotaxis channel, and the output from the lower network is diverted to the waste. Closing valves #2 and #4 while opening the valves #1 and #3 will direct the output form the lower network to the chemotaxis channel, enabling the fast switching of the chemical gradients within 5 to 10 seconds. Valve #5 is opened only during the loading of whole blood into the chemotaxis channel. During the chemotaxis assay valve #5 is always maintained in closed position.

before starting the experiment. This process of bubble removal from PDMS channels is called priming and is achieved by gradually pressurizing the device from inside.

1. Prepare four 1-foot-long tubing pieces with connecting needles at the end of each tube.
2. Then take four 1-mL syringes and load two of them with $500\,\mu L$ of HBSS buffer in each while $500\,\mu L$ of IL8 and fMLP solutions are loaded in the other two syringes.
3. Connect syringes and needle/tubing assembly together and then tap the syringes with needle facing down to remove any trapped air inside the needle.
4. Hold the open end of each piece of tubing with a pair of forceps (about 1–2 mm away from the end) and insert the tubing into the inlet ports of the device. The two syringes are connected to the device in an order such that chemokine always forms a gradient from top (100%) to bottom (0%) in the main channel to maintain consistency. For example, if the IL8 and FMLP syringes are connected to the inlet ports 2 and 4, then the buffer syringes are connected to ports 1 and 3 so that the laminar flow will always keep the chemokine layer at the top of the cell chamber.
5. Connect similar lengths of tubing to the cell chamber and waste outlet ports and place the syringes on a multi-rack syringe pump.
6. Actuate all the valves to open position.
7. Turn on the pump at a flow rate of about $5\,\mu L/min$. Once the fluidic network is filled with fluids and air bubbles, block all the outlet ports by clamping the outlet tubing using ordinary binder clips while leaving the pump on. Since all the outlets are blocked, the pump starts pressurizing the device by continuously infusing the fluids and pushing air bubbles to diffuse through PDMS matrix. The priming process should be carefully performed by continuously monitoring the channels under the microscope to avoid overpressure buildup. As soon as all the bubbles disappear, the pump should be turned off and outlet tubing unclamped to release the pressure.

8.4.3.3 Blood sample loading and neutrophil isolation

Neutrophils are isolated directly from whole blood on the P selectin coated surfaces. The selectivity of capture is 80% to 90%.

1. Close valves #1 through #4 and open valve #5 (Figure 8.4).
2. Take a 2-mL BD vacutainer containing sodium heparin and shake well after filling with 2 mL HBSS.
3. Then take a 5-cm piece of tubing with needle on one side and inject 70 µL of heparin solution into the open end of the needle. Break and remove any visible air bubbles with the pipette tip.
4. Using a 1.25-mm lancet, gently prick the finger and collect $10\,\mu L$ of whole blood with a pipette. Immediately inject the blood in the needle containing heparin solution and mix well by applying suction/compression to the pipette several times.
5. Gradually apply slight pressure to the open end of the needle using index finger to push the blood sample through the tubing until a small drop of fluid appears on the other end of the tubing.
6. To avoid entrapment of air bubbles while connecting the tubing to the PDMS device, a drop of HBSS should always be present in excess at the connecting ports of the

device. Hold the loose end of the sample tubing with forceps and carefully insert in the PDMS port.

7. Once a sealed connection is established, again apply pressure to the loading needle to push more sample through the flow chamber (Figures 8.5, 8.6). The inlet and

Figure 8.5 A syringe needle filled with blood is connected to the blood inlet port of the microfluidic device. The needle is filled with 10-μL whole blood mixed with 70-μL heparin solution in Hank's buffer. A 5-cm-long piece of Tygon tubing is used to connect the needle to the inlet port. The same type of tubing is used to bring different solutions to the chip and for applying pressure and vacuum to control the microfluidic valves.

100 μ

Figure 8.6 Magnified image of the chemotaxis channel filled with whole blood. The proximal region of the channel with the inlet channel and the output channels from the gradient generator networks are shown. Most cells in this image are red blood cells. They will be removed from the chemotaxis channel once the flow is started leaving the captured neutrophils behind into the channel.

outlet tubings are then clamped using binder clips for 5 to 10 minutes allowing cells to bind to P-selectin.

8.4.3.4 Device operation for neutrophil migration

The four valves are actuated in an order (two valves simultaneously) such that when valves #1 and #3 are open, the valves #2 and #4 are closed and vice versa (Figure 8.4). This actuation scheme allows fluids from one of the gradient networks to enter the chemotaxis channel while diverting the fluids from the other network to the waste chamber. When the valve patterns are reversed, fluids from the second network are diverted back to the cell chamber while opening the first network to the waste chamber. This is done by actuating the alternate valves simultaneously and the switching process takes only about 10 seconds.

After loading the cells and waiting for 10 minutes, open all the outlet tubes those were previously clamped except the cell chamber outlet. Turn the syringe pump on at the flow rate of 0.2 to 0.5 μL/min. Open valve #5 to remove blood cells from the channel connecting valve 5 to the chemotaxis channel. Also open the pair of valves #2 and #4. This will establish a uniform steady state flow inside the gradient networks before the gradient enters the main cell chamber. The flow will immediately remove most of the red blood cells and allow the visualization of the captured neutrophils (Figure 8.7(a)).

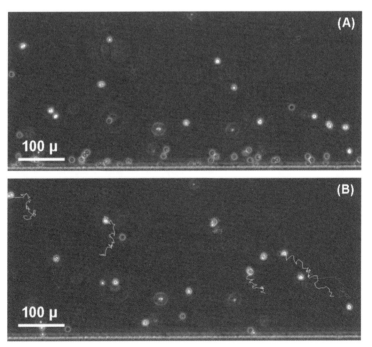

Figure 8.7 Neutrophils isolated in the chemotaxis channel from whole blood. (a) Immediately after applying the gradient, neutrophils have a round shape, indicating the lack of activation during the isolation on the P-selectin coated surface. Some remaining red blood cells, smaller and less bright, are being gradually washed away. (b) At 45 minutes after the application of a linear gradient of fMLP (0 to 20 nM across the 450-μm-wide channel) most of the neutrophils have migrated towards the higher concentration. Migration tracks are overlaid. Concentration gradient is from top to bottom across the channel while the flow is from right to left.

1. After few minutes, open the cell chamber outlet, and close valve #5 immediately. This will divert streams from the top gradient network to enter the cell chamber immediately forming a linear gradient. Run for the desired experimental duration, usually 15 to 30 minutes.

2. Actuate control syringes to open valves #1 and #3 and immediately close valves-#2 and #4. This will switch the gradient in the chemotaxis channel allowing the gradient generated in the bottom network to enter the cell chamber, while diverting the top streams to the waste channel. Again, run for 15 to 20 minutes and repeat the switches if required.

8.5   Data acquisition

A Pixelink digital camera is used to capture time-lapse images. As soon as the gradient is established, a suitable section of the channel is focused under microscope and time-lapse images are captured in phase contrast mode every 6 seconds for simplicity to yield 10 images every minute. Captured images are analyzed and individual cells are tracked using Metamorph imaging software. Elapsed time and migration distance from origin in the direction transversal to the channel are derived for each cell and average displacements can be calculated for the entire cell population in consideration. Eventually, the migration curves representing average distance traveled in the direction of increasing chemokine concentration against time are plotted and compared (Figure 8.7(b)).

8.6   Troubleshooting tips

1. It is extremely important to keep excess PBS or HBSS at the port interface while inserting tubing connections as well as while switching the connections if required. This will avoid air bubbles to sneak into the channel network while making or breaking tube connections.

2. The device priming can be done significantly faster by using a higher flowrate of up to 20 μL/min; however, the process should be constantly monitored under the microscope or tubing connections may pop out of the device due to overpressure. Also, at higher flowrates, pump should be intermittently turned on and off to provide sufficient time for air to diffuse through the PDMS.

3. During the priming step, part of the connected tubing may be pushed out by the pressure buildup inside the device; therefore after the priming, tubing connections should be secured tight by pushing the tubing ends back into the connection ports.

4. Before blocking the blood sample, pushing the sample volume at a high flow rate helps capture higher population of neutrophils.

Appendix 8A

MATLAB code for calculating the position of the dividers for the universal gradient generator.

```
%
% MATLAB code for calculating the position of the dividers for the uni-
versal gradient generator
%

% the program uses the output gradient profile to calculate back
% the position of the dividers for generating such gradient
% starting with two solutions
% with concentrations equal to the maximum and minimum of the target
output gradient

% concentrations are normalized in the range 0 to 1.


clear all
% N = number of concentrations on the last row
N=10;

% n = number of walls including the side walls
n=N+1;

% C = matrix of concentrations for all levels from the entrance to the
output
C=zeros(N);

% W matrix of the positions of the walls relative to the left side of
the channel (O concentration)
W=zeros(N);

% M = median position of the all channels at different levels
M=zeros(N);

% alpha = splitting ratios for flow from row to next row
alpha=-1*ones(N);

% delta = thickens of the dividers [in microns]
delta=10

% L = total width of the channel
L=1000

% P = position of the left side of the delta width dividers
P=zeros(N-1);

%****************************************
%%%DESCRIPTION OF THE LAST ROW

%walls (N-1) walls equally spaced, not including the left sidewall
%W(1,:)=[(1:N)/N];
```

221

```
% or walls spaced such that equal jumps in concentrations between output
channels
W(1,:)=1/2*(log((1:N)/(N).*(exp(2)-1)+1));

%M(r,n) is the position of the center of a channel r at level n
r=1;
j=1;
M(r,1)=W(r,j)/2;
j=2;
while ((j<=N)&(W(r,j)<=1))
    M(r,j)=(W(r,j-1)+W(r,j))/2;
    j=j+1;
end

% normalized concentrations in (n-1) output channels
% based on the desired function and precision
% C(1,:)=1/(N-1)*(0:N-1); %example of linear function
C(1,:)=((exp(2*M(1,:)))-1)/(exp(2)-1); %exponential function

%C(1,:)=(0:N-1).^5/(N-1)^5; % power function
%C(1,:)=((0:N-1)/(N-1)).^(1/3); % sqrt function
%C(1,:)=erf((0:N-1)/(N-1))/erf(1); % error function
%C(1,:)=((0:N-1)/(N-1)).^(5); % power function

%DESCRIPTION OF THE NEXT SET OF ROWS "r"

for r=2:n-2
    alpha(r,:)=[1-(1:(N-r-1))/(N-r) -1*(1:r+1)];
    % alpha is an arbitrary parameter
    % one choice for alpha : split ratio = number of walls -1
    % other choices for alpha could be a random number that changes :
alpha(r,:)=[rand(1,N-r) -1*(1:r)];
    % alpha(n-4,:)=[0.5 0.5 -1*(1:N-2)]

    C(r,1)=0;
for j=2:(N-r)
    C(r,j)=alpha(r,j-1)*C(r-1,j)+(1-alpha(r,j-1))*C(r-1,j+1);
end
    C(r,N-r+1)=1;


    for j=1:n-2
        if (C(r,j)-C(r,j+1))<0

W(r,j)=(C(r-1,j+1)*(W(r-1,j+1)-W(r-1,j))+C(r,j)*W(r-1,j)-C(r,j+1)*W(r-1,
j+1))/(C(r,j)-C(r,j+1));
        else
            W(r,j)=1; % fills in one for extreme right and inexistent walls
        end
    end
end
```

```
%W

for r=1:N-1
    for j=1:N-1
        if W(r,j)<1
            P(r,j)=W(r,j)*(L-(N-r)*delta)+(j-1)*delta;
        end
    end
end


% matrix P gives the position of the left side of dividers at different
levels
P
```

References

[1]   Luster, A. D., "The Role of Chemokines in Linking Innate and Adaptive Immunity," *Curr Opin Immunol*, Vol. 14, No. 1, 2002, pp. 129–135.

[2]   Grellner, W., "Time-Dependent Immunohistochemical Detection of Proinflammatory Cytokines (IL-1beta, IL-6, TNF-alpha) in Human Skin Wounds," *Forensic Sci Int*, Vol. 130, No. 2–3, 2002, pp. 90–96.

[3]   Reinhardt, P. H., and P. Kubes, "Differential Leukocyte Recruitment from Whole Blood via Endothelial Adhesion Molecules under Shear Conditions," *Blood*, Vol. 92, No. 12, 1998, pp. 4691–4699.

[4]   Kobayashi, Y., "The Role of Chemokines in Neutrophil Biology," *Front Biosci*, Vol. 13, 2008, pp. 2400–2407.

[5]   Petri, B., M. Phillipson, and P. Kubes, "The Physiology of Leukocyte Recruitment: An In Vivo Perspective," *J Immunol*, Vol. 180, No. 10, 2008, pp. 6439–6446.

[6]   Keenan, T. M., and A. Folch, "Biomolecular Gradients in Cell Culture Systems," *Lab Chip*, Vol. 8, No. 1, 2008, pp. 34–57.

[7]   Li Jeon, N., H. Baskaran, and S. K. Dertinger, et al., "Neutrophil Chemotaxis in Linear and Complex Gradients of Interleukin-8 Formed in a Microfabricated Device," *Nat Biotechnol*, Vol. 20, No. 8, 2002, pp. 826–830.

[8]   Walker, G. M., J. Sai, and A. Richmond, et al., "Effects of Flow and Diffusion on Chemotaxis Studies in a Microfabricated Gradient Generator," *Lab Chip*, Vol. 5, No. 6, 2005, pp. 611–618.

[9]   Herzmark, P., K. Campbell, and F. Wang, et al., "Bound Attractant at the Leading vs. the Trailing Edge Determines Chemotactic Prowess," *Proc Natl Acad Sci USA*, Vol. 104, No. 33, 2007, pp. 13349–13354.

[10]   Lin, F., C. M. Nguyen, and S. J. Wang, et al., "Neutrophil Migration in Opposing Chemoattractant Gradients Using Microfluidic Chemotaxis Devices," *Ann Biomed Eng*, Vol. 33, No. 4, 2005, pp. 475–482.

[11]   Boyden, S. "Chemotactic Effect of Mixtures of Antibody and Antigen on Polymorphonuclear Leucocytes," *Journal of Experimental Medicine*, Vol. 115, No. 3, 1962, pp. 453–466.

[12]   Brown, A. F., "Neutrophil Granulocytes-Adhesion and Locomotion on Collagen Substrata and in Collagen Matrices," *Journal of Cell Science*, Vol. 58(DEC), 1982, pp. 455–467.

[13]   Nelson, R. D., P. G. Quie, and R. L. Simmons, "Chemotaxis Under Agarose—New and Simple Method for Measuring Chemotaxis and Spontaneous Migration of Human Polymorphonuclear Leukocytes and Monocytes," *Journal of Immunology*, Vol. 115, No. 6, 1975, pp. 1650–1656.

[14]   Zigmond, S. H.., "Ability of Polymorphonuclear Leukocytes to Orient in Gradients of Chemotactic Factors," *Journal of Cell Biology*, Vol. 75, No. 2, 1977, pp. 606–616.

[15]   Cheng, X., D. Irimia, and M. Dixon, et al., "A Microfluidic Device for Practical Label-Free CD4(+) T Cell Counting of HIV-Infected Subjects," *Lab Chip*, Vol. 7, No. 2, 2007, pp. 170–178.

[16]   Nagrath, S., L. V. Sequist, and S. Maheswaran, et al., "Isolation of Rare Circulating Tumour Cells in Cancer Patients by Microchip Technology," *Nature*, Vol. 450, No. 7173, 2007, pp. 1235–1239.

[17]   Agrawal, N., M. Toner, and D. Irimia, "Neutrophil Migration Assay from a Drop of Blood," *Lab Chip*, Vol. 8, 2008, pp.2054–2061.

[18] Irimia, D., S. Y. Liu, and W. G. Tharp, et al., "Microfluidic System for Measuring Neutrophil Migratory Responses to Fast Switches of Chemical Gradients," *Lab Chip,* Vol. 6, No. 2, 2006, pp. 191–198.

[19] Irimia, D., D. A. Geba, and M. Toner, "Universal Microfluidic Gradient Generator," *Anal Chem,* Vol. 78, No. 10, 2006, pp. 3472–3477.

Microfluidic Immunoassays

Dean Y. Stevens, Kjell E. Nelson, Elain Fu, Jennifer O. Foley, and Paul Yager
Department of Bioengineering, University of Washington, Seattle, WA

Abstract

Advances in the fields of microfluidics and microfabrication have enabled the development of microfluidic immunoassays with significant advantages over their conventional bench-top counterparts. These advantages include reduction in sample and reagent volumes, reduction in the amount of waste generated, and the use of inexpensive, disposable polymeric devices. Implementation in a microfluidic format is also compatible with the integration of additional micro-processing and analysis features within the same device. These qualities make microfluidic immunoassays attractive for a multitude of applications, especially those at the point of care. In this chapter, the key design and operating parameters of two recently developed heterogeneous microfluidic immunoassays are presented.

Key terms	heterogeneous assay
	microfluidics
	surface binding
	flow-through membrane immunoassay
	concentration gradient immunoassay
	polymeric laminate device
	gold colloid label

9.1   Introduction

An immunoassay is an analytical method that uses an antibody-antigen recognition event to detect an analyte, which may be the antigen or the antibody. Antibodies are proteins produced by the immune system that recognize a variety of antigens including proteins, peptides, and small organic molecules [1, 2]. The ability of antibodies to recognize a range of molecules, coupled with the commercial mass-production of antibodies, has enabled immunoassays to become a platform diagnostic technology with an important role in research, medicine, forensics, and biodefense. Immunoassays generally fall into two classes—homogeneous assays, in which all components remain in solution, and heterogeneous assays, in which there is at least one solid surface to which reagents are immobilized. Because of its versatility and accuracy, the enzyme-linked immunosorbant assay (ELISA) is among the best of the heterogeneous immunoassays and has dominated all others for the last four decades [3, 4]. However, ELISAs typically have long incubations times, from several hours to overnight, and the assay is conducted at pseudoequilibrium time points. Advances in the field of microfluidics have enabled the development of microfluidic immunoassays with significant advantages over their conventional bench-top counterparts. These advantages include reduction in sample and reagent volumes, reduction in the amount of waste generated, and the use of inexpensive, microfabricated disposable polymeric devices. Implementation in a microfluidic format is also compatible with the integration of additional microprocessing and analysis features within the same device. These qualities make microfluidic immunoassays attractive for a multitude of applications, especially those for patient diagnosis at the point of care.

There has been much recent activity in the development of microfluidic immunoassays (see Henares et al. [5] for a recent review). Microfluidic immunoassays have been developed to detect a variety of clinically relevant targets, such as C reactive protein [6, 7], prostate specific antigen [6], and ferritin [6]; larger targets, such as the bacteria *Escherichia coli* [8] and *Helicobacter pylori* [9]; and toxins, such as *Staphylococcus* enterotoxin B [10] and cholera toxin [11]. Microfluidic assay analysis times of minutes [12, 13] and detection in the attomolar range [14] have been achieved.

One approach for achieving a high performance microfluidic immunoassay at a low per-test cost has been to combine a disposable card that incorporates all sample-contacting elements with a portable, permanent detector or "reader" [12, 15, 16]. Sample and waste are retained on the card, and never contact the electrical and optical components of the reader, thereby eliminating the need for cleaning the reader between samples to prevent cross-contamination. Advanced control capabilities may be implemented in the permanent reader, bridging the gap between expensive and sophisticated bench-top assays and disposable dipstick assays.

This laboratory initially focused on development of homogeneous microfluidic immunoassays in the form of the diffusion immunoassay [17–23], and much development was made in other laboratories of homogeneous immunoassays in capillary electrophoresis-on-a-chip formats [24–26]. Such assays can now be performed in commercially-available instruments. In this chapter, the key design and operating parameters of two more recently-developed and complementary heterogeneous microfluidic immunoassays based on the disposable-plus-reader model will be presented.

9.1.1 Microfluidic immunoassay design/operation considerations

The choice of materials for use in the disposable card is a key factor in the design, as well as implementation, of a microfluidic assay. The application will often require the disposable card to be compatible with a specific surface modification process, detection method, or generation of a minimum feature size. This will set requirements on the chemical and physical properties of the material, such as optical transparency, level of autofluorescence, and chemical resistance to a given substance. For example, polydimethylsiloxane (PDMS) is commonly used for immunoassay development [5] because of its surface modification properties and its ability to produce submicrometer features. However, it is permeable to many small and hydrophobic molecules, is not ideal for the fabrication of more complex devices with a large number of layers, and is not well suited to higher-throughput production. The choice of material should also balance the application requirements with the cost of the material plus the processing costs [16]. Practical devices may require the use of a combination of materials. Lamination of laser-cut polymer sheets allows for inexpensive and extremely rapid prototyping when features smaller than 100 μm are not required [27–29].

The flow rate is another key parameter in the operation of a microfluidic assay. The choice of flow rate will be a balance between restrictions on sample/reagent consumption and any requirements on the time to result for the assay. Higher flow rates increase the flux of molecules of interest to the capture surface, and thus reduce the time to result. However, this occurs at the expense of sample/reagents and capture efficiency. Higher flow rates also increase the pressure in the device, requiring a greater degree of structural integrity. Lower flow rates can increase the time to result for the same magnitude signal, but reduce the volume of sample and other reagents required [30–32].

The choice of assay parameters is also dependent on the quantification method; frequently, assay systems measure either the binding rate or the signal at a single time point (e.g., when the number of bound molecules has reached a steady state with unbound molecules in the solution, as based on the chemical equilibrium or pseudoequilibrium). Both methods require precisely timed delivery of the sample volume to the capture surface. Measurement of the binding rate may reduce the total assay time, but is linear only at relatively low fractional surface coverage and depends strongly on the amount and reactivity of immobilized ligand. Quantification using the binding rate also requires reproducible sample introduction to the capture surface, especially with regard to leading-edge Taylor dispersion [33–35].

The application will also set the requirements for sensitivity, limit of detection (LOD), speed of the assay, and the minimum sample volume. Achieving a certain sensitivity or LOD may require a longer time to result or a greater input sample volume; the set of optimal parameters used in the operation of the assay should balance all the application requirements.

These key considerations will be revisited in the context of the two example immunoassays detailed in this chapter.

9.1.2 Example microfluidic immunoassay formats

This chapter discusses microfluidic immunoassays in the context of two specific assays: the flow-though membrane immunoassay and the concentration gradient immunoassay. Each assay format is compatible with a number of reagent systems and detection methods that would enable detection of a variety of targets. We present a review of the detection of a midsized protein using the flow-through membrane immunoassay with simple optical detection of a gold colloid label (originally published in [15]) and the concentration measurement of a clinically relevant small molecule drug using the concentration gradient immunoassay coupled with surface plasmon resonance (SPR) imaging (originally published in [36]).

9.1.2.1 Microfluidic flow-through membrane immunoassay

The microfluidic flow-through membrane immunoassay (FMIA) is a rapid (less than 10 minutes) microfluidic immunoassay based on a flow-through membrane. A porous membrane is patterned with capture molecules and positioned within a microfluidic channel. For a sandwich assay, shown in Figure 9.1, a sample is passed through the membrane to allow capture of sample analyte, followed by a buffer wash to remove unbound sample. Buffer is also passed over a fibrous pad that contains a labeling reagent dried in a preservative matrix. The rehydrated label is then passed through the assay membrane to allow binding of the label to the captured analyte, followed by a wash to remove unbound label. For visible labels, an image of membrane-bound label can be used to quantify the amount of captured analyte. The assay is implemented on a laminate card containing stable, anhydrous labeling reagent and the membrane, with external hardware handling the fluid pumping and optical readout.

The example analyte is the malarial antigen *Plasmodium falciparum* histidine-rich protein II (PfHRP2), a ~30 kDa water-soluble protein produced by the *Plasmodium falciparum* strain of malarial parasites that induces heme polymerization in erythrocyte hosts [37–39]. In this system, antibody nonspecifically bound to the assay membrane is the capture molecule and gold-antibody conjugate is the label that generates a visible increase in redness proportional to the concentration of analyte present.

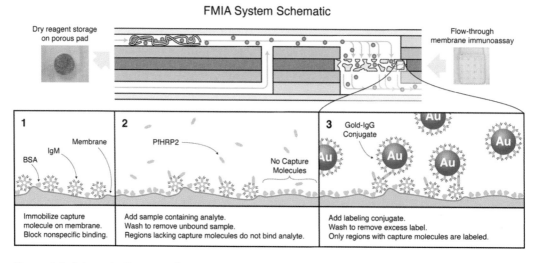

Figure 9.1 Schematic illustrating the FMIA format [15]. The central element of the assay is the porous membrane pictured in the upper right of the image. As can be seen in the upper part of the schematic, reagent flows within sealed channels through the membrane, dissolving dry reagents in the process. The lower panels illustrate the steps of the sandwich assay for PfHRP2 described in the text.

Use of the porous membrane provides two significant advantages over planar assay substrates: (1) increased surface area for sample capture, and (2) decreased distances for diffusion of the sample to the capture surface. The results are shorter assay times for the same level of capture or increased signal strength for the same interaction time. The fibrous pad provides large surface area for rehydration, and the sugar matrix acts to stabilize protein structure and thus preserve function [40]. Implementation of these components into a microfluidic format with external fluid actuation provides additional advantages: sequential reagent addition, variable flow rates, automated timing, and parallel assay multiplexing via spatial separation of different capture molecules on the membrane.

9.1.2.2 Microfluidic concentration gradient immunoassay

The microfluidic concentration gradient immunoassay (CGIA) is a rapid (less than 10 minutes) competition immunoassay for the quantitative detection of low molecular weight analytes. It allows for simultaneous controls and multianalyte detection and does not require mixing (other than by diffusion) or labeled reagents. The operation of the CGIA is based on laminar fluid flow in microchannels, where the mass transport of solutes between miscible fluids occurs only by diffusion and is governed by Fick's laws [41–43].

The operation of the assay is schematically shown in Figure 9.2.

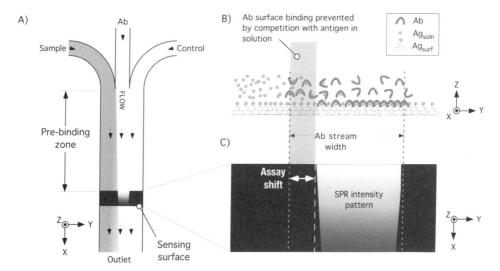

Figure 9.2 Schematic illustrating the CGIA format. (a) The device has the following features: (1) a microfluidic channel consisting of three inlets that converge into a common duct; (2) a "prebinding" zone functionalized with a PEG-terminated alkylthiol self-assembled monolayer to resist nonspecific adsorption from species in the bulk to the surface; and (3) a capture region that has been functionalized with an analog of the target analyte. The following solutions are flowed into one of the three device inlets: (1) a sample, containing an unknown quantity of a relatively rapidly diffusing analyte, (2) a solution of antibody that binds specifically to the analyte, and (3) a reference solution, containing a known concentration of analyte or other reference species. (b) Analyte from the competitor stream diffuses into the antibody stream and binds antibody in the bulk, thus preventing antibody binding to the capture surface. (c) Distribution of antibody binding to the capture surface, measured as an "assay shift," is monitored with SPR imaging. (Reprinted with permission from [36]. Copyright 2007 American Chemical Society.)

The mass flux across the fluid interfaces is determined by the diffusion coefficients of the solutes, their concentrations, and the flow rate used (which determines the time allowed for diffusion). Since the diffusion time for a small molecule analyte is an order of magnitude smaller than that of an IgG molecule [45], mass transfer between the fluids is dominated by the movement of the analyte from the sample and/or reference streams into the antibody stream. At a constant flow rate, a steady-state gradient of analyte concentration is established perpendicular to convective flow, from the sample and reagent streams into the antibody stream. Binding between antibody and analyte leads to a gradient of occupied antibody binding sites. The gradient of occupied binding sites will broaden transverse to flow (along the y-axis) with increasing channel distance downstream of the fluid entry points (along x). Variables controlling the distribution of this gradient are the flow rate, channel length, and the concentrations of antibodies and analytes, while the relative diffusion coefficients and reaction rates of the binding partners used are experimental parameters determined by the particular reagents employed. Based on these parameters, a distance downstream is selected at which point the solutions encounter the capture region coated with a surface-bound analog of the analyte. Here, antibodies having at least one available binding site may bind to the capture surface. The rate of antibody binding to the sensor surface near the interface with the sample stream containing analyte will be smaller than the rate of binding observed further from the interface where the analyte concentration is lower, due both to the increased diffusion distance and capture of analyte by antibody near the interface. If the analyte concentration is high relative to the concentration of antibody, antibody binding near the interface with the sample stream will be zero. Thus, the presence of analyte in the sample stream modulates the "width" of the sensor surface area to which antibody binds; a greater concentration of analyte results in a smaller width, or greater assay shift. For a given antibody concentration, the assay shift can be related to the concentration of the analyte in the sample solution since, by Fick's law, the diffusive flux is proportional to the concentration gradient of a solute. Accurate concentration determinations can be made by comparing the assay shift caused by the sample to the assay shift caused by the presence of a known concentration of analyte in a reference stream.

We demonstrate the application of this method for the quantitative analysis of phenytoin, a ~250 kDa small molecule drug used in the treatment of epilepsy.

9.2 Materials

9.2.1 Microfluidic device

Adhesive-backed Mylar and poly(methyl methacrylate) (PMMA) for fabrication of flowcell layers were purchased from Fraylock, Inc. (San Carlos, CA). Autocad LT (Autodesk, San Rafael, CA) was used to create the designs, and the polymer layers were cut with a 35W CO_2 laser, M-360 from Universal Laser Systems Inc. (Scottsdale, AZ).

9.2.1.1 FMIA

The nitrocellulose membrane was Whatman (Maidstone, Kent, UK) Protran® with 0.45-μm pore size. Recombinant PfHRP2 (53 kDa with GST tag used for purification) was purchased from Immunology Consultants Laboratory (Newberg, OR). Anti-PfHRP2 IgM (clone PTL-3) and IgG (clone C1-13) were purchased from National Bioproducts Institute (Pinetown, South Africa). Membranes were cut to size using the CO_2 laser and were patterned with capture antibodies. The membrane was then soaked in Zymed Membrane Blocking Solution (Invitrogen, Carlsbad, CA) to decrease nonspecific binding to the membrane. Gold colloid with a 40-nm diameter was produced by reduction of tetrachloroauric acid with trisodium citrate [46]. Immunogold conjugates were formed by sequential incubation of the colloid with anti-PfHRP2 IgG and BSA [47], followed by centrifugation and resuspension at $OD_{534}=10$ in tris-buffered saline with 1% BSA, 10% sucrose, and 5% trehalose. Conjugate was filtered through a 0.2-μm cellulose acetate filter (Whatman FP 30/0,2 CA-S), and 20 μL was dried at 35°C on BSA-treated, laser-cut polyester conjugate pads measuring 0.25 to 0.27 inches in diameter (Ahlstrom, Holly Springs, PA).

9.2.1.2   CGIA

To create the SPR sensing surface, soda lime glass microscope slides ($25\times75\times1$ mm^3) were purchased from Fisher Scientific (Fair Lawn, NJ) and subsequently coated with 1-nm chromium and 45-nm gold (99.999% purity) by electron beam evaporation. A mixture of a PEG-terminated alkylthiol (Prochimia, Sopot, Poland) and a custom-synthesized phenytoin-alkylthiol conjugate (Asemblon, WA) (98% PEG, 2% Phenytoin-alkylthiol, in ethanol, 1 mM total thiol concentration) was deposited on gold in the capture region such that a transition zone (or "edge") of the patterned area was established at a specified location downstream of the points where the inlets converge into a common channel (22 mm in this case, see Figure 9.1). This interface may be produced by a number of methods, including microcontact printing, capillary wetting under a mask placed in close contact with the surface, or noncontact (e.g., piezoelectric) printing. PEG-terminated alkylthiol was used to create nonfouling regions upstream of the capture region by flooding the slide with a 1-mM ethanolic solution following surface patterning.

9.2.2 Pumps and interconnections

Positive displacement syringe pumps were used to drive fluid flow (Microflow, Micronics, Redmond WA). Tubing and interconnects were purchased from Upchurch Scientific (Oak Harbor, WA) or from McMaster-Carr (Atlanta, GA) for the ethylene propylene diene monomer (EPDM) rubber O-rings.

9.3 Methods

9.3.1 Fabrication of flowcells

9.3.1.1   FMIA

The microfluidic card design, shown in Figure 9.3, consists of a chamber containing the assay membrane, three upstream fluid lines (sample line, conjugate pad line, and bubble venting line) connecting to syringe pumps, and a downstream waste line. The card is valveless and hence requires three pumps, although simpler fluid actuation systems will work.

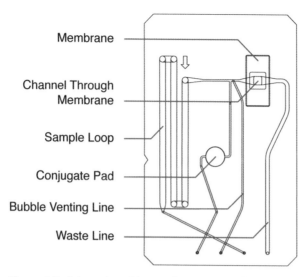 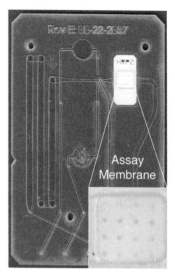

Figure 9.3 Schematic and image of an FMIA card [15]. Many of the main components of the card are labeled. The port located near the bottom-right edge of the card connects to a waste reservoir, and the remaining three ports connect to pumps. The image shows a card, 83× 52× 2.3 mm³, before use, while the inset shows capture areas on the membrane after assay completion.

A piezoelectric spotting system (Microfab Technologies, Plano, TX) was used to pattern the nitrocellulose membrane with a 4×4 grid of capture spots, each measuring 120 μm in diameter and containing 0.15 μL of 0.25 mg/mL anti-PfHRP2 IgM. This patterning approach produces small spots with a high density of capture antibody and is amenable to multiplexing. Another patterning method suited to large-scale production is striping, in which narrow tubes are drawn over the membrane while fluid is dispensed from the end of the tube. Striping is a faster method, but gives lower-resolution features. Hand-pipetting of microliter volumes is a reasonable patterning approach for initial testing. The patterned membrane was allowed to dry for 20 minutes at room temperature and then was blocked for 30 minutes in the membrane blocking solution, followed by drying at 20°C and storage in a desiccator. The patterned membrane was then placed in a pocket cut through a polymer layer, to be sandwiched between two other layers.

Cards were assembled from laser-cut layers of adhesive-backed Mylar and PMMA. Channel surfaces were immersed in 10% bovine serum albumin (BSA) for 30 minutes, followed by rinsing with deionized water and baking at 35°C for 30 minutes, to prevent nonspecific binding to the membrane. During final card assembly, a conjugate pad containing 20 μL of anti-PfHRP2 gold conjugate was installed.

9.3.1.2 CGIA

The microfluidic cards used for the CGIA were assembled from laser-cut Mylar and PMMA laminated together. The cards consisted of the following layers: (1) a gold-coated glass substrate patterned with mixed phenytoin-terminated/PEGylated alkylthiol SAM (described below), (2) a 3.6-mm wide×45-mm long channel composed of 62 μm-thick Mylar plus double-sided adhesive, (3) a channel cap, composed of 50 μm thick Mylar plus adhesive, that contains vias for fluid flow positioned over the inlets and outlets of layer 2, (4) an O-ring seat, composed of 2.5-mm thick PMMA layer, that contains holes over the inlets and outlet to hold ethylene propylene diene monomer (EPDM) O-rings, and (5) an O-ring retainer, composed of 50-μm thick Mylar plus adhesive, that contains holes to hold the connection tubing to the off-card valves and pumps. The card layers were assembled on an alignment jig and then pressed at 2,000 psi for ~10 seconds with a hydraulic press (International Crystal Laboratories, Garfield, NJ). Surface patterning was completed by a simple two-step method. First, a suitable mask was prepared using Mylar or another material that provided an opening where the binding area was to be patterned on the substrate. This mask was placed in contact with the cleaned gold slide, and a 2-mM ethanolic solution of functionalized alkylthiols was sprayed onto the slide using a commercially available artist's airbrush. Once the solution had dried, the mask was removed and the substrate was flooded with a 2-mM ethanolic solution of PEG thiol to render the remaining surface resistant to nonspecific protein adsorption. After a 30 minute incubation, the slide was rinsed with pure ethanol and blown dry under an N2 stream. Patterned substrates could be stored in the dark under N2 for up to 24 hours prior to use.

9.3.2 Sample and reagent delivery

9.3.2.1 FMIA

Recombinant PfHRP2 in fetal bovine serum was injected into the sample line of the card (~185 μL volume). A microFlow system provided positive-displacement pumping and software control for the assays. Pump reservoirs were filled with phosphate-buffered saline (PBS) with 0.1% Tween 20 (PBST). The membrane flow-through area was 7.6 mm^2. An automated script performed the following steps: (1) PBST was pumped into the conjugate pad to rehydrate the dry conjugate (19 μL at 4.0 μL/sec) (Figure 9.4); (2) sample was pushed slowly through the membrane to allow capture (120 μL at 0.5 μL/sec); (3) PBST was pumped quickly through the membrane to remove unbound sample (300 μL at 4.0 μL/sec); (4) PBST was pumped through the conjugate pad while drawing negative pressure on the bubble vent line to prevent bubble formation near the membrane (9 μL at 4 μL/sec); (5) PBST was pumped through the conjugate pad to drive the conjugate slowly through the membrane (12 μL at 0.1 μL/sec); and (6) PBST was pumped through the membrane to remove unbound conjugate (180 μL at 4.0 μL/sec).

9.3.2.2 CGIA

A microFlow™ system (with 64-μl pump barrels) provided positive-displacement pumping for these assays, although other stepper-motor controlled pumping systems may also be used (e.g., Kloehn, Las Vegas, NV). A common upstream reservoir for the pumps was filled with PBS. The outlet of each pump was connected to a manual six-port injection valve; the injection valve was connected to sample loops of 100 μl volume and the card. Tubing connecting the pumps to the injection valves was 1/16" OD, 0.03" ID PEEK, while the tubing connecting the valves to the device were 1/16" OD, 0.005" ID PEEK. Tubing ID and length between the valves and the device were chosen to reduce dispersion and the time required to introduce the antibody solution into the device during the experiment. A schematic of the system is shown in Figure 9.5. The following steps were performed to conduct the assay: (1) the channel was filled with buffer and positioned on the SPR imager; (2) the sample loops were loaded with a sample

Conjugate release on assay card

Conjugate release profiles under constant lateral flow, 0.5 μL s⁻¹

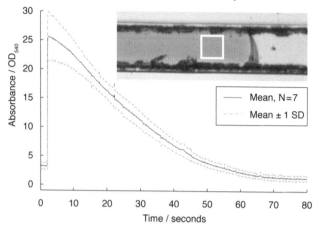

Figure 9.4 Image sequence and concentration profile of reagent release from a conjugate pad (0.25 inches in diameter) at a constant flow rate of 0.5 μL/sec and $N=7$ (15). The inset shows the solution front downstream of the pad with the measurement area outlined in white. Note that while this plot demonstrates repeatable release of reagent into fluid advancing in a dry channel, the concentration of reagent at the assay membrane of the PfHRP2 assay card would differ. The assay card stops fluid flow over the pad prior to use, and the rehydrated reagent displaces buffer rather than air when it is introduced to the assay membrane; both of these factors will affect these concentration measurements at the assay membrane.

containing phenytoin, anti-phenytoin monoclonal antibody in PBS, and PBS containing a known amount of phenytoin; (3) the pumps were operated to flood the solutions, except the antibody solution, into the tubing downstream of the sample injection valve; (4) the tubing was connected to the device; (5) the pumps were operated at a flow rate of 30 nL/sec for all fluid streams. This method resulted in the establishment of steady-state concentration gradients in the device prior to the arrival of the antibody solution (which required additional time to enter the device from the sample holding loop). Flow continued until adequate signal was obtained from the imaging equipment to calculate a result, as described below. The experiment may be stopped at the earliest time point at which adequate signal is available, though the assay measurement is not very sensitive to the total elapsed time of the experiment.

The assay signal depends on the locations of antibody accumulation relative to the locations of the fluid interfaces within a given device. Thus, it is necessary to accurately determine the positions of the fluid interfaces. This can be accomplished by including an inert tracer molecule that produces a bulk refractive index change to the solution such that the distribution of the tracer species in the reference region (i.e., upstream of the capture surface) can be monitored. By definition, the interface position is located

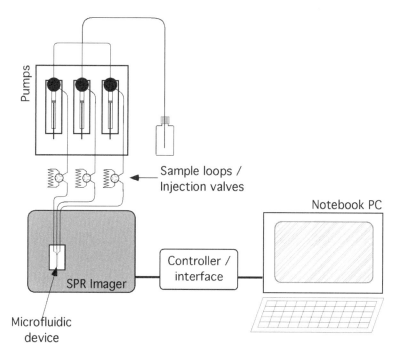

Figure 9.5 Schematic of the experimental system. A common upstream reservoir for the microFlow system pumps was filled with PBS. The outlet of each pump was connected to a manual six-port injection valve; the injection valve was connected to sample loops of 100 μl volume and the card. Tubing connecting the pumps to the valves was 1/16″ OD, 0.03″ ID PEEK, while the tubing connecting the valves to the device were 0.005″ ID. The card was coupled to a custom built SPR imager [48]. (Reprinted with permission from [36]. Copyright 2007 American Chemical Society.)

where the concentration of the tracer molecule is 50% of its maximum intensity, the latter being located far from the putative interface location (e.g., near the channel wall). Such a measurement can be determined for a given device by conducting a calibration run (i.e., flowing buffer, water, and buffer in the three streams from left to right, respectively). An alternative (and preferred) method is to add a tracer species to either the sample and control streams or just the antibody stream during the assay run. The ideal tracer species will have the diffusivity of the antibody, will not bind competitor, and will not foul the capture surface. Dextran has been demonstrated to be an acceptable tracer species for a typical IgG, although other smaller compounds such as dextrose have been used successfully to identify interface positions during the experiment by conducting analogous intensity measurements to those described above, but in parallel with the assay measurement using intensity data in the reference region immediately upstream of the binding surface.

As the retention time is a critical factor in determining the magnitude of the assay shift, the velocity of the fluids in the channel must be carefully controlled. However, variations in channel cross-section as a result of manufacturing variations may change the channel volume significantly. To control for these variations and provide for accurate device calibration, it is best to include a reference solution with a known concentration of analyte opposite the sample stream. The concentration of analyte in the sample can then be estimated from the ratio of the sample and reference assay shifts using a prevalidated calibration model.

9.4 Data acquisition and results

9.4.1 FMIA

Images of the cards were captured on either a flatbed scanner (ScanMaker i900, MicroTek International, Inc., Cerritos, CA) in 48-bit RGB at a resolution of 2,400 ppi or a low-cost USB camera (AM211 Dino-Lite, AnMo Electronics Corp., Hsinchu, Taiwan). The image data was processed using the following steps using ImageJ [48]: (1) selection of regions of interest (ROI) inside and outside of each visible spot; (2) measurement of the mean green-channel pixel intensity of each ROI; (3) calculation of the difference between the ROI means inside and outside of each spot; and (4) report of the mean of these differences.

The signal obtained from the FMIA is a visible increase in optical density, especially in the green part of the electromagnetic spectrum, when 40 nm gold conjugates are used as labels. The capture regions show signal increases that are a function of the analyte concentration due to specific binding of analyte and label, and regions lacking capture molecules remain unchanged from the beginning of the assay, except as a result of non-specific interactions. The pre- and post-assay membranes are pictured at the 0- and 140-second time points in Figure 9.6, showing a clear pattern of capture spots used for end-point measurement. These spot locations can be used either for replicates (where the same capture molecule is used) or for assay multiplexing with additional analytes or references.

During signal development, the conjugate in solution obscures reading of the membrane, preventing measurement of binding rates. Additionally, the nonuniform exposure of gold conjugate to the membrane can be seen during conjugate introduction, indicating the need for replicate sampling across the membrane. These exposure gradients are caused by displacement of PBST by the denser conjugate solution, containing a high concentration of sugars, resulting in gravitationally induced advection and mixing.

The response of the assay is shown in Figure 9.7. The microfluidic FMIA has demonstrated detection of subnanomolar concentrations of PfHRP2 (~10 ng/mL), in less than 10 minutes using an automated protocol. This is comparable to the limit of detection of a well-based ELISA assay for PfHRP2 [49], which requires two hours to give a detection limit of 3.91 ng/mL of PfHRP2 in PBS, and requires such materials as micropipettes, clean containers, a humidity chamber, and a spectrophotometer.

As presented here, the assay response measures changes in the amount of light transmitted back towards the detector. This value changes as photons scattered by the porous

Spot development in FMIA assay for PfHRP2

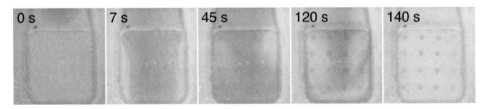

Figure 9.6   Image sequence of the FMIA [15]. The addition of gold-antibody conjugate results in the development of red spots at PfHRP2 capture antibody regions.

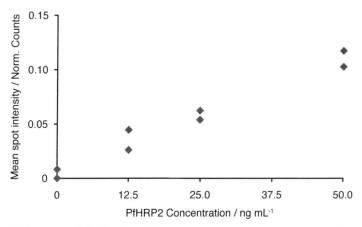

Figure 9.7 FMIA response [15]. The signal (shown as mean spot intensity) versus analyte concentration for eight samples of PfHRP2 in fetal bovine serum.

membrane encounter labels and are absorbed, an event whose likelihood is increased by greater concentrations of analyte and label. The relationship between the amount of light transmitted to the detector and the concentration of label should not be linear. (The absorbance, not the transmittance, would be the measurement that is linear with label concentration if the Beer-Lambert law held, which it does not for scattering systems.) There is however a near-linear range below about 50 ng/mL, with the dynamic range spanning at least 10 to 400 ng/mL. Strong nonlinearity at high concentrations requires generation of a standard curve or further signal transformation to achieve reasonable quantification at these levels. The dynamic range will also depend strongly on the binding capacity of the capture surface and the affinity of the antibodies used for capture and labeling.

9.4.2 CGIA

Composite images, created by averaging 30 20-millisecond exposures, were collected using a custom-built miniature SPR imaging instrument [50]. An initial image was taken before the start of the assay and was used to calculate the change in reflectivity for each subsequent image. Line profiles were taken across the channel perpendicular to fluid flow. The assay shift was measured as the difference between the location of the fluid interface and the edge of the antibody band on the sample side of the channel.

An SPR difference image from the CGIA is shown in Figure 9.8. In this example, the edge of the antibody band on the buffer side (right) matches up with the fluid interface as measured in the calibration run, while the edge of the antibody band on the competitor side (left) is significantly shifted from the fluid interface as measured in the calibration run. Line profiles for both the sample and calibration runs are shown in Figure 9.9.

The assay shift, measured over a range of competitor concentrations against 100-nM antibody, is shown in Figure 9.10. The assay shift is approximately linear over the lower end of the measured range of competitor concentrations, 0 to 300 nM. A three-dimen-

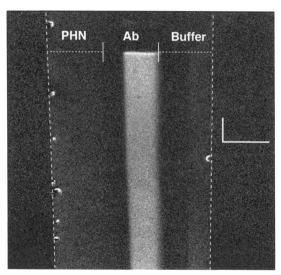

Figure 9.8 CGIA SPR difference image. The streams contain 400-nM phenytoin in PBS, 75-nM anti-phenytoin antibody in PBS, and PBS alone, from left to right, respectively. Antibody bound to the capture surface over a 10-minute period appears as a bright band in the image. The locations of the fluid interfaces, based on the buffer calibration streams, are noted in the image. The scale bars are 1 mm. The contrast in the image has been adjusted for display. (Reprinted with permission from [36]. Copyright 2007, American Chemical Society.)

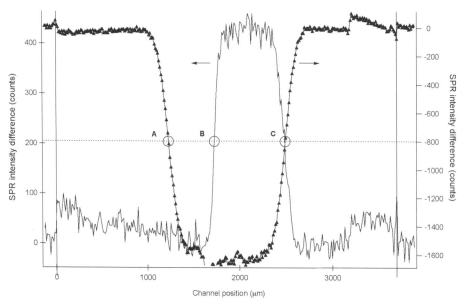

Figure 9.9 Line profiles, perpendicular to flow, for the sample (-) and calibration (^) runs. The edge of the antibody band on the buffer side (right) matches up with the fluid interface as measured in the calibration run (C), while the edge of the antibody band on the competitor side (left) is significantly shifted from the fluid interface (B) as measured in the calibration run (A). The assay shift, the difference between the edge of the antibody band, and the fluid interface on the phenytoin side of the channel (B-A), is ~ 500 μm. (Reprinted with permission from [36]. Copyright 2007 American Chemical Society.)

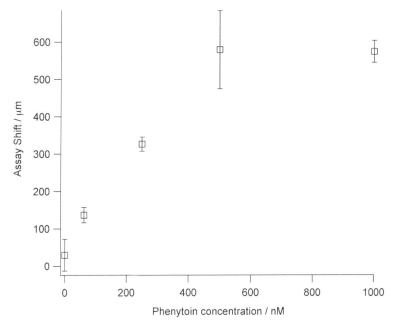

Figure 9.10   CGIA response. Assay shift vs. analyte concentration. Each point is an average of at least three measurements and the error bars represent the standard deviation. The assay shift was measured for 100-nM anti-phenytoin antibody over a range of competitor concentrations. (Reprinted with permission from [36]. Copyright 2007 American Chemical Society.)

sional computational model of the CGIA shows excellent agreement with this experimental data [36, 51].

9.5   Discussion

9.5.1   FMIA

Multiple parameters can be controlled to modulate the signal generated in the FMIA. These include, among other factors, the interaction time of the reagents with the analyte and the concentration of capture and labeling reagents.

In general, assay signal can be increased for a given analyte concentration by lengthening the interaction time between the analyte and the molecules binding it. The longer the contact time between the analyte and the capture molecules, the more binding events occur, and the relatively low off-rates of antibody-antigen binding allow greater accumulation of analyte in regions patterned with capture molecules than in those without. The same holds true for the contact time with the secondary label, which needs time to bind to the analyte. Although maximum signal would be obtained by running the system to equilibrium, a rapid diagnostic device requires a balance between short assay times and sufficiently high binding to achieve the desired performance characteristics. Two factors that will affect the interaction time for a given sample volume and fluid flux through the membrane are the volumetric flow rate and the cross sectional area of the membrane. For example, using a constant fluid flux for the PfHRP2 assay, a decrease in the flow-through area of the membrane from 140 to 7.6 mm^2 produced an

18-fold increase in the time the sample and secondary label were exposed to the membrane, yielding higher signals for longer exposure times.

The concentrations of the capture and labeling reagents also have a strong impact on the signal strength. Generally, functionalizing the membrane with higher concentrations of capture molecules results in a higher density of these molecules on the membrane surface, thereby allowing faster and denser accumulation of analyte at the membrane surface. At high enough capture density, however, additional increases in concentration will predominantly result in an increase in the area of the patterned region, not an increase in capture molecule density. For the PfHRP2 assay, a spotting concentration was chosen to be high enough to give measurable responses at low analyte concentrations and low enough to prevent excessive waste of the capture reagent during spotting. The signal was also improved by allowing at least three days of drying after membrane preparation and prior to use; the effect is presumed to be due to increased binding between the capture molecules and the membrane during the drying period, resulting in decreased wash-off of the capture molecules during reagent introduction, and thus preserving a high capture molecule density. Using a higher concentration of the secondary label results in increased binding between the label and the analyte, and is thus another way to increase assay signal. For a given analyte concentration, however, there is a concentration, volume, and exposure time of secondary label at which most of the analyte is labeled, and higher concentrations will not result in a significant increase in signal. This may be an ideal situation, but it should be noted that its reality may be limited by design choices on factors such as reagent cost and assay time. Additionally, increasing label concentration can result in increased *nonspecific* interactions that can raise the background signal. For the PfHRP2 FMIA, higher label concentrations were shown to give higher assay signals in a bench-top format, prompting a change to the fluid actuation protocols used in the on-card format. As seen in Figure 9.4, the leading edge of the labeling reagent has a higher concentration of label than the trailing edge. For a given exposure time of secondary label, an increased signal was generated by limiting that exposure to the high-concentration leading edge of the reagent plug, instead of using a larger volume of the reagent that included lower-concentrations of label.

9.5.2 CGIA

The magnitude of the CGIA assay response is determined in part by the relative concentrations of analyte and antibody. The antibody concentration must be sufficiently low relative to the expected analyte concentration so that analyte may occupy a majority fraction of antibody binding sites near the sample/antibody interface and thereby affect the distribution of antibody binding near the interface. Lowering the relative antibody concentration may increase the assay shift for a given analyte, and may also increase the response of the assay to analytes with relatively slow diffusivities, though this will also reduce the available sensor signal obtained over a given period. Increasing the relative antibody concentration will increase the dynamic range of the assay and the available assay signal for a given elapsed time, with concomitant reduction in analyte sensitivity. Optimal assay conditions will therefore depend on the expected range of analyte concentrations for a given sample. In some cases, analyte concentrations may be high enough, and detectors sensitive enough, to permit dilution of the sample so that lower concentrations of antibody may be used and adequate sensitivity may be obtained over a broader range of starting analyte concentrations. Due to the nonlinearity of the diffusion-based concentration at distances far from the fluid interfaces, the assay response plateaus at high analyte concentrations relative to the selected antibody concentration. Uncertainties in flow rate for a given experimental run (caused, for example, by variations in pump performance or channel dimensions) can reduce precision of repeated measurements. Using a reference solution in the control stream can improve reliability of quantification when using low-cost pumps or device manufacturing techniques.

9.5.3 Challenges of analyzing complex samples

The analysis of complex samples presents several challenges. Complex samples often contain substances that interfere with the transport/binding of species of interest in the

Troubleshooting Table

Problem	Explanation	Potential Solution
FMIA low assay signal.	Interaction time between analyte and capture molecules is too short.	Increase the density of capture molecules when spotting and/or increase the drying time of the spotted membrane prior to use; decrease the flow-through area of the membrane for a fixed sample volume and fluid flux; decrease the rate at which sample flows through the membrane.
	Interaction time between analyte and secondary label is too short.	Increase the secondary label concentration; decrease the flow-through area of the membrane for a fixed reagent volume and fluid flux; decrease the rate at which secondary label flows through the membrane.
FMIA high background signal.	Nonspecific binding of secondary label.	Decrease the concentration of the secondary label; adjust surface chemistry by changing blockers, reagent pH, or cosolutes; increase wash duration and/or stringency.
CGIA non-steady fluidic interface.	Bubble in system.	Flush system with fluid to remove bubble.
	Pump malfunction.	Fix or replace pump.
CGIA limit of detection is too high for target analyte concentration.	Antibody concentration is too high.	Reduce antibody concentration.
	Retention time is too short.	Increase retention time by reducing the flow rate or increasing the upstream channel length.
CGIA dynamic range is too small.	Antibody concentration is too low for the target analyte concentration range.	Increase the antibody concentration or increase the relative volumetric flow rate of antibody solution to increase the width of the antibody stream.
CGIA poor sensitivity.	Antibody concentration is too high.	Reduce antibody concentration.
CGIA poor interface resolution.	Tracer species diffuses too quickly.	Select an alternate tracer species to sharpen the transition between fluid streams.
	Surface patterning defects.	Improve uniformity of surface patterns (and definition of boundaries).
	Image noise.	Identify and correct image artifacts.

bulk or at a capture surface. For example, nonspecific adsorption of substances in the sample to the capture surface increases the background signal, while interactions between substances in the bulk sample with the analyte, such that transport or binding of the analyte to the capture surface is impeded, will decrease the signal. These effects may be problematic, especially for the implementation of a quantitative assay. One potential solution is the use of a blocking of solution, as described above in the FMIA protocol. Another potential solution is the use of reference channel compensation to "subtract out" the signal due to nonspecific adsorption. For example, Navratilova et al. have described the successful use of this method in a competition assay using SPR detection for urine samples (the sample channel contained urine samples with human serum albumin (HSA) and anti-HSA, while the reference channel contained urine samples with HSA only [52]). In some cases, neither blocking nor compensation alone is effective and processing of the sample to remove interfering substances before delivery to the capture surface is required. For example, a two-stage preprocessing protocol for human saliva samples [16, 53], based on the microfluidic H-filter [54, 55] has enabled quantitative operation of a SPR-based competition immunoassay for the detection of a small molecule analyte in saliva [53]. Herr et al. [12, 56] have also reported development of a preprocessing method for saliva, compatible with a downstream electrophoretic immunoassay, that has been fully integrated into a hand-held device.

Acknowledgments

The authors would like to thank Yager Laboratory members past and present, particularly Benedict Y. Hui for his skillful assistance in the development and validation of the CGIA assay.

Support for research on the FMIA was provided through funding from The Bill and Melinda Gates Foundation's Grand Challenges in Global Health Initiative under grant number 37884, "A Point-of-Care Diagnostic System for the Developing World." Support for the CGIA was provided by NIDCR grant number 1 U01 DE14971-01. The views expressed by the authors do not necessarily reflect the views of the funding agencies. Note that one of us (Yager) has financial interest in Micronics, Inc.

The authors would like to thank Yager Laboratory members past and present, particularly Benedict Y. Hui for his skillful assistance in the development and validation of the CGIA assay. The authors would also like to thank other members of the DxBox team led by Patrick Stayton of the University of Washington Department of Bioengineering, Fred Battrell of Micronics, Walt Mahoney of Epoch Biosciences, Inc., and Gonzalo Domingo of PATH.

References

[1]   Castellion M. 2002. *Fundamentals of General, Organic, and Biological Chemistry*, New York: Prentice Hall.
[2]   Glick BR, Pasternack JJ. 1998. *Molecular Biotechnolgy: Principles and Applications of Recombinant DNA*. Washington D.C.: ASM Press.
[3]   Engvall E, Perlmann P. 1971. *Immunochemistry* 8: 871.
[4]   Hennion MC, Barcelo D. 1998. *Analytica Chimica Acta* 362: 3–34.
[5]   Henares TG, Mizutani F, Hisamoto H. 2008. *Analytica Chimica Acta* 611: 17–30.

[6]   Kartalov EP, Zhong JF, Scherer A, Quake SR, Taylor CR, Anderson WF. 2006. *Biotechniques* 40: 85–90.

[7]   Laib S, MacCriath BD. 2007. *Analytical Chemistry* 79: 6264–70.

[8]   Xiang Q, Hu G, Gao Y, Li D. 2006. *Biosensors and Bioelectronics* 21: 2006–9.

[9]   Gao Y, Lin FYH, Hu G, Sherman PM. 2005. *Analytica Chimica Acta* 543: 109–16.

[10]   Haes AJ, Terray A, Collins GE. 2006. *Analytical Chemistry* 78: 8412–20.

[11]   Phillips KS, Dong Y, Carter D, Cheng Q. 2005. *Analytical Chemistry* 77: 2960–5.

[12]   Herr AE, Hatch AV, Throckmorton DJ, Tran HM, Brennan JS, et al. 2007. *PNAS* 104: 5268–73.

[13]   Kakuta M, Takahashi H, Kazuno S, Murayama K, Ueno T, Tokeshi M. 2006. *Measurement Science and Technology* 17: 3189–94.

[14]   Goluch ED, Nam J-M, Georganopoulou DG, Chiesl TN, Shaika KA, et al. 2006. *Lab on a Chip* 6: 1293–9.

[15]   Stevens DY, Petri CR, Osborn JL, Spicar-Mihalic P, McKenzie KG, Yager P. 2008. *Lab on a Chip* 8: 2038–45.

[16]   Yager P, Edwards T, Fu E, Helton K, Nelson K, et al. 2006. *Nature* 442: 412–8.

[17]   Hatch A, Garcia E, Yager P. 2004. *Proceedings of the IEEE* 92(1): 126–39.

[18]   Hatch A, Kamholz AE, Hawkins KR, Munson MS, Schilling EA, et al. 2001. *Nature Biotechnology* 19: 461– 5.

[19]   Hatch A, Yager P. 2001. *Diffusion immunoassay in polyacrylamide hydrogels*. Presented at Micro Total Analysis Systems 2001.

[20]   Hatch AV. 2004. *Diffusion Based Analysis of Molecular Binding Reactions in Microfluidic Devices*. Ph.D. thesis. University of Washington, Seattle. 224 pp..

[21]   Hawkins KR, Hatch A, Chang H, Yager P. 2002. *Diffusion immunoassay for protein analytes*. Presented at Microtechnologies in Medicine and Biology, Madison, WI.

[22]   Hawkins KR, Yager P. 2004. *The aggregation of multivalent immune complexes expands the useful analyte size range of the diffusion immunoassay*. Presented at Micro Total Analysis Systems 2004, Malmo, Sweden.

[23]   Yager P, Cabrera C, Hatch A, Hawkins K, Holl M, et al. 2000. *Analytical devices based on transverse transport in microchannels*. Presented at Micro Total Analysis Systems 2000, University of Twente, the Netherlands.

[24]   Cheng SB, Skinner CD, Taylor J, Attiya S, Lee WE, et al. 2001. *Analytical Chemistry* 73: 1472–9.

[25]   Chiem N, Harrison DJ. 1997. *Anal Chem* 69: 373–8.

[26]   Chiem NH, Harrison DJ. 1998. *Clin. Chem.* 44: 591–8.

[27]   Weigl BH, Bardell RL, Cabrera C, R. 2003. *Advanced Drug Delivery Reviews* 55: 349–77.

[28]   Yager P, Bell D, Brody JP, Qin D, Cabrera C, et al. 1998. *Applying microfluidic chemical analytical systems to imperfect samples*. Presented at Micro Total Analysis Systems, Banff, Canada.

[29]   Becker H, Locascio L. 2002. *Talanta* 56: 267–87.

[30]   Lionello A, Josserand J, Jensen H, Girault HH. 2005. *Lab on a Chip* 5: 1096–103.

[31]   Lionello A, Josserand J, Jensen H, Girault HH. 2005. *Lab on a Chip* 5: 254–60.

[32]   Zimmermann M, Delamarche E, Wolf M, Hunziker P. 2005. *Biomedical Microdevices* 7: 99–110.

[33]   Bancaud A, Wagner G, Dorfman KD, Viovy J. 2005. *Analytical Chemistry* 77: 833–9.

[34]   Ruzicka J, Hansen HE. 1988. *Flow Injection Analysis*. New York: J. Wiley. 498 pp..

[35]   Dutta D, Ramachandran A, Leighton D. 2006. *Microfluidics and Nanofluidics* 2: 275–90.

[36]   Nelson KE, Foley JO, Yager P. 2007. *Analytical Chemistry* 79: 3542–8.

[37]   Howard RJ, Uni S, Aikawa M, Aley SB, Leech JH, et al. 1986. *Journal of Cellular Biology* 103: 1269–77.

[38]   Huy NT, Serada S, Trang DT, Takano R, Kondo Y, et al. 2003. *Journal of Biochemistry* 133: 693–8.

[39]   Papalexis V, Siomos MA, Campanale N, Guo X, Kocak G, et al. 2001. *Molecular and Biochemical and Parasitology* 115: 77–86.

[40]   Crowe JH, Carpenter JF, Crowe LM. 1998. *Annual Reviews in Physiology* 60: 73–103.

[41]   Hatch AV, Kamholz AE, Hawkins K, Munson MS, Shilling EA, et al. 2001. *Nature Biotechnology* 19: 461–5.

[42]   Kamholz AE, Weigl BH, Finlayson BA, Yager P. 1999. *Analytical Chemistry* 71: 5340–7.

[43]   Stone HA, Stroock AD, Ajdari A. 2004. *Annual Review of Fluid Mechanics* 36: 381–411.

[44]   Prime KL, Whitesides GM. 1993. *Journal of the American Cancer Society* 115: 10714–21.

[45]   Creighton TE. 1993. *Proteins: structures and molecular properties*. New York: W.H. Freeman.

[46]   Frens G. 1973. *Nature-Physical Science* 241: 20–2.

[47]   Deroe C, Courtoy PJ, Baudhuin P. 1987. *Journal of Histochemistry and Cytochemistry* 35: 1191–8.

[48]   Rasband WS. 1997–2007. Image J, U. S. National Institutes of Health, Bethesda, Maryland, USA, http://rsb.info.nih.gov/ij/. .

[49]   Kifude CM, Rajasekariah HG, Sullivan DJ, Stewart VA, Angov E, et al. 2008. *Clinical Vaccine and Immunology* 15: 1012–8.

[50]   Chinowsky T, Johnston K, Edwards T, Nelson K, Fu E, Yager P. 2007. *Biosensors and Bioelectronics* 22: 2208–15.

[51]   Foley JO, Nelson KE, Mashadi-Hossein A, Finlayson BA, Yager P. 2007. 79: 3549–53.

[52]   Navratilova I, Skladal P. 2003. *Supramolecular Chemistry* 15: 109–15.

[53]   Helton KL, Nelson KE, Fu E, Yager P. 2008. *Lab on a Chip*, 8: 1847–5.

[54]   Brody J, Yager P. 1997. *Sensors and Actuators A* 58: 13–8 .

[55]   Brody J, Yager P, Goldstein R, Austin R. 1996. *Biophysical Journal* 71: 3430–41.

[56]   Herr AE, Hatch AV, Giannobile WV, Throckmorton DJ, Tran HM, et al. 2007. *Integrated Microfluidic Platform for Oral Diagnostics*. Presented at Oral-Based Diagnostics, Atlanta.

Droplet Based Microfluidics by Shear-Driven Microemulsions

Robert Lin[1], Shia-Yen Teh[1], Abraham P. Lee[1, 2, 3]

[1]Department of Biomedical Engineering, University of California at Irvine
[2]Department of Mechanical & Aerospace Engineering, Department of Mechanical &
 Aerospace Engineering
[3]Director, Micro/Nano Fluidics Fundamentals Focus (MF3) Center, University of
 California at Irvine

Abstract

Droplet based microfluidics presents a platform with features that enable both novel experimental methods and solutions to traditional challenges for biological and chemical analyses. With the ability to accommodate a wide array of materials ranging from gases, oils, and aqueous solutions to organic solvents, it has been proven as a versatile and robust system for a diverse field of bioengineering applications. Using immiscible phases to discretize continuous fluid streams, individual nano- to femtoliter-sized reactors can be generated at rates up to thousands per seconds. Particles with extremely narrow size distribution can be created while controllably encapsulating cells, proteins and nanoparticles. In addition to the unique properties offered by the platform, a variety of manipulation techniques have been developed. This chapter will highlight the various techniques and methods required to carry out the basic droplet microfluidic operations.

Key terms microfluidics
droplets
high-throughput assay
digital microfluidics
microreactor
encapsulation
microparticles

10.1 Introduction

Droplet microfluidics utilizes immiscible phases to create discrete volumes of one fluid inside another. The droplets can be created at high rates, with high fidelity and low polydispersity. A variety of materials can be used with the platform including oils, aqueous solutions, and gases; enabling applications in fields including, biological detection, particle synthesis, tissue engineering, and drug delivery.[1, 2] A wide array of methodologies has been developed to create droplets using microfluidic systems including shear-focusing [3], flow-focusing [4], T-junction [5, 6], diaelectrophoretic [7, 8], and electrowetting-on-dielectric (EWOD) [9, 10]. In addition to the different generation methods, other manipulation techniques have also been developed including droplet fusion, fission, and sorting [1]. Although the theory behind droplet generation will not be discussed in detail in this chapter, those who are interested can refer to recent reviews which provide excellent insights [11, 12].

10.1.1 Advantages of droplet-based microfluidics

As discussed in previous chapters, microfluidic systems take advantage of specific behaviors and phenomenon at small scales to enable novel technologies. In addition to the benefits offered by continuous flow systems, droplet-based platforms offer numerous additional advantages including homogeneous reactions, rapid mixing, fast generation rates and monodispersity.

10.1.1.1 Rapid mixing

Due to laminar flow characteristics, mixing inside microfluidic devices is challenging. Although the lack of turbulent mixing in microchannels can be taken advantaged of for specific applications [13], many applications require mixing of multiple components. Various geometric designs have been demonstrated to increase the rate of mixing, but combining various components effectively is still a challenge [14–16].

Droplet-based systems offer a unique solution to the mixing problem. Although flow inside of microchannels is laminar; fluidic behavior inside of droplets follows a different pattern. It has been shown that a recirculating flow occurs inside of droplets that significantly enhance mixing inside of droplets [17]. Figure 10.1 shows the recirculating pattern inside a droplet. When two distinct components are used to form droplets, the recirculating flow begins to enhance mixing of the components almost immediately after the formation of a droplet; whereas laminar flow keeps the two components separate prior to the rapid mixing.

In addition to utilizing the recirculating flow, channel geometries can be used to enhance this behavior. Winding or serpentine channels have been utilized to further mix the contents as shown in Figure 10.1. Due to the differential distances traveled by the contents inside a droplet when going around a corner, the contents shift unequally as the droplets make the turn. After exiting the corner, the recirculating flow mixes the components again, allowing the entire droplet to be quickly homogenized. With optimized design, the contents of a droplet can be completely mixed within tens of milliseconds [18].

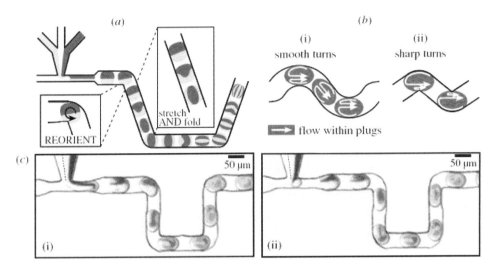

Figure 10.1 Schematic showing the recirculating pattern as droplets move around corners. Note the pattern changes orientation allowing the contents to be effectively mixed [17].

10.1.1.2 Large numbers

One main attraction of microfluidic systems is the ability to miniaturize experimental apparatus and carry out experiments with significantly less reagents. An upside of this is the potential to perform many more parallel experiments while consuming the same amount of reagents and taking up less space. With its ability to generate large number of droplets, droplet-based systems take the concept one step further.

With the ability to continuously generate droplets at a high rate, a large number of droplets could be made in a short amount of time. Each could be thought of as discrete reaction vessel capable of performing individual experiments without contamination from other droplets. This enables thousands or more parallel experiments to be performed simultaneously while using very little amount of reagents and very limited amount of space. Droplet generation rates have been reported to be as high as twenty thousand per second and can be controlled with device geometry and flowrates [19].

10.1.1.3 Monodispersity

Generating large numbers of droplets without control would render the platform virtually useless. Fortunately, droplet-based microfluidic platforms also offer precise control and consistent generation of droplets. Monodispersity, or the consistency in droplet size has been reported to be as low as 1 % [20] and the mixing ratio inside droplets can be easily controlled by relative flowrates of components [21].

The high fidelity of droplet generation is a critical feature of the droplet-based systems as it allows each droplet to be treated as an identical reaction vessel to produce statistically significant results. Also, it allows highly monodisperse particles to be synthesized using this platform.

In addition to the inherent advantages of working on the micrometer scale, droplet-based microfluidics possesses unique properties including rapid mixing, high generation rate, and monodispersity that make it highly versatile and adaptable to a variety of applications in a number of fields.

10.2 Biomedical applications of droplet microfluidics

In typical macroscale reactions, one major time-limiting factor is the distance with which the components of the reaction must travel in order to initiate the reaction. Mechanical agitation can be utilized to shorten this distance and improve reactant interaction, but the effect is limited when compared to reactions performed at a more miniaturized scale, where the diffusion distance is dramatically shorter. Droplet-based microfluidic systems achieve this by generating micrometer-sized droplets that can serve as small reaction vessels. In addition to providing a large supply of uniform picoliter-sized droplets, droplet microfluidics presents a platform to manipulate droplets effectively and efficiently; two characteristics ideal for performing high throughput reactions, bioassays, and particle formation. We only provide a brief overview of these applications as more detailed information can be found elsewhere in the literature [1].

10.2.1 Bioassays

One of the primary motivations behind the field of microfluidics is to develop lab-on-a-chip platforms to replace traditional, bulky, reagent-consuming lab equipment. This will enable the development of small, portable diagnostic tools that can be effective in both a hospital setting or in the field. The first step to achieve this goal is to develop a platform that can perform all the functions of a biological assay, such as sample preparation, detection, and analysis. Droplet microfluidics has advantages over continuous-flow microfluidics in that samples are contained in protective droplets that prevent unwanted adsorption of analytes onto the channel surface. In addition, since each droplet is in itself a single reaction chamber, and there are thousands of droplets made in a given period of time, multiple conditions and analytes can be tested simultaneously within the device.

Srinivasan et al. developed a chip capable of detecting glucose from saliva, serum, plasma, and urine using an electrowetting on dielectric (EWOD) droplet-based platform [22, 23]. On-chip electrodes are used to transport, fuse, and mix droplets containing the sample and assay reagents. The glucose concentration is then calculated by a colorimetric assay, and measured with a LED and photodiode. Multiple samples and reagents can be introduced into the system to perform sample preparation and testing on one chip. Recent efforts are being focused on implementing and configuring these miniaturized assay systems, pumps, and valves onto a single portable chip.

A wide array of microfluidic devices have demonstrated processing of biological materials such as DNA, protein, and cells [24–26]. Droplet microfluidic chips have also been used for more specific applications such as the improvement of real-time polymerase chain reactions, analysis of DNA, and for the screening of protein crystallization conditions. Beer et al. performed PCR reactions in picoliter sized droplets on microfluidic platform [27]. Due to the size of the droplets, single-copy PCR could be performed in significantly fewer cycles than that of commercial PCR instruments, and with much higher sensitivity. Droplet microfluidics has also successfully been applied for screening optimal conditions for protein crystallization. Li et al. was able to complete nearly 1,300 protein crystallization trials in less than 20 minutes [28]. Thus, droplet microfluidics has streamlined the parallelization of multiple experiments, by simultaneously increasing productivity while decreasing experiment time.

10.2.2   Particle formation

Particles composed of polymeric or biological materials such as alginate or chitosan are important in therapeutic delivery and tissue engineering applications, however current methods of production result in particles of large size variation, and limited encapsulation control. Since formation of monodispersed droplets can be readily achieved in droplet microfluidic platforms, many groups have taken advantage of this feature to form stable gels beads and polymer particles with narrow size distribution [29–31]. Additionally, nonspherical yet monodispersed particles are difficult to achieve using macroscopic methods, since it is entropically favorable for droplets to conform into a spherical shape. However with microfluidic methods, the shape of the droplets can be controlled by creating microchannels that physically confine the droplets to define their shape.

There are three primary methods used to alter the phase of a droplet from liquid to gel or solid: ultraviolet (UV), chemical catalyzed polymerization, and solvent extraction/evaporation. In the first method, UV light activates photo-initiators within the droplet to cause monomers to link together and solidify the droplet. The second method relies on polymerization that is initiated with the addition of chemical agents. In the third method, the solvent is removed by either evaporation or extraction to concentrate and polymerize the remaining monomers. Channel design becomes more important for the two latter methods because regions for the introduction and mixing of chemical catalyst or extracting solution need to be incorporated onto the platform without creating instability in the overall system.

10.2.2.1   Photo-initiated polymerization

Photo initiators are required for UV-activated polymerization. UV light causes the photo initiators to become reactive radicals that link monomers in the solution together, and cause the droplet to solidify into a gel or particle. Irradiation of UV can be done outside the microfluidic platform in a collection vial or array [32, 33], or in situ, immediately after the droplet is formed [34]. A couple challenges of on-chip photo polymerization are preventing the droplets from premature polymerization, yet ensuring that the droplets have sufficient exposure to UV activation. One method demonstrated by Jeong et al. prevents unwanted UV exposure by covering the entire device with aluminum foil except for the desired illumination section [35].

More complicated and functional particles have also been formed using UV-initiated polymerization. For instance polymer microcapsules have been formed by photo-initiated polymerization. Utada et al. formed water-in-oil-in-water (W/O/W) emulsions with external and internal phases consisting of water, and a middle phase of Norland Optical Adhesive (NOA) [36]. As shown in Figure 10.2, NOA hardens after it has been exposed to UV light for 10 seconds, turning the W/O/W emulsion into a microcapsule.

By forming the double emulsions using droplet microfluidic techniques, one has control over the amount of substance encapsulated, thickness of the polymer shell, and size of the capsule.

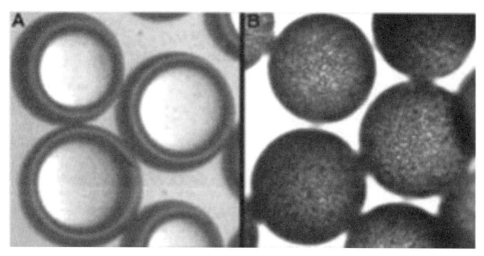

Figure 10.2 Photomicrograph of W/O/W double emulsions (a) prior, and (b) after photopolymerization into microcapsules [36].

10.2.2.2 Chemical-catalyzed polymerization

Chemical species such as ions can be used to trigger crosslinking, which causes the droplet to solidify or to gel. There are two main techniques for introducing the crosslinking agent to prepolymerized droplets. The first is to dissolve the chemical catalyst in the continuous phase, surrounding the droplets. The catalyst passively diffuses into the droplets to initiate polymerization [37, 38]. Zhang et al. used this method to produce droplets made from solutions of biopolymers such as alginate, kappa-carrageenan, and carboxymethylcellulose, and polymerized them into microcapsules with Ca^{2+} ions [39]. The group controlled the depth of ion diffusion by limiting the amount of time the droplets were exposed to the ions, and by varying the ion concentration.

The second technique is to produce droplets containing both the monomer and the chemical catalyst in separate dispersed phases. With careful design, the two droplets can be fused downstream, allowing more control over when the reaction is initiated. Liu et al. formed calcium alginate particles using this method [30]. Alginate and $CaCl_2$ droplets are formed at separate shear-focusing junctions and then led into a fusing chamber. The fusing chamber is a region of the channel that has a wider diameter than the rest of the channel. The widening of the channel slows the droplets' velocities, bringing them into contact. Fusion of the droplets allows calcium ions to interact with the alginate to harden it into an alginate gel bead.

10.2.2.3 Solvent extraction/evaporation polymerization

Unlike UV or chemical-catalyzed polymerization, where external agents are required for cross-linking, solvent extraction and evaporation involves the removal of the solvent. As the solvent is depleted, a more concentrated monomer solution that polymerizes over time is left behind. In the extraction method, also known as liquid-liquid extraction, separation of compounds is based on the relative solubilities of the monomer's solvent to the immiscible phase, which usually consists of an aqueous solution and organic solvent. Hung et al. formed nanometer to micrometer sized PLGA particles using the solvent extraction method [40]. As shown in Figure 10.3, oil is first used to shear a solution of PLGA dissolved in dimethyl sulfoxide (DMSO) into monodispersed droplets. Downstream, the DMSO/PLGA droplets are fused with water droplets. Due to the high solubility of DMSO and low solubility of PLGA in water, PLGA is desolvated immediately after fusion with the water droplet. Removal of DMSO causes PLGA to precipitate out into ~100 nanometer-sized particles that can be used to encapsulate materials for drug delivery applications.

The solvent evaporation method requires the use of a solvent that can be quickly removed by evaporation. An important property of solvents is the boiling point, which determines the rate of evaporation. Solvents with low boiling points such as acetone or dichloromethane tend to have higher rates of evaporation. This method has three primary steps: (1) dissolve the polymer into an organic solvent, (2) form monodispersed polymer/solvent droplets by shearing with an immiscible liquid, and (3) evaporate away solvent and collect the resulting particles. Kobayashi et al. created lipid microspheres using a cross-flow microchannel platform [41]. As shown in Figure 10.4, monodispersed 11-μm droplets of Tripalmitin dissolved in hexane were formed in a continuous phase consisting of water. The droplets were collected in vial where evaporation of hexane resulted in 2-μm particles.

Microcapsules can also be made using the solvent evaporation method. Hayward et al. [42] and Lorenceau et al. [43] formed diblock polymerosomes by creating water/oil/water double emulsions in a glass microcapillary emulsifier. The diblock polymers are dissolved in a mixture of the water-immiscible solvents, toluene, and tetrahydrofuran, respectively, which serve as the "oil" phase. The solvents dissolve into the surrounding aqueous phase and eventually evaporate, causing the polymer shell to become thinner. Because the polymer is amphiphilic, it self-assembles to form a bilayer shell similar to that of phospholipids in a liposome. It is advantageous to form

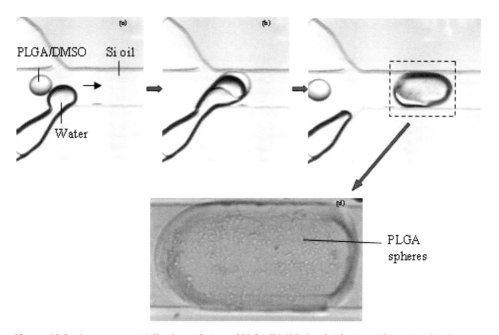

Figure 10.3 Image sequentially shows fusion of PLGA/DMSO droplet fusing with a water droplet (a–c). The resulting droplet acts as a microreactor that forms many PLGA nanoparticles (d) [40].

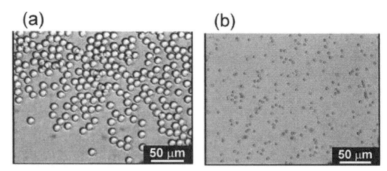

Figure 10.4 1% tripalmatin in hexane droplets (a) prior, and (b) after solvent evaporation [41].

microcapsules in droplet microfluidic devices because the encapsulation efficiency can be better controlled. The materials encapsulated can be kept separate from the outer phase, offering more flexibility in encapsulation choice and protection from unwanted outside interactions.

10.2.3 Therapeutic delivery

Using droplet microfluidics, a wide range of biocompatible materials can be used to form size-controlled and surface tunable monodispersed particles for therapeutic delivery applications. These particles can be customized to alter drug release profiles, improve site directed targeting, and affect the absorption rate of the therapeutic entity. Polymer nanoparticles synthesized in microfluidic platforms can be used as an effective means of delivering therapeutic agents due to their small size (<200 nm) and monodispersity. Encapsulation of polymer beads, metal particles, and dyes has been achieved in microdroplets [44–47]. Furthermore, cell encapsulation has been demonstrated in both aqueous, lipid, and gel droplets with potential applications in both tissue engineering and delivery of cells for implantation [48–53]. Figure 10.5(a) shows a collection of alginate particles containing cells. The live and dead cells can be easily distinguished using a live dead stain as shown in Figure 10.5(b). Cell viability in droplets can be maintained for several weeks since cell culture media is encapsulated with the cell, and since the droplets allow diffusion of nutrients from the outside environment.

Protein and drug compounds have also been encapsulated inside of droplets [54]. Yang et al. encapsulated Ampicillin within chitosan particles and monitored the drug release profile [55]. Variation in the drug release rate could be achieved by changing the size of the particle. Since the size is highly controllable and there is a small size distribution, this device has the potential to form particles for controlled drug delivery. Microfluidic systems offer quick droplet generation and photopolymerization processing times, reducing the risk of optical or chemical damage to enzymes or biological substances encapsulated within the particles.

There is also interest in developing droplet microfluidic platforms for the generation of multifunctional particles for drug delivery. Hettiarachchi et al. formed microbubbles with a therapeutic-containing lipid shell that can be decorated with targeting ligands [56]. The microbubbles can be used as contrast agents to image the target region, while offering the capability to release drug compounds simultaneously.

While there is still much work to be done to automate and integrate components of microfluidic systems together, droplet microfluidics technology has tremendous poten-

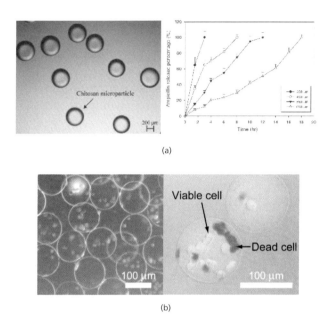

(a)

(b)

Figure 10.5 (a) Alginate hydrogel microbeads encapsulating Jurkat cells, (b) encapsulated cells are stained with trypan blue to determine cell viability. Blue color indicates dead cells [53].

tial to positively impact many biomedical applications. Experiments can be streamlined, therapeutics can be optimized, and the functionality and performance of current technology platforms can be improved.

10.3 Materials

A variety of materials have been used by different research groups in a wide array of droplet experiments. The materials listed is meant to serve as a starting point that would allow the readers to follow the protocols and begin preliminary work in the field and is not intended to be an exhaustive list.

Device fabrication

SU-8 photoresist from Microchem (Newton, MA) is used to fabricate the mold for soft lithography. Silicon wafers (vendor location) is the substrate used to pattern the SU-8. Polydimethylsiloxane (PDMS) in the form of Sylgard 184 from Dow Corning (Midland, MI) is the primary material used for fabricating the devices. Precleaned 75x25-mm glass slide from Corning (Corning, NY) are used as the substrate for the bonding of the PDMS devices. Other materials including SU-8, silicon, and thiolene have also been used for forming the channels for droplet generation. Solvent compatibility is often a limitation in the generation of polymeric particles and several groups have attempted to address this issue [57–61].

Trichlorosilane (tridecafluoro-1, 1, 2, 2-tetrahydrooctyl) from Gelest Inc. (Morrisville, PA) and Teflon AF from DuPont (Wilmington, DE) are used for hydrophobic coating of the devices. Polyvinyl alcohol from Sigma-Aldrich (St. Louis, MO) is used for hydrophilic coating of devices.

Fluid manipulation

Fluid injection and pumping is done with syringe pumps Pico Plus from Harvard Apparatus (Holliston, MA). A computer interface to control multiple pumps simultaneously is created in LabVIEW from National Instruments (Austin, TX).

Gas flow is controlled using a combination of the pressure gauge Cole-Palmer EW-68022-02 from Cole Palmer Instrument Company (Vernon Hills, IL) and valve control using Upchurch P-445 from Upchurch Scientific (Oak Harbor WA).

Oils

Many different fluids can be used for generation of aqueous droplets including oils and PFCs (perfluorocarbons). Here we provide a partial list of available reagents.

Silicon oil from Fisher (Pittsburgh, PA) and Dow Corning Fluid 200 from Dow Corning (Midland, MI) are both silicone based oils used for droplet generation. Fluid 200 comes in a variety of viscosities ranging from 1 to 10,000 cst and can be selected for specific needs. In general viscosities from 5 to 50 cst can be used as a starting point.

Heavy mineral oil from Fisher Scientific (Pittsburgh, PA) and light mineral oil from Sigma-Aldrich (St. Louis, MO) can also be used. Oleic Acid from Fisher Scientific (Pittsburgh, PA) is another oil that can be used to generate aqueous droplets.

Aqueous reagents

Virtually all aqueous-based reagents and solutions can be used with droplet microfluidics. One added advantage of the platform is that a thin oil layer coats the channel walls and prevent protein and DNA adsorption onto the walls.

Water used is from MilliQ A10 from Millipore (Billerica, MA). Cell culture media and reagents are purchased from Invitrogen (Carlsbad, CA). Sodium alginate and calcium chloride is purchased from Sigma Aldrich (St. Louis, MO).

Gas

Nitrogen gas is purchased from Airgas (Radnor, PA).

Surfactants

Surfactants play an important role in droplet systems as it increases the stability of the dispersed phase by decreasing the interfacial energy between the two immiscible phases and greatly reduce unwanted fusion of the droplets. It can be put into either the dispersed or the continuous phase or both. However, it is important to consider potential adverse effects the surfactants might have on sensitive materials such as cells or proteins.

Oil phase surfactants include SPAN 80 from Sigma Aldrich (St. Louis, MO) and ABIL EM 90 from Evonik (Essen, Germany). Water phase surfactants include PVA, Tween 20 and Pluronic F-68; all from Sigma Aldrich (St. Louis, MO).

Solvents

Solvents including DMSOS, chloroform, ethyl acetate, and toluene have all been used in droplet experiments.

10.4 Methods

10.4.1 Channel surface modification

Channel surface properties play an important role in the formation and operation of droplets. Namely, the hydrophilicity or hydrophobicity of the channel surface determines the ease in which an oil or aqueous droplet can be formed. Since the majority of droplet microfluidic systems are fabricated using PDMS, we will cover methods that have been demonstrated to control the wetting properties of PDMS-based channel walls.

10.4.2 Hydrophilic surface treatment

Cured PDMS is naturally hydrophobic, with a contact angle of $90°$ or higher [62]. Hydrophilic processing of the channel allows one to better wet an aqueous continuous phase, and also to prevent unwanted protein or cell adsorption. Two methods of hydrophilic modification are discussed in this section: plasma surface oxidation and polyvinyl alcohol coating.

10.4.2.1 Plasma surface oxidation

The hydrophobicity of PDMS is due to the repeating $OSi(CH_3)_2O$- units that compose the PDMS structure [63]. Air or oxygen plasma oxidation replaces methyl groups ($Si-CH_3$) with silanol groups ($Si-OH$), thus converting the units into a hydrophilic structure [64]. The contact angle of PDMS post-plasma treatment ranges from $30°$ to $60°$ [65] (Figure 10.6).

Materials

- Device components (PDMS channels, glass slide, or other material used to seal PDMS channels).
- Plasma Oven Cleaner from Harrick (Ossining, NY).
- KJLC 205BM Thermocouple Gauge Controller.

Method

1. Place device components (treatment-side up) for plasma bonding and surface treatment into plasma oven chamber. Completely close the needle valve.
2. Turn on the vacuum pump.
3. When the pressure gauge reaches 200 mTorr or lower, switch on the plasma.
4. Adjust the needle valve carefully until the vacuum pressure drops to around 400 mTorr.
5. After approximately 15 seconds, the plasma should be a bright pink color.
6. Maintain the pink plasma color for 2–3 minutes, then turn off the vacuum and plasma.
7. Open the needle valve, remove device components, bond together, and use immediately.

Figure 10.6   Image showing the interface of water and PDMS (a) before and (b) after plasma treatment.

Air plasma-treated PDMS surfaces begin to lose hydrophilicity within a few hours of exposure to air and revert to being hydrophobic by the following day. To retain PDMS hydrophilicity for multiple days, keep the channels exposed to a polar liquid such as water until the device is needed [66].

10.4.2.2   Poly(vinyl alcohol) coating

Since plasma surface treatment renders the PDMS surface hydrophilic for only a short period of time, numerous methods have been developed to more permanently modify the PDMS surface. One such method uses a highly hydrophilic polymer called poly(vinyl alcohol) (PVA) as a coating material. Kozlov et al. demonstrated that PVA could be irreversibly adsorbed onto hydrophobic polymer surfaces [67], and it was shown by Wu et al. that a repetitive adsorption/drying and heat immobilization cycle of PVA onto PDMS results in a hydrophilic treatment that lasts for multiple weeks [68].

Materials

- Polyvinyl alcohol from Sigma Aldrich (St. Louis, MO).
- DI water.
- Thin polypropylene tubing connected to vacuum line or vacuum pump.
- Oven or hotplate set at 110°C.

Method

1. Make a 1% PVA solution by dissolving the PVA powder into DI water
2. After plasma bonding the PDMS channels, add 1% PVA to the channel reservoir until all channels are filled with solution. (The channels may be filled automatically by capillary action or manually with a syringe.)
3. Incubate for 5 minutes at room temperature then remove the solution by vacuum.
4. Dry the channels by heating to 110°C for 10–15 minutes.

If the PDMS channels are not hydrophilic enough, the above method can be repeated two or more times to better immobilize PVA onto the surface. This method can also be applied for the formation of double or multiple emulsions by selectively coating channel surfaces with hydrophilic PVA.

10.4.3 Hydrophobic surface treatment

During fabrication of PDMS-based devices, channels are often sealed through air or oxygen plasma bonding. However, plasma bonding renders the channels hydrophilic, leaving it susceptible to unwanted sticking of hydrophobic species or preventing the channel to be used for water-in-oil emulsion formation. In this section we will cover a common PDMS hydrophobic modification method known as silane vapor deposition.

10.4.3.1 Silane-based vapor deposition

Although PDMS surfaces begin to lose its hydrophilicity within a few hours after plasma bonding, it does not return fully to its hydrophobic state in one day. If a hydrophobic surface is needed, one would need to wait a couple days before being able to use the device. To circumvent this waiting period, techniques have been developed to allow same-day use of plasma bonded devices. Silanes react with the hydroxyl groups on oxygenated PDMS to form covalent bonds on the surface [69]. After silane vapor deposition, the treated PDMS has a contact angle near its value prior to plasma exposure.

Materials

- Fluorosilane such as tridecafluoro-1, 1, 2, 2-tetrahydrooctyl)trichlorosilane from Gelest (Morrisville, PA).
- Desiccator connected to a vaccum pump or vacuum line.
- Small container such as a 15-mL conical tube.

Method

1. After plasma bonding, place PDMS device into vacuum desiccator.
2. Place one drop of silane (about 5 μL) into a conical tube, and place the tube inside vacuum dessicator along with the PDMS device.
3. Seal the desiccator and turn on vacuum line. Continue pumping for one hour, then shut off the vacuum. The device should be ready for use.

10.4.4 Solution preparation

Interfacing with the microfluidic devices is done through tubing and syringes. The tubing provides the physical link to the device and the syringes serve as both a reservoir and the means to pump the fluids. Although most syringes will work for this purpose, there are some important points to consider when choosing syringes. First and foremost is that the syringe needs to be able to be properly actuated by the pump. This includes both syringe size and the desired flow rates. Since most syringe pumps use a plunger mechanism, they depend on velocity of the plunger as the means to control flow rate. Therefore, for very low flow rates, it is necessary to use small syringes while high flow rates would require larger ones depending on the capabilities of the specific pump. For example, the Pico Plus will be able to reach around 25 μl/min using a 1-ml syringe but for higher flow rates a larger syringe will need to be used. The second issue is the dead volume. Although the increments on a 1ml syringe goes down to 100 μl and below, a small sample size will not be able to be pumped effectively. In general a 1ml syringe will require at least 250 μl of sample to operate effectively. This covers both the dead volume inside the syringe as well as inside the tubing. If the sample size needs to be kept small, the sample could be loaded into a smaller syringe or only into the tubing to avoid dead volumes. Last, although not critical with oil/water experiments, the material of the syringe comes into play when using solvents. It is important to use chemical resistant materials such as the Norm-Ject syringes from HSW (Tuttlingen, Germany), which has a barrel made of polypropylene and plunger made from polyethylene. Using syringes such as this one will prevent the syringe from contaminating the solvent sample. Another choice to use with solvents is glass syringe, which also provide excellent solvent resistance.

Chemical resistance and reagent adsorption are two critical issues concerning tubing selection. Although Tygon tubing provides a cheap and effective method of connecting the syringes to the devices, it works well only with nonreactive reagents such as oils and aqueous solutions. Teflon tubing will be effective in both withstanding solvents and preventing precious reagents such as proteins and DNA from being lost due to adsorption.

Depending on the amount of reagents used, they can be either loaded into the syringe directly using suction or drawn into the syringe after attaching the needle. Tubing attachment is usually done after loading of the reagents since it is sometimes difficult to manipulate the tip of the tubing into small containers such as an Eppendorf tube. After loading the reagents into the syringe, be sure to hold the syringe vertically with the needle pointing up and tap the syringe gentle to move all the air bubbles from the syringe. After all the bubbles have reached the top of the syringe, slowly push the plunger until all of the air is removed.

Slide the tubing carefully along the needle without puncturing the tubing. Any puncture could lead to leakage during experiments causing inaccurate flow control and loss of reagents. After connecting the tubing, move the fluid front to the tip of the tubing before inserting into the device. It is important to remove all air bubbles because air is compressible. When pumping, the compressibility of the air will cause the downstream flow rates to differ from the intended flow rates. The compressibility will also cause problems when attempting to balance the pressure/flow between different inputs as air bubbles will "cushion" the applied pressure and hinder the ability to balance the flows.

10.4.5 Generation of droplets

After inserting the tubing begin flowing the continuous phase through the device. It is important to allow the continuous phase to wet the surfaces of the channels, especially the shear junction to ensure successful droplet generation. If the dispersed phase were to stick to the channel walls, the generation process would be unstable and may fail to generate droplets.

After allowing the continuous phase to wet the walls, the dispersed phase can be allowed to flow into the channels. Although some back flow of the continuous phase into the disperse phase channel(s) might occur, this will not affect the generation process. Initially, some air bubbles will be formed from the air that originally occupies the fluidic channels, but fluids should enter the channels shortly as the air is displaced.

At the beginning of the experiment, lower flow rates should be used on both the continuous and disperse phase to reduce excess back flow. Usually equal flow rates could be used to begin an experiment. After the droplet generation process has stabilized the flow rates can be adjusted to change the droplet sizes. In general, increasing the disperse phase flow rates or decreasing continuous phase flow rates will increase droplet size. Conversely, decreasing disperse phase flow rates or increasing continuous phase flow rates will decrease the droplet size [3, 4, 36]. For a given geometry, higher flow rates will result in higher generation rates and lower flow rates will produce lower frequencies. Generally speaking, droplet sizes depend on the ratio of the flow rates, whereas the generation frequency depends on the combination of the flow rates.

10.5   Data acquisition

Visualization and the capture of images and movies are important tools in droplet microfluidic experiments. The following sections will discuss various methods that can be employed for the visualization of droplets.

10.5.1   Droplet arrays

A large number of droplets can be observed simultaneously in the array format. A large array takes advantage of the ability of droplet systems to produce large numbers of identical droplets that function as individual vessels. By monitoring the array, many experiments can be examined at the same time, dramatically reducing the amount of time needed to perform multiple experiments. Droplet arrays can be generated using two different channel geometries. The first utilizes winding or serpentine channels to create a linear array of droplets. The long winding channels are placed after the generation junction so that the droplets pass through the winding channels before reaching the outlet as shown in Figure 10.7. When fluid flows are stopped, the droplets are "frozen" inside the winding channels presenting a linear array with the droplets generated first located towards the outlet and the droplets generated later towards the generation junction. The advantage of this method is that the location of each droplet in the array/line contains temporal information that can be useful for applications such as the study of protein binding kinetics. Combining information such as droplet generation rates and velocity of droplets inside the channels, this type of array could provide "frame-by-frame" images of the observed experiment [70].

A recent development in droplet array was demonstrated by Shi et al., utilizing specifically designed channels to capture and localize droplets as they float by [72]. The group demonstrated culturing of the worm *C. elegans* inside the droplets for up to two hours. The use of channel geometry allows one to trap droplets without having to stop fluid flow and interrupt droplet generation. Figure 10.8 shows droplets encapsulating *C. elegans* in the array.

Another type of array is one that is created with a reservoir or very large microfluidic channel, as shown in Figure 10.9. Although temporal information is not preserved precisely as in the winding channel designs, a much higher density of droplets can be

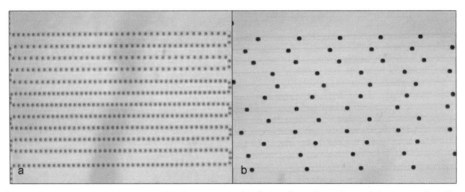

Figure 10.7 Two images showing droplet array inside identical winding channel designs with different densities. Aqueous droplets are colored with dye for enhanced imaging [71].

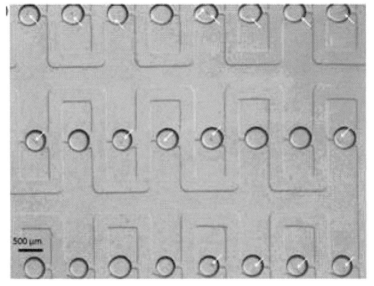

Figure 10.8 Droplets captured by the array. White arrows indicate droplets encapsulating the worms [72].

observed in the same area. The high density of droplets is the result of allowing the droplets to self-pack, which results in close to optimal orientation inside the reservoirs. It is worth noting that although no accurate relative time information is preserved; in slower reactions a difference can still be easily observed. The earlier droplets will be located closer to the outlet but the precise order that they were generated will be difficult to elucidate. Another advantage of this design is that the speed of the droplets is dramatically reduced when they reach the reservoir. Due to the fact that the reservoir channel is much wider than the channels directing the droplets there, the overall flow velocity is reduced. This allows the droplets to be monitored as more are generated despite high generation rates. The reservoir design is good for experiments that either have a much longer completion time compared to its generation rate or experiments that are triggered after an array is formed.

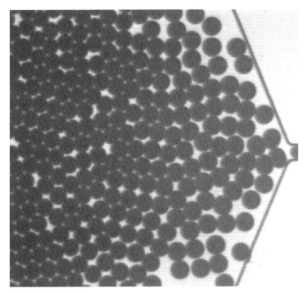

Figure 10.9   Image of droplet array in a reservoir. Droplets enter from right into the reservoir with the outlet hold downstream to the left [71].

10.5.2   High-speed cameras

Since droplets are often generated at rates up to 20,000 per second [19], specialized high-speed cameras are needed to observe the generation process. Monitoring the generation process serves two important purposes: to accurately measure the generation rate, and to observe the break-off process. Figure 10.10 shows a series of images captured using a high-speed camera of the generation of a single droplet. Measuring generation rates ensures precise quantification of the experiment and allows better control over the generation process by allowing the users to correlate generation rates with different experimental conditions. Observing the breakoff process will allow the users to ensure

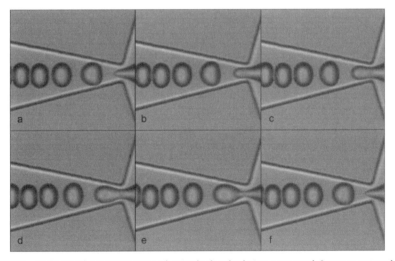

Figure 10.10   High-speed camera images of a single droplet being generated. Images captured at 1,000 fps [71].

stability in the generation process. Successful droplet experiments rely on the stable generation of highly monodispersed droplets. Thus it is important to monitor and fine tune how factors such as channel geometry or surface chemistry affect the generation process. Furthermore, mixing inside droplets and phase transitions such as from liquid-to-gel or liquid-to-solid can be captured and observed using high-speed cameras. For instance the Photron (San Diego, CA) PCI series of PC-based high-speed cameras with speeds up to 100,000 fps can be mounted onto microscopes for monitoring droplet generation.

10.5.3 Conventional imaging methods

In addition to the use of high-speed cameras, droplet experiments can be observed using other imaging methods as well. High-speed cameras, due to the fact that they are built for speed, generally do not have the sensitivity or resolution needed for certain applications. For instance it is difficult to detect fluorescent signals using a high-speed camera due to the lack of sensitivity and inability to utilize long exposure times. On the other hand, the high velocity of the droplets makes it difficult to capture images on a camera normally used for static fluorescence imaging. Although each type of camera has its limitations, their capabilities compliment each other and together allow the monitoring of droplets at all stages of the experiment.

By collecting and gathering droplets in an array, conventional bright-field microscopy and fluorescence microscopy could be performed on droplet experiments. As mentioned above, arrays dramatically reduce the velocity of droplets, allowing them to be interrogated using methods that are compatible with any other microfluidic devices. For instance, Hatakeyama et al. observed protein crystallization inside droplets that are kept inside winding channels [73]. Fluorescent images can be also taken in an array using long exposure times as with other microfluidic experiments. Furthermore, droplets, particles and beads synthesized using droplet devices can be collected after they are removed from the device, allowing them to be examined in the same manner as with products made using other methods. For instance cells encapsulated in beads can be kept in media and monitored over periods of several days [74]. Particles can be examined using SEM to precisely measure their size and distribution.

10.6 Discussion and commentary

It is worth spending a few words discussing basic device design. In the basic shear-focusing design, two streams of the continuous phase intersect the disperse phase stream at the shearing junction and the three streams flow towards the outlet through a narrow orifice [3]. The narrow orifice creates a point of highest shear and the single point of highest shear enables consistent droplet breakoff. Many variations and modifications can be made to the design to suit specific applications. Additionally, change in channel geometry can also be used to adjust the droplet generation process. For instance, a smaller orifice will produce smaller droplets at the same flow rates.

It is also important to note that droplet generation is a dynamic process and the process will be affected by channel geometry, surface chemistry, as well as the properties of the continuous and disperse phases. Although it seems that the numerous factors make

Troubleshooting Table

Problem	Explanation	Potential Solutions
Disperse phase does not shear into droplets and seems to stick to the channel walls.	The disperse phase is flowing along the channel walls due to favorable fluid-surface interaction.	1. Flow the continuous phase into the shear junction before the dispersed phase, to prevent contact between the disperse phase and the walls.
		2. Utilize surface treatment methods to ensure the surface interaction is more favorable for the continuous phase. See "Surface Treatment" section.
Disperse phase does not shear but there is no stiction to the walls.	This is most likely due to high viscosity of the disperse phase.	1. Increase the viscosity of the continuous phase by either using a different material or additives that would increase viscosity. For example adding glycerol into an aqueous continuous phase.
		2. Decrease the width of the shear orifice. This will increase the shear rate and potentially improve shearing.
The continuous phase flows into the disperse phase channel.	The relative flow rate difference is too high.	1. Change flow rates slowly when adjusting the values. Large changes should be done incrementally to allow stabilization and prevent back flow.
		2. Fluidic resistances such as winding channels could be added to the dispersed phase inlet channel to decrease the likelihood of backflow.
The device is generating unusually small droplets for a set of given flow rates.	The shearing orifice is clogged by debris and the actual width of the orifice is drastically decreased.	1. The solutions should be filtered prior to use to eliminate the presence of debris. For instance, syringe filters could be used.
		2. Channel filters can be designed into the device to prevent debris from reaching the shear orifice.

it a difficult process to optimize; in fact it is the ability to tune so many factors that allows droplet based microfluidic platforms to be so versatile.

Acknowledgments

The authors would like to thank the Micro/Nano Fluidics Fundamentals Focus (MF3) Center for the generous support.

References

[1] Teh, S.-Y., Lin, R., Hung, L.-H. and Lee, A. P., "Droplet microfluidics," Vol. 8, No. 2, 2008, pp. 198-220.

[2] Song, H., Chen, D. L. and Ismagilov, R. F., "Reactions in droplets in microflulidic channels," *Angewandte Chemie, International Edition*, Vol. 45, no. 44, 2006, pp. 7336-56.

[3] Tan, Y. C., Cristini, V. and Lee, A. P., "Monodispersed microfluidic droplet generation by shear focusing microfluidic device," *Sensors and Actuators B-Chemical*, Vol. 114, No. 1, 2006, pp. 350-6.

[4] Anna, S. L., Bontoux, N. and Stone, H. A., "Formation of dispersions using "flow focusing" in microchannels," *Applied Physics Letters*, Vol. 82, No. 3, 2003, pp. 364-6.

[5] Thorsen, T., Roberts, R. W., Arnold, F. H. and Quake, S. R., "Dynamic pattern formation in a vesicle-generating microfluidic device," *Physical Review Letters*, Vol. 86, No. 18, 2001, pp. 4163-6.

[6]   Zheng, B. and Ismagilov, R. F., "A microfluidic approach for screening submicroliter volumes against multiple reagents by using preformed arrays of nanoliter plugs in a three-phase liquid/liquid/gas flow," *Angewandte Chemie, International Edition*, Vol. 44, No. 17, 2005, pp. 2520-3.

[7]   Ahmed, R. and Jones, T. B., "Dispensing picoliter droplets on substrates using dielectrophoresis," *Journal of Electrostatics*, Vol. 64, No. 7-9, 2006, pp. 543-9.

[8]   Jones, T. B., "Liquid dielectrophoresis on the microscale," *Journal of Electrostatics*, Vol. 51, 2001, pp. 290-9.

[9]   Pollack, M. G., Shenderov, A. D. and Fair, R. B., "Electrowetting-based actuation of droplets for integrated microfluidics," *Lab Chip*, Vol. 2, No. 2, 2002, pp. 96-101.

[10]   Lee, J., Moon, H., Fowler, J., Schoellhammer, T. and Kim, C. J., "Electrowetting and electrowetting-on-dielectric for microscale liquid handling," *Sensors and Actuators a-Physical*, Vol. 95, No. 2-3, 2002, pp. 259-68.

[11]   Garstecki, P., Fuerstman, M. J., Stone, H. A. and Whitesides, G. M., "Formation of droplets and bubbles in a microfluidic T-junction - scaling and mechanism of break-up," *Lab Chip*, Vol. 6, No. 3, 2006, pp. 437-46.

[12]   Li, W., Nie, Z., Zhang, H., Paquet, C., Seo, M., Garstecki, P. and Kumacheva, E., "Screening of the Effect of Surface Energy of Microchannels on Microfluidic Emulsification," *Langmuir*, Vol. 23, No. 15, 2007, pp. 8010-4.

[13]   Brody, J. P., Yager, P., Goldstein, R. E. and Austin, R. H., "Biotechnology at low Reynolds numbers," *Biophysical Journal*, 1996, pp. 3430-41.

[14]   Stroock, A. D., Dertinger, S. K. W., Ajdari, A., Mezic, I., Stone, H. A. and Whitesides, G. M., "Chaotic Mixer for Microchannels," *Science,* Vol. 295, No. 5555, January 25, 2002, pp. 647-51.

[15]   Stone, H. A., Stroock, A. D. and Ajdari, A., "Engineering flows in small devices: Microfluidics toward a lab-on-a-chip," *Annual Review of Fluid Mechanics*, Vol. 36, 2004, pp. 381-411.

[16]   Chang, C. C. and Yang, R. J., "Electrokinetic mixing in microfluidic systems," *Microfluidics and Nanofluidics*, Vol. 3, No. 5, Oct 2007, pp. 501-25.

[17]   Bringer, M. R., Gerdts, C. J., Song, H., Tice, J. D. and Ismagilov, R. F., "Microfluidic systems for chemical kinetics that rely on chaotic mixing in droplets," *Philosophical Transactions of the Royal Society of London Series a-Mathematical Physical and Engineering Sciences*, Vol. 362, No. 1818, 2004, pp. 1087-104.

[18]   Sarrazin, F., Prat, L., Di Miceli, N., Cristobal, G., Link, D. R. and Weitz, D. A., "Mixing characterization inside microdroplets engineered on a microcoalescer," *Chemical Engineering Science*, Vol. 62, No. 4, 2007, pp. 1042-8.

[19]   Kobayashi, I., Uemura, K. and Nakajima, M., "Formulation of monodisperse emulsions using submicron-channel arrays," *Colloids and Surfaces A—Physicochemical And Engineering Aspects*, Vol. 296, No. 1-3, 2007, pp. 285-9.

[20]   Nisisako, T., Torii, T., Takahashi, T. and Takizawa, Y., "Synthesis of monodisperse bicolored janus particles with electrical anisotropy using a microfluidic co-flow system," *Advanced Materials*, Vol. 18, No. 9, 2006, pp. 1152-6.

[21]   Tan, Y. C., Fisher, J. S., Lee, A. I., Cristini, V. and Lee, A. P., "Design of microfluidic channel geometries for the control of droplet volume, chemical concentration, and sorting," *Lab Chip*, Vol. 4, No. 4, 2004, pp. 292-8.

[22]   Srinivasan, V., Pamula, V. K. and Fair, R. B., "An integrated digital microfluidic lab-on-a-chip for clinical diagnostics on human physiological fluids," *Lab Chip*, Vol. 4, No. 4, 2004, pp. 310-5.

[23]   Srinivasan, V., Pamula, V. K. and Fair, R. B., "Droplet-based microfluidic lab-on-a-chip for glucose detection," *Analytica Chimica Acta*, Vol. 507, No. 1, 2004, pp. 145-50.

[24]   Sims, C. E. and Allbritton, N. L., "Analysis of single mammalian cells on-chip," *Lab Chip*, Vol. 7, No. 4, 2007, pp. 423-40.

[25]   Burns, M. A., Johnson, B. N., Brahmasandra, S. N., Handique, K., Webster, J. R., Krishnan, M., Sammarco, T. S., Man, P. M., Jones, D., Heldsinger, D., Mastrangelo, C. H. and Burke, D. T., "An integrated nanoliter DNA analysis device," *Science*, Vol. 282, 1998, pp. 484-7.

[26]   Haeberle, S. and Zengerle, R., "Microfluidic platforms for lab-on-a-chip applications," *Lab Chip*, Vol. 7, No. 9, 2007, pp. 1094-110.

[27]   Beer, N. R., Hindson, B. J., Wheeler, E. K., Hall, S. B., Rose, K. A., Kennedy, I. M. and Colston, B. W., "On-Chip, Real-Time, Single-Copy Polymerase Chain Reaction in Picoliter Droplets," *Anal Chem*, Vol. 79, 2007, pp. 8471-5.

[28]   Li, L., Mustafi, D., Fu, Q., Tereshko, V., Chen, D. L. L., Tice, J. D. and Ismagilov, R. F., "Nanoliter microfluidic hybrid method for simultaneous screening and optimization validated with crystallization of membrane proteins," *Proc. Natl. Acad. Sci. USA*, Vol. 103, 2006, pp. 19243-8.

[29]   Hung, L. H. and Lee, A. P., "Microfluidic Devices for the Synthesis of Nanoparticles and Biomaterials," *Journal of Medical and Biological Engineering*, Vol. 27, No. 1, 2007, pp. 1-6.

[30] Liu, K., Ding, H. J., Liu, J., Chen, Y. and Zhao, X. Z., "Shape-controlled production of biodegradable calcium alginate gel microparticles using a novel microfluidic device," *Langmuir*, Vol. 22, No. 22, 2006, pp. 9453-7.

[31] Hsieh, A. T.-H., Pan, J.-H., Pinasco, P. G., Fisher, J., Hung, L.-H. and Lee, A. P., "Polymer Microsphere Mass Production using 128-Channel Digital Fluidic," in *The Eleventh International Conference on Miniaturized Systems for Chemistry and Life Sciences*. Paris, France, 2007, pp. 346-8.

[32] De Geest, B. G., Urbanski, J. P., Thorsen, T., Demeester, J. and De Smedt, S. C., "Synthesis of monodisperse biodegradable microgels in microfluidic devices," *Langmuir*, Vol. 21, No. 23, 2005, pp. 10275-9.

[33] Ikkai, F., Iwamoto, S., Adachi, E. and Nakajima, M., "New method of producing mono-sized polymer gel particles using microchannel emulsification and UV irradiation," *Colloid and Polymer Science*, Vol. 283, No. 10, 2005, pp. 1149-53.

[34] Seo, M., Nie, Z. H., Xu, S. Q., Mok, M., Lewis, P. C., Graham, R. and Kumacheva, E., "Continuous microfluidic reactors for polymer particles," *Langmuir*, Vol. 21, No. 25, 2005, pp. 11614-22.

[35] Jeong, W. J., Kim, J. Y., Choo, J., Lee, E. K., Han, C. S., Beebe, D. J., Seong, G. H. and Lee, S. H., "Continuous fabrication of biocatalyst immobilized microparticles using photopolymerization and immiscible liquids in microfluidic systems," *Langmuir*, Vol. 21, No. 9, 2005, pp. 3738-41.

[36] Utada, A. S., Lorenceau, E., Link, D. R., Kaplan, P. D., Stone, H. A. and Weitz, D. A., "Monodisperse double emulsions generated from a microcapillary device," *Science*, Vol. 308, No. 5721, 2005, pp. 537-41.

[37] Quevedo, E., Steinbacher, J. and McQuade, D. T., "Interfacial polymerization within a simplified microfluidic device: Capturing capsules," *Journal of the American Chemical Society*, Vol. 127, No. 30, 2005, pp. 10498-9.

[38] Huang, K. S., Lai, T. H. and Lin, Y. C., "Using a microfluidic chip and internal gelation reaction for monodisperse calcium alginate microparticles generation," *Frontiers in Bioscience*, Vol. 12, 2007, pp. 3061-7.

[39] Zhang, H., Tumarkin, E., Sullan, R. M. A., Walker, G. C. and Kumacheva, E., "Exploring microfluidic routes to microgels of biological polymers," *Macromolecular Rapid Communications*, Vol. 28, No. 5, 2007, pp. 527-38.

[40] Hung, L. H. and Lee, A. P., "PLGA Micro/Nanosphere Synthesis by Microfluidic Droplet Evaporation/Extraction Approaches," submitted.

[41] Kobayashi, I., Iitaka, Y., Iwamoto, S., Kimura, S. and Nakajima, M., "Preparation characteristics of lipid microspheres using microchannel emulsification and solvent evaporation methods," *Journal of Chemical Engineering of Japan*, Vol. 36, No. 8, 2003, pp. 996-1000.

[42] Hayward, R. C., Utada, A. S., Dan, N. and Weitz, D. A., "Dewetting instability during the formation of polymersomes from block-copolymer-stabilized double emulsions," *Langmuir*, Vol. 22, No. 10, 2006, pp. 4457-61.

[43] Lorenceau, E., Utada, A. S., Link, D. R., Cristobal, G., Joanicot, M. and Weitz, D. A., "Generation of polymerosomes from double-emulsions," *Langmuir*, Vol. 21, No. 20, 2005, pp. 9183-6.

[44] Zhang, H., Tumarkin, E., Peerani, R., Nie, Z., Sullan, R. M. A., Walker, G. C. and Kumacheva, E., "Microfluidic production of biopolymer microcapsules with controlled morphology," *Journal of the American Chemical Society*, Vol. 128, No. 37, 2006, pp. 12205-10.

[45] Abraham, S., Jeong, E. H., Arakawa, T., Shoji, S., Kim, K. C., Kim, I. and Go, J. S., "Microfluidics assisted synthesis of well-defined spherical polymeric microcapsules and their utilization as potential encapsulants," *Lab Chip*, Vol. 6, No. 6, 2006, pp. 752-6.

[46] Huang, K. S., Lai, T. H. and Lin, Y. C., "Manipulating the generation of Ca-alginate microspheres using microfluidic channels as a carrier of gold nanoparticles," *Lab Chip*, Vol. 6, No. 7, 2006, pp. 954-7.

[47] Kim, J. W., Utada, A. S., Fernandez-Nieves, A., Hu, Z. B. and Weitz, D. A., "Fabrication of monodisperse gel shells and functional microgels in microfluidic devices," *Angewandte Chemie, International Edition*, Vol. 46, No. 11, 2007, pp. 1819-22.

[48] Choi, C.-H., Jung, J.-H., Rhee, Y., Kim, D.-P., Shim, S.-E. and Lee, C.-S., "Generation of monodisperse alginate microbeads and in situ encapsulation of cell in microfluidic device," *Biomedical Microdevices*, Vol. 9, 2007, pp. 855-62.

[49] Oh, H. J., Kim, S. H., Baek, J. Y., Seong, G. H. and Lee, S. H., "Hydrodynamic micro-encapsulation of aqueous fluids and cells via 'on the fly' photopolymerization," *Journal of Micromechanics and Microengineering*, Vol. 16, No. 2, 2006, pp. 285-91.

[50] Tan, Y. C., Hettiarachchi, K., Siu, M., Pan, Y. P. and Lee, A. P., "Controlled microfluidic encapsulation of cells, proteins, and microbeads in lipid vesicles," *Journal of the American Chemical Society*, Vol. 128, No. 17, 2006, pp. 5656-8.

[51] Edd, J. F., Carlo, D. D., Humphry, K. J., Koster, S., Irimia, D., Weitz, D. A. and Toner, M., "Controlled encapsulation of single-cells into monodisperse picolitre drops," *Lab Chip*, 2008, pp. 1262-4.

[52] Mazumder, M. A. J., Shen, F., Burke, N. A. D., Potter, M. A., Sto, x and ver, H. D. H., "Self-Cross-Linking Polyelectrolyte Complexes for Therapeutic Cell Encapsulation," *Biomacromolecules,* 2008, pp. 2292-300.

[53] Tan, W.-H. and Takeuchi, S., "Monodisperse Alginate Hydrogel Microbeads for Cell Encapsulation," *Advanced Materials,* Vol. 19, No. 18, 2007, pp. 2696-701.

[54] Lewis, P. C., Graham, R. R., Nie, Z. H., Xu, S. Q., Seo, M. and Kumacheva, E., "Continuous synthesis of copolymer particles in microfluidic reactors," *Macromolecules,* Vol. 38, No. 10, 2005, pp. 4536-8.

[55] Yang, C. H., Huang, K. S. and Chang, J. Y., "Manufacturing monodisperse chitosan microparticles containing ampicillin using a microchannel chip," *Biomedical Microdevices,* Vol. 9, No. 2, 2007, pp. 253-9.

[56] Hettiarachchi, K., Talu, E., Longo, M. L., Dayton, P. A. and Lee, A. P., "On-chip generation of microbubbles as a practical technology for manufacturing contrast agents for ultrasonic imaging," *Lab Chip,* Vol. 7, No. 4, 2007, pp. 463-8.

[57] Lee, J. N., Park, C. and Whitesides, G. M., "Solvent Compatibility of Poly(dimethylsiloxane)-Based Microfluidic Devices," *Anal. Chem.,* 2003, pp. 6544-54.

[58] Hung, L.-H., Lin, R. and Lee, A. P., "Rapid microfabrication of solvent-resistant biocompatible microfluidic devices," *Lab Chip,* 2008, pp. 983-7.

[59] Abate, A. R., Lee, D., Do, T., Holtze, C. and Weitz, D. A., "Glass coating for PDMS microfluidic channels by sol-gel methods," *Lab Chip,* 2008, pp. 516-8.

[60] Rolland, J. P., VanDam, R. M., Schorzman, D. A., Quake, S. R. and DeSimone, J. M., "Solvent-Resistant Photocurable 'Liquid Teflon' for Microfluidic Device Fabrication," *Am. Chem. Soc.,* 2004, pp. 2322-3.

[61] Yoon, T.-H., Park, S.-H., Min, K.-I., Zhang, X., Haswell, S. J. and Kim, D.-P., "Novel inorganic polymer derived microreactors for organic microchemistry applications", *Lab Chip,* 2008, pp. 1454-9.

[62] Bodas, D. and Khan-Malek, C., "Formation of more stable hydrophilic surfaces of PDMS by plasma and chemical treatments," *Microelectronic Engineering,* Vol. 83, No. 4-9, 2006, pp. 1277-9.

[63] Duffy, D. C., McDonald, J. C., Schueller, O. J. A. and Whitesides, G. M., "Rapid Prototyping of Microfluidic Systems in Poly(dimethylsiloxane)," Vol. 70, No. 23, 1998, pp. 4974-84.

[64] McDonald, J. C., Duffy, D. C., Anderson, J. R., Chiu, D. T., Wu, H. K., Schueller, O. J. A. and Whitesides, G. M., "Fabrication of microfluidic systems in poly(dimethylsiloxane)," *Electrophoresis,* Vol. 21, No. 1, Jan 2000, pp. 27-40.

[65] Martin, I. T., Dressen, B., Boggs, M., Liu, Y., Henry, C. S. and Fisher, E. R., "Plasma modification of PDMS microfluidic devices for control of electroosmotic flow," *Plasma Processes and Polymers,* Vol. 4, No. 4, May 2007, pp. 414-24.

[66] Kim, B., Peterson, E. T. K. and Papautsky, I., 'Long-term stability of plasma oxidized PDMS surfaces', In: *Engineering in Medicine and Biology Society, 2004. IEMBS '04.* 26th Annual International Conference of the IEEE, 2004, Vol. 7, pp. 5013-6.

[67] Kozlov, M., Quarmyne, M., Chen, W. and McCarthy, T. J., "Adsorption of poly(vinyl alcohol) onto hydrophobic substrates. A general approach for hydrophilizing and chemically activating surfaces," *Macromolecules,* Vol. 36, No. 16, Aug 2003, pp. 6054-9.

[68] Wu, D. P., Luo, Y., Zhou, X. M., Dai, Z. P. and Lin, B. C., "Multilayer poly(vinyl alcohol)-adsorbed coating on poly(dimethylsiloxane) microfluidic chips for biopolymer separation," *Electrophoresis,* Vol. 26, No. 1, Jan 2005, pp. 211-8.

[69] Bhushan, B., Hansford, D. and Lee, K. K., "Surface modification of silicon and polydimethylsiloxane surfaces with vapor-phase-deposited ultrathin fluorosilane films for biomedical nanodevices," *Journal of Vacuum Science & Technology A,* Vol. 24, No. 4, Jul-Aug 2006, pp. 1197-202.

[70] Cygan, Z. T., Cabral, J. T., Beers, K. L. and Amis, E. J., "Microfluidic platform for the generation of organic-phase microreactors," *Langmuir,* Vol. 21, No. 8, 2005, pp. 3629-34.

[71] Lin, R. and Lee, A. P., *unpublished.*

[72] Shi, W., Qin, J., Ye, N. and Lin, B., "Droplet-based microfluidic system for individual Caenorhabditis elegans assay," *Lab Chip,* Vol. 8, No. 9, 2008, pp. 1432-5.

[73] Hatakeyama, T., Chen, D. L. L. and Ismagilov, R. F., "Microgram-scale testing of reaction conditions in solution using nanoliter plugs in microfluidics with detection by MALDI-MS," *Journal of the American Chemical Society,* Vol. 128, No. 8, 2006, pp. 2518-9.

[74] Sugiura, S., Oda, T., Aoyagi, Y., Matsuo, R., Enomoto, T., Matsumoto, K., Nakamura, T., Satake, M., Ochiai, A., Ohkohchi, N. and Nakajima, M., "Microfabricated airflow nozzle for microencapsulation of living cells into 150 micrometer microcapsules," *Biomedical Microdevices,* Vol. 9, No. 1, 2007, pp. 91-9.

MicroFACS System

Myoung Gon Kim[1], Yang Jun Kang[1], Sang Youl Yoon[2], and Sung Yang[1, 2, 3]

[1]School of Information and Mechatronics, [2]Graduate Program of Medical System Engineering, [3]Department of Nanobio Materials and Electronics, Gwangju Institute of Science and Technology (GIST), Gwangju, Republic of Korea

Corresponding author: Sung Yang, School of Information and Mechatronics, GIST, School of Medical System Engineering, GIST, Room 406, Mechatronics Building, 1 Oryong-dong, Buk-gu, Gwangju, 500-712, Republic of Korea, phone: 82-62-970-2407, e-mail: syang@gist.ac.kr

Abstract

The main tasks of FACS system are to count and sort cells in mixed cellular populations. A miniaturized FACS system, so called microFACS (¥iFACS) system, could be easily integrated with other devices for conducting additional tasks. The overall process of microFACS system is similar with conventional FACS system, namely, sample preparation, transportation, focusing, sensing, sorting, and analysis. In this chapter, first, particle types used in development of microFACS system are briefly reviewed based on literature surveys conducted from 1999 to 2006. Second, fundamental characteristics of scattered and absorbed lights by a particle are related with size and granularity of a particle. Third, sorting mechanisms used in microFACS systems, such as hydrodynamic, pneumatic, electrokinetic, and optical methods, are compared in terms of throughput, purity, recovery rate, and switching time. And last, statistical methods to identify cells via fluorescent and scattered light characteristics (FCS, SSC) are briefly discussed.

Key terms	MicroFACS (microfluorescence-activated cell sorting)
	fluorescence
	actuation mechanism
	cell sorting
	throughput
	statistical analysis

11.1 Introduction

Fluorescence-activated cell sorting (FACS) has been known as one of the common techniques used in investigation of cell populations from mixed cells in liquid. At the beginning of the 1940s, a flow cytometry was introduced for counting and sorting blood cells suspended in liquid medium at laboratories [1]. In 1972, a conventional FACS system, which consisted of specific fluorescent markers, a laser source for fluorescent excitation, and electrostatic (± 1,000V, DC) charged plates for sorting had been released with a sorting rate of 2,000 cells/sec [2]. As shown in Figure 11.1, a conventional FACS system is composed of four main parts: cell focusing, specific fluorescence sensing, target cells sorting, and optical signal analysis. At present, a general FACS system is capable of detecting, analyzing, and sorting a great number of cells in one second [3]. For example, a commercially available FACS machine (Becton Dickinson FACSAria, United States) could analyze and sort target cells at a rate of 25,000 cells per second and 70,000 cells per second, respectively.

In preparation of cells for FACS analysis, high concentrations of cells, which have specific antigenic surface markers, are generally incubated in cell culture media containing the fluorochrome-labeled antibodies. This incubation leads antibodies, which are conjugated with fluorochromes, to be bound with antigenic surfaces of target cells. Then, fluorochrome-labeled target cells are mixed with carrier buffer fluid and introduced into a FACS machine. Once the target cells are confined within the narrow stream produced by both sheath flows, the target cells pass across a focal region of a laser beam. The target cells are then excited by a laser beam and emit a specific wavelength of fluorescent light. The emitted fluorescent light is detected by optical sensors such as photodiode and a photonic multiplier tube (PMT). An electronic system coupled to the

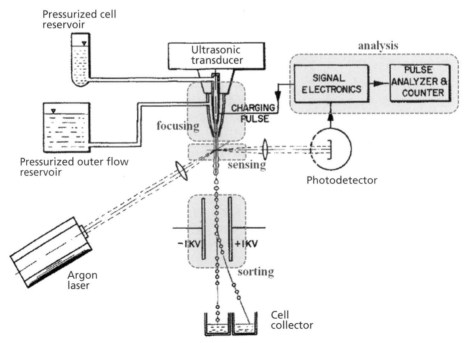

Figure 11.1 Schematic drawing of conventional FACS system (fluorescence-activated cell sorting system). (From [2].

optical signal detector transfers light signals into electric voltages, with which target cells are identified. After finishing the identification of the target cells, an ultrasonic transducer breaks the narrow sample stream into small droplets containing only one target cell per droplet. Each droplet identified is negatively charged by applying DC electric pulse. The degree of charge density on each droplet varies depending on the identification [2]. The identified target cell contained in a droplet passes through a transverse electric field and is deflected by an electrostatic force. Since the degree of deflection is influenced by the degree of charge density that each droplet holds, the target cells could be directed into a designated cell collector [2, 4, 5].

Although the commercially available FACS system is capable of counting and sorting of target cells with outstanding performance, variation in illumination and detection might occur frequently due to clogging of the orifice, particle adsorption, and contamination in the tubing [6]. In addition, relatively large amount of sample volume (~minimum 0.5 ml) is necessary to conduct appropriate analysis. Moreover, it is hard to integrate with other functional devices due to the complexity of the system.

Recently, as new technologies have been introduced in diverse applications including biomedical sample preparation and sample analysis. FACS systems have also been miniaturized and integrated based on microelectromechanical system (MEMS). Since Fu et al. [6] reported a microFACS system, which has electroosmosis flow channels for cell introduction, an optical detector for cell recognition, and a novel algorithm for cell identification, various methods have been utilized to realize microFACS systems. In the early stage, electric or pneumatics were predominant methods in microFACS systems. As new technologies have been introduced, optical and hydrodynamic methods were applied for microFACS systems. The following chapter is composed of five parts: materials, methods, results, discussion of pitfall, and statistical analysis.

11.2 Materials

Between 1999 and 2006, many representative target cells have been used in demonstration of microFACS systems. Since fluorescent particles provide relatively easy evaluation of the system due to high fluorescent intensity of fluorescent particles, the most frequent target cell used was fluorescent particles (~40%) [7–13] in the beginning of the study. As technologies mature, live cells, Escherichia coli (*E. coli.*) [6, 11, 12, 14], blood cells [9, 15, 16], cancer cells [8, 10], mammalian cells [17], and yeast cells [18], were used.

In demonstration of microFACS systems using both particles and live cells, the process called "fluorescence" is commonly utilized. Fluorescence depends on the quantum properties of fluorochromes inside particles or tagged on target cells. For instance, once the incident beam excites fluorochromes by photon energy, the electrons in the ground energy level move to the higher energy level. The excited electrons at the higher energy level quickly return downward to slightly the lower energy level by releasing energy as heat (internal conversion process). Then, the electronic state moves from the higher energy level to ground level, and simultaneously release the excess energy as fluorescent light. According to Planck's law, $E = hv = h\dfrac{c}{\lambda}$ where $h = 6.63 \times 10^{-34}$ J-s (Planck's constant), v is a frequency, and λ is a wavelength. Since the energy is inversely proportional to the wavelength, the wavelength of the emitted fluorescent light is longer than that of the excitation light. This energy transition is called as fluorescence.

Representative commercial fluorochromes for specific cells or particles are summarized in Table 11.1. The commonly used excitation and emission wavelengths of fluorochromes are ranged from 460 to 580 nm. Each fluorochrome has its own excitation and emission wavelengths, which produce the maximum excitation and emission efficiencies, respectively (Ex and Em in Table 11.1). Thus, appropriate optical cutoff filters should be carefully equipped in the system.

11.3 Methods

A MicroFACS system roughly consists of a microfluidic platform, an optical module, an electronic control system, an actuation module as shown in Figure 11.2. Among these components, the optical module is the most important one because it provides information for identification of the cells examined.

In optical module, when the incident laser beam strikes the target cell, the incident light is scattered into specific directions as well as absorbed by the target cells. These light scattering and absorption are used in identification of the target cells as shown in Figure 11.3. Thus, there are two methods in the identification process of the cells.

One is a fluorescent labeling method utilizing the absorbed light to the target cells for which the specific target cells are labeled with appropriate fluorochromes (fluorescent molecules) by antigen-antibody reaction. Since fluorochromes tagged onto the specific target cells absorb the incident light and emit a fluorescent light with characteristic wavelength, it is possible to discriminate the specific target cells from the other cell types. The labeling method is the most popular one used in microFACS systems because it allows relatively easy detection and simple identification algorithm [6, 8, 9, 20].

The other is a label free method, which utilizes intrinsic optical scattering properties of each cell type. In the light scattering based identification without any labeling process, the direction of the scattered light depends on the physical properties of the cells. In other words, cell size as well as membranes, nucleus, and granular materials inside the cells exert influence upon the direction of the scattered light. The scattered light is composed of two components; forward-scattered (FSC) light and side-scattered (SSC)

Table 11.1 Representative Commercial Fluorescence Dyes

Fluorescent Dye	Ex* (nm)	Em† (nm)	Cells or Particles	Company	Reference
Blue fluorescence	350	440	Fluorescent particles	Invitrogen	[6, 9, 11–13, 15, 19]
Green fluorescence	505	515			
Red fluorescence	580	605			
GFP‡	395	509	E. coli	—	[6, 11, 12, 14]
Alexa Fluor 488	497	519	Blood cells (RBCs, WBCs)	Invitrogen	[9, 15, 16]
SYTO80	531	545			
TO-PRO-3	632	663	Mammalian cells (HL-60 cell)	Invitrogen	[17, 20]
Calcein-AM	490	515			
MitoTracker Red 580	588	644	Cancer cells (lung, hepatoma)	—	[8, 10]
Bengal rose B	550	568	Yeast cells	Fluka	[18]

*EX: Maximum excitation wavelength.
†EM: Maximum emission wavelength.
‡GFP: Green fluorescent protein.

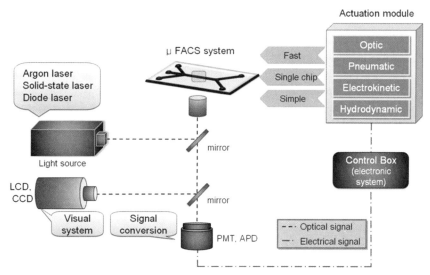

Figure 11.2 Schematic drawing of microFACS system. (Adapted from [6].)

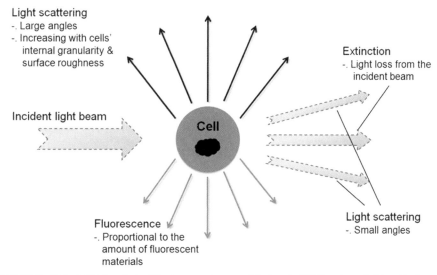

Figure 11.3 Schematic drawing of the two measurement factors in FACS system; light scattering, and fluorescence in all directions. (Adapted from [1].)

light. The FSC light is the diffracted light and changes with cell size. The FSC light is detected at the offset (less than 10 degree) from the axis of the incident laser beam in the forward direction. The SSC light is the refracted/reflected lights and occur at any interface of the target cell. The SSC light is a strong function of the granularity of the target cells. The SSC light is collected by an optical lens at the 90 degree to the incident beam and then redirected into an optical detector such as PMT or photodiode.

Another important module in microFACS system is the actuation module for cell sorting. Unlike conventional FACS systems taking advantages of charged droplet in sorting, microFACS systems utilize various methods in order to sort the target cells including electroosmotics, pneumatics, hydrodynamics, optical forces, and so forth.

Electroosmotic flow (EOF) is a liquid flow induced by an applied electrical DC potential across a microfluidic channel. In general, capillary electrophoresis adapts EOF to separate charged small particles in pH-controlled liquid solutions. Owing to relatively simple operation principle, the EOF-based sorting methods have also been tested [6, 12, 15, 17–20] in microFACS systems. Fu et al. [15] have used EOF for cell sorting and flow focusing as well as flow generation as shown in Figure 11.4. The system is composed of electroosmotic sample focusing, optical identification, and electroosmotic sorting zones. In both sample focusing and sorting zones, the degree of focusing and the flow switching time for cell sorting depend on the magnitude of electrical potentials applied. The flow direction at the sorting zone could be modulated by switching applied potentials at each electrode placed in three outlets. Other combinations of hydrodynamic focusing and electroosmotic flow sorting methods can be found in Table 11.2.

Another simple way to manipulate cells in microFACS systems is to utilize hydrodynamic focusing and switching. Figure 11.5(a) shows a schematic drawing of a hydrodynamic focusing and sorting device proposed by Jan Kruger et al.[13]. The cell focusing is realized by forming sheath flows (Figure 11.5(a)). In order to obtain fast and robust sorting function, the cross flow to the sample flow is introduced (Figure 11.5(b). The direction of target cells, however, is often disrupted by backpressure in the collection channel under the dynamic condition where a syringe pump operates at 200-ms pulse (Figure 11.5(c)). To minimize this problem, precise fluidic control should be made.

Focusing and sorting could be also realized by only using pneumatics as reported by Sung-Yi Yang et al. [10]. They have integrated micropneumatic pumps and valves onto a polymer-based chip device. Hydrodynamic focusing as well as sample introduction is achieved by three pneumatic/peristaltic micropumps as shown in Figure 11.6. Once cells are identified by their fluorescent property, sorting is performed using three pneumatic

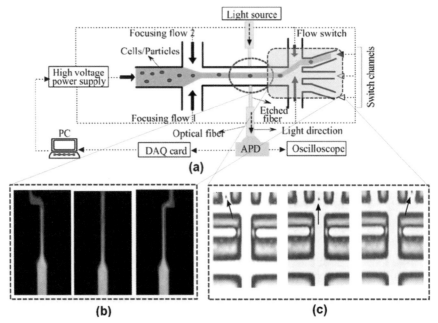

Figure 11.4 (a) Schematic illustration of experimental setup and operational principle for the microFACS system. (b) Switched medium and (c) sorted cell with respect to ratio of applied voltage. (Adapted from [15].)

Table 11.2 Operating Methods of MicroFACS Systems: Pumping, Focusing, Optical Sensing, and Sorting

Pumping	Focusing	Optical Sensing	Sorting	Reference
Hydrodynamic	Hydrodynamic	488-nm laser, PMT	Optical force (IR laser, AOM)	[7]
Electroosmotic	Non-focusing	Microscopy, video camera	Hydrodynamic and electroosmotic	[18]
Hydrodynamic	Hydrodynamic	Diode laser, PMT an APD	Hydrodynamic (rotary valve)	[13]
Hydrodynamic	Hydrodynamic	532-nm laser, PMT	Hydrodynamic (high-speed check valve)	[16]
Hydrodynamic (peristaltic pump)	Hydrodynamic	Argon-ion laser, PMT	Pneumatic (pneumatic valve)	[14]
Hydrodynamic	Hydrodynamic	532-nm laser, APD	Pneumatic (pneumatic valve)	[10]
Capillary	Nonfocusing	488-nm laser, PMT	Electroosmotic	[6]
Electroosmotic	Electroosmotic	488-nm laser, PMT	Electroosmotic	[19]
Hydrodynamic	Hydrodynamic	Confocal microscopy, 488-nm laser, APD	Electroosmotic	[12]
Electroosmotic	Electroosmotic	Optical fiber, He-Ne laser, APD	Electroosmotic	[15]
Hydrodynamic	Nonfocusing (DEP trapping)	Microscopy, CCD	DEP Trapping	[17]
Gravity	Hydrodynamic	635-nm diode laser, PMT	Electroosmotic	[20]
Hydrodynamic	Nonfocusing	Microscopy, CCD	Physical (T-shaped structure)	[8]
Hydrodynamic	Hydrodynamic	488-nm laser, CCD	Passive structure (sorting slit)	[9]
Hydrodynamic	Nonfocusing	Solid-state laser, PMT	Optical (IR lser, TGP; thermo-reversible gelation polymer)	[11]

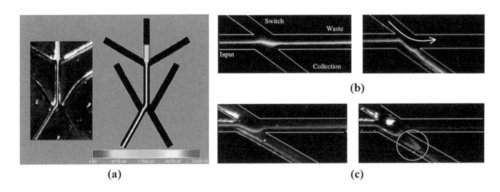

Figure 11.5 (a) Schematic illustration of a prototype device in operation and computational model-ling of hydrodynamic cell sorting device. (b) Changed path of target cells at X-junction by a clear buffer fluid. (c) A path of target cells in dynamic mode (200-ms pulse) of syringe pumps. (Adapted from [13].)

microvalves positioned at the sorting zone. When the identified target cells approach to the sorting region, each pneumatic microvalve is on/off controlled independently so that each identified target cell is directed into a designated outlet depending on identifi-cation obtained.

Other cell sorting methods described in previous discussion not only require com-plex fluidic connection but also cause significant microflow disturbance in sorting pro-cess so that require relatively long switching time. On the other hand, Mark M. Wang et al. [7] have reported a unique optical sorting mechanism that influences only on a target cell movement with minimal microflow disturbance and complexity in fluidic connec-

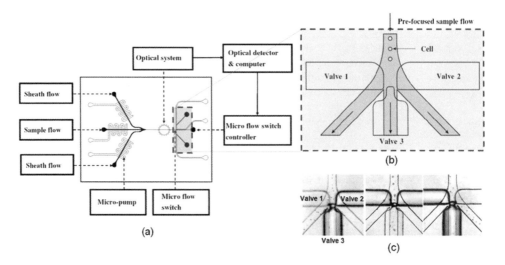

Figure 11.6 (a) Schematic illustration of purely pneumatics driven cell focusing and sorting device. (b) Zoom-in view of the sorting zone consisting of three microvalves and three outlets (c) Photographs showing three combinations of microvalve actuation resulting in the target cell sorting into a designated outlet. (From [10].)

tion (Figure 11.7(a)) [7]. Once a target cell is identified, the acousto-optic modulator (AOM) is activated so that the laser spot moves across the microfluidic channel. In other words, radiation pressure forces push or pull the identified target cells depending on optical power and the relative optical properties of the cells as well as surrounding medium (Figure 11.7(b)). Thus, the lateral movement of the laser beam then influences on the motion of the identified target cells.

Although major attention for the development of cell sorting mechanism have given to electroosmotics, hydrodynamics, pneumatics, and optics, various other methods have also been tried. Some of those methods include air/liquid two-phase flows utilizing electrowetting [21], thermo-reversible sol-gel polymer [11], and bubble actuation force generated by an electrochemical force [8]. Table 11.2 summarizes operating methods of microFACS systems.

11.4 Results

Performances of microFACS systems are generally characterized by three factors: throughput, purity, and switching time. The throughput is defined as the number of cells that can pass through identification or sorting zones under the given unit time. The purity is a ratio of the number of the correctly identified target cells to the total number of cells in the designated outlet. The switching time is defined as a minimum time required for appropriate sorting of the identified target cells each time. According to reports discussed, an optical sorting provides the highest sorting rate up to 100 cells/sec, while other methods including electroosmostics, pneumatics, hydrodynamics only allow the sorting rate up to 10 cells/sec. However, in all methods, the purity varies significantly from 70% to 99% depending on the system operation conditions. However, in

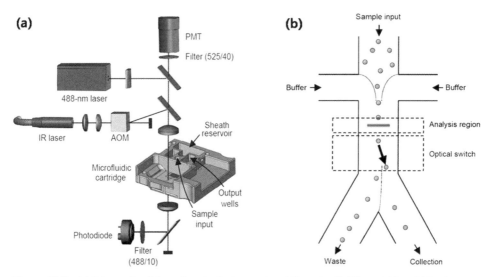

Figure 11.7   (a) Schematic of the cell sorter instrument and the microfluidic cartridge. (b) Sorting principle using a radiation pressure force. (From [7].)

general, the shorter switching time a microFACS system has, the higher throughput could be achieved. On the other hand, the higher throughput a microFACS system has, the lower the purity could be obtained. Table 11.3 summarizes the performance of the microFACS systems reported from 1999 to 2006.

11.5   Discussion of pitfalls

Comprehensive comparison between various sorting methods discussed in this chapter is summarized in Table 11.4. Although there is no ideal method for cell sorting with high throughput, sorting rate, and purity, further improvement might be possible either by efficiently combining these methods or developing a noble sorting method for microFACS systems.

Although conventional FACS systems currently achieve higher throughput than that of microFACS systems, microFACS systems are still promising since they provide

Table 11.3   Experiment Results of Proposed MicroFACS System

Pumping	Sorting	Throughput	Purity	Recovery rate	Switching time	Reference
Capillary	Electroosmotic	10 beads/s; 16.7 cells/s	84%~95%; 31%	—	—	[6]
Hydrodynamic (peristaltic pump)	Pneumatic (pneumatic valve)	1.3~44 cells/s	—	16%~50%	5 ms	[14]
Hydrodynamic	Optical force (IR laser, AOM)	23~106 cells/s	83%~99%	74%~95%	2, 4 ms	[7]
Hydrodynamic	Pneumatic (pneumatic valve)	2 cells/s	—	—	67 ms	[10]
Hydrodynamic	Optical (IR laser, TGP)	0.24~5 cells/s	66%~97%	64%~86%	3 ms	[11]
Hydrodynamic	Electroosmotic	0.38~0.79 beads/s	85%~95%	—	—	[12]
Gravity	Electroosmotic	1.2~1.7 cells/s	55%~80%	—	—	[20]

Table 11.4 Comprehensive Comparisons Between Various Sorting Methods

| | Cell Sorting Method | | | |
	Electroosmotic	Hydrodynamic	Pneumatic	Optical
The complexity of the total system	Simple	Simple	Simple	Complex
The complexity of the sorting component	Moderate	Complex (many fluidic connection)	Complex (multilayer fabrication, many pneumatic connections)	Simple
Cell damage during cell sorting process	Yes (high electric voltage)	No	No	Minimal
Operation principle	Simple	Simple	Simple	Simple
Sorting rate	Slow	Slow	Relatively fast	Relatively fast
Possibility of the parallelization	Possible	Very difficult	Difficult	Very difficult

the possibility of integration with other functional devices. In other words, other chemical or enzymatic reactions, such as cell lysis, DNA purification, or polymerase chain reaction (PCR), can also be carried out downstream immediately after cell sorting. The complete microFACS system, however, still asks approximately $15,000 to build up due to high costs of external devices; optics, detectors, illumination equipment, and so forth, although most researchers insist that microFACS system has considerable advantages in cost saving due to the simple fabrication process and inexpensive materials. Thus, considerable cost savings should be realized by integrating external devices on the chip.

11.6 Statistical analysis

The goals of the statistical analysis are to extract meaningful information from its raw data. The obtainable information includes number of cells, concentrations, cell size, DNA contents, and identification of the cell type in mixed populations. The basic procedure in statistical analysis of FACS data consists of displaying data acquired from electric system, selecting data region of interest, and analyzing the statistical parameters such as mean, standard deviation, correlation coefficient, and regression analysis. In the displaying process of the statistical analysis, once data is saved through data acquisition process, cell populations are displayed in various plot types such as histogram, dot plot, and contour plot as shown in Figure 11.8. Among them, histogram and dot plot are often used in displaying and selecting area of interest. Single parameter histogram is a direct graphical representation of the number of events occurring for the counts against intensity.

A single parameter such as forward-scatter light (FSC), side-scatter light (SSC), and fluorescent intensities, is often displayed in a single parameter histogram where the x and y axes represent single parameter values and the number of the events, respectively. By plotting FSC or SSC histograms, the statistical distributions of the cell size and granularity could be estimated (Figure 11.8(a, b)). Dot plot is another representation of the two independent parameters and provides a correlation of two parameters. Thus, by dot plotting FSC versus SSC, the correlation between cell size and granularity could be obtained (Figure 11.8(c)). For instance, the subpopulations of leukocytes, namely, monocyte (M), lymphocyte (L), and granulocyte (G), could be discriminated from a heterogeneous cell population by plotting FSC versus SSC on the x-y plane (Figure 11.9). In

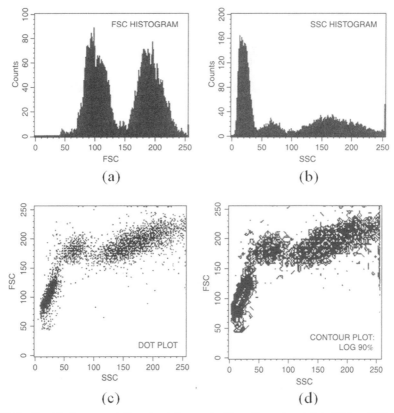

Figure 11.8 Graphical representations to plot one set of two-dimensional data (FSC vs. SSC) where two separate histograms (a, b), a dot plot (c), and contour plots (d) according to two different plotting algorithms. (From [23].)

order to conduct a thorough study on the correlation between FSC and SSC, relatively higher population regions are often investigated. In order to investigate specific region of interest, a "gate process" is utilized. A gate is a boundary drawn around a subpopulation to isolate events for analysis. Representative gate [1] shapes used in selection of the region of interest include rectangular, ellipsoidal, and polygonal (Figure 11.10(a)). Data for events within a gate can then be displayed in subsequent plots (Figure 11.10(b, c)). As a final procedure, statistical dependent parameters such as mean, standard deviation, correlation coefficient, and regression are obtained using well known statistical formulae. Commercially available software packages, MiniTAB, SPSS, and SAS, may be used in statistical analyses.

11.7 Application notes

Currently, microFACS systems are used in separation and sort of various target cells such as *E. coli*, blood cells, cancer cells, and yeast cells with a high sensitivity and low costs. In addition these, microFACS systems could be integrated with several functional modules to construct a total analysis platform. For instance, a sophisticated single-cell microarray could be integrated with a microFACS system in order to study proteomic and genomic functions of the target cell.

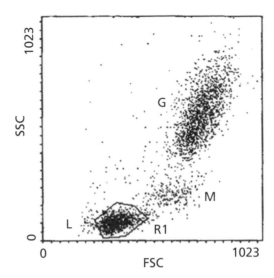

Figure 11.9 A cytogram or dot plot depicting a sample of human peripheral blood leukocytes. The intensity of side scatter light (SSC) is plotted against the intensity of forward scatter light (FSC) for each cell. L= lymphocyte cluster; M= monocyte cluster; G= granulocyte cluster, R12= region of interest encircling the lymphocyte cluster. (From [22].)

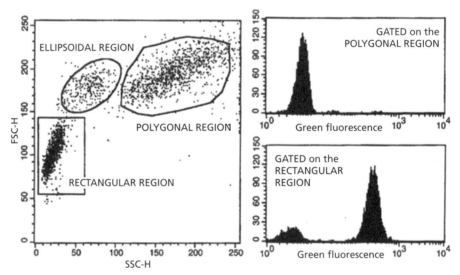

Figure 11.10 Regions can be drawn to define clusters of cells. Regions can then be used to form gates for restricting subsequent analysis to certain groups of cells. (From [23].)

11.8 Summary points

The goal of microFACS system is to conduct cell counting and sorting with high throughput, purity, and recovery of live target cells. Major working principles for cell counting and sorting include hydrodynamic, pneumatic, electrokinetic, and optical methods. Optical counting and sorting methods seem to provide the best performance in terms of throughput. MicroFACS systems demonstrated that most cell types including *E. coli*, blood cells, cancer cells, and yeast cells could be successfully sorted with sorting

Troubleshooting Table

Problem	Explanation	Potential Solution
Irregular shape of target cells.	No isotonic condition (hypertonic, hypotonic solution).	Adjust the concentration: osmolality.
Lower fluorescent intensity.	Improper fluorescent-labeling process or death of cell under the cell culture process.	Check the cell culture conditions (time, temperature, pH).
		Check the cell viability by fluorescein diacetate/ethidium bromide (FDA/EtBr) [8].
Lower optical signal sensitivity.	Converted electric signals below specific threshold condition.	Check defects of optical sensing parts : lens or fiber [22].
		Add negative feedback to maintain a constant light output to stabilize signal fluctuations of laser sources [22].
Saturation of converted optical signal in the PMT or APD.	Improper experimental conditions	Add neutral-density filters or control power outputs from the laser beam sources [22].
Bubble generation in the electroosmotic method.	Local joule heating by electric source.	Decrease the operation voltage or adjust the conductivity of suspending medium.

rate of 20~100 cell/sec depending on sorting mechanisms. Although the performance of microFACS systems falls way behind in performance compared with conventional FACS systems, it is still promising that a microFACS system could be the most efficient one for the development of an integrated micro total analysis system, which would allow us to study various proteomic or genomic functions of the target cells on a chip.

Acknowledgments

This work was partially supported by the Korea Research Foundation (KRF-D00583), the Ministry of Education (World Class University Program, project number R31-20008-000-10026-0), and the Institute of Medical System Engineering (iMSE), GIST, Republic of Korea.

References

[1]   Shapiro, H. M., *Practical Flow Cytometry,* New York: Wiley-Liss, 1995.

[2]   Bonner, W. A., H. R. Hulett, and R. G. Sweet, et al., "Fluorescence Activated Cell Sorting," *The. Rev. Sci. Ins.,* Vol. 43, No. 3, 1972, pp. 404–409.

[3]   Herzenberg, L. A., D. Parks, and B. Sahaf, et al., "The History and Future of the Fluorescence Activated Cell Sorter and Flow Cytometry: A View from Stanford," *Clin. Chem.,* Vol. 48, No. 10, 2002, pp. 1819–1827.

[4]   Fulwyer, M. J., "Electronic Separation of Biological Cells by Volume," *Science,* Vol. 150, 1965, pp. 910–911.

[5]   Richard, G. S., "High Frequency Recording with Electrostatically Deflected Ink Jets," *Rev. Sci. Instrum.,* Vol. 36, No. 2, 1965, pp. 131–136.

[6]   Fu, A. Y., C. Spence, and A. Scherer, et al., "A Microfabricated Fluorescence-Activated Cell Sorter," *Nature Biotechnology,* Vol. 17, 1999, pp. 1109–1111.

[7]   Wang, M. M., E. Tu, and D. E., Raymond et al., "Microfluidic Sorting of Mammalian Cells by Optical Force Switching," *Nature Biotech.,* Vol. 23, No. 1, 2005, pp. 83–87.

[8]   Ho, C.-T., R.-Z. Lin, and H.-Y. Chang et al., "Micromachined Electrochemical T-Switches for Cell Sorting Applications," *Lab. Chip,* Vol. 5, 2005, pp. 1248–1258.

[9]   Lancaster, C., M. Kokoris, and M. Nabavi, et al., "Rare Cancer Cell Analyzer for Whole Blood Applications: Microcytometer Cell Counting and Sorting Subcircuits," *Method*, Vol. 37, 2005, pp. 120–127.

[10]   Yang, S.-Y., S.-K. Hsiung, and Y.-C. Hung, et al., "A Cell Counting/Sorting Systems Incorporated with a Microfabricated Flow Cytometer Chip," *Meas. Sci. Technol.*, Vol. 17, 2006, pp. 2001–2009.

[11]   Shirasaki, Y., J. Tanaka, and H. Makazu, et al., "On-Chip Cell Sorting Systems Using Laser-Induced Heating of a Thermoreversible Gelation Polymer to Control Flow," *Anal. Chem.*, Vol. 78, 2006, pp. 695–701.

[12]   Dittrich, P. S., and P. Schwille, "An Integrated Microfluidic Systems for Reaction, High-Sensitivity Detection, and Sorting of Fluorescent Cells and Particles," *Anal. Chem.*, Vol. 75, 2003, pp. 5767–5774.

[13]   Kruger, J., K. Singh and A. O'Neil, et al., "Development of a Microfluidic Device for Fluorescence Activated Cell Sorting," *J. Micrometh. Microeng.*, Vol. 12, 2002, pp. 486–494.

[14]   Fu, A. Y., H.-P. Chou, and C. Spence, et al., "An Integrated Microfabricated Cell Sorter," *Anal. Chem.*, Vol. 74, 2002, pp. 2451–2457.

[15]   Fu, L.-M., R.-J. Yang, and C.-H. Lin, et al., "Electrokinetically Driven Micro Flow Cytometers with Integrated Fiber Optics for On-Line Cell/Particle Detection," *Ana. Chim. Acta.*, Vol. 507, 2004, pp. 163–169.

[16]   Bang, H., C. Chung, and J. K. Kim, et al., "Microfabricated Fluorescence-Activated Cell Sorter Through Hydrodynamic Flow Manipulation," *Microsyst. Technol.*, Vol. 12, 2006, pp. 746–753.

[17]   Voldam, J., M. L. Gray, and M. Toner, et al., "A Microfabrication-Based Dynamic Array Cytometer," *Anal. Chem.*, Vol. 74, 2002, pp. 3984–3990.

[18]   Johann, R., and P. Renaud, "A Simple Mechanism for Reliable Particle Sorting in a Microdevice with Combined Electroosmotic and Pressure-Driven Flow," *Electrophoresis*, Vol. 25, 2004, pp. 3720–3729.

[19]   Kunst, B. H., A. Schots and A. J. W. G. Visser, "Design of Confocal Microfluidic Particle Sorter Using Fluorescent Photon Burst Detection," *Rev. Sci. Instrum.*, Vol. 75, No. 9, 2004, pp. 852–857.

[20]   Yao, B., G.-a. Luo, and X. Feng, et al., "A Microfluidic Device Based on Gravity and Electric Force Driving for Flow Cytometry and Fluorescence Activated Cell Sorting," *Lab. Chip*, Vol. 4, 2004, pp. 1248–1258.

[21]   Huh, D., A. H. Tkaxayk, and J. H. Bahng, et al., "Reversible Switching of High-Speed Air-Liquid Two-Phase Flows Using Electrowetting-Assisted Flow-Pattern Change," *J. Am. Chem. Soc.*, Vol. 125, No. 48, 2003, pp. 14678–14679.

[22]   Ormerod, M. G., *Flow Cytometry*, 3rd edition, UK: Oxford, 2000.

[23]   Alice, L. G., *Flow Cytometry: First Principles*, 2nd edition, New York: Wiley-Liss, 2001.

Optical Flow Characterization—Microparticle Image Velocimetry (μPIV)

Han-Sheng Chuang, Aloke Kumar, and Steven T. Wereley*

Birck Nanotechnology Center and School of Mechanical Engineering, Purdue University, West Lafayette, IN 47907,*e-mail: wereley@purdue.edu

Abstract

Microfluidics occupies an important position in biological and biochemical analyses due to the rapid development of lab-on-a-chip (LoC) systems. Characterizing fluidic behavior at the microscale often becomes essential. Among other flow visualization tools, micro particle image velocimetry (μPIV) was proposed in recent years and has been widely adopted for many microfluidic applications. μPIV features noninvasive and quantitative two-dimensional (2D) measurements of microfluidic velocity fields. Also, it significantly improves the spatial resolution as compared to its macroscopic counterpart, thus bridging the gap between the macroscopic and the microscopic domains. This chapter aims to introduce the fundamentals of μPIV from a beginner's point of view and then guides readers from theory to practice step by step. Representative applications and handy troubleshooting tables are provided at the end of this chapter.

Key terms	microfluidics
	PIV
	particle
	velocimetry
	microscope
	optical diagnostics
	microchip

12.1 Introduction

μPIV has become a powerful and prevailing tool for a wide variety of microfluidic diagnostics [1–4] since it was first proposed by Santiago et al. [5]. This invention provides researchers an intuitive perspective of flow visualization as well as the ability to quantitatively assess fluid behavior in microscopic domains. Another popular technique based on optical diagnostics is laser Doppler velocimetry (LDV) [66] which is inherently a point-wise measurement technique. The strengths of LDV are high accuracy and an excellent capability of handling high Reynolds number flows. Integrating LDV with a microscope, a novel laser Doppler microscope (LDM) has been developed for microfluidic measurements [6, 7]. However, time-dependent measurement over a large flow field and limited spatial resolution constrain its extent of application. Hot-wire anemometry is another mature technique that has been developed for fluid dynamic measurements over the past few decades [8, 9]. By miniaturizing the hot-wire with a polysilicon strip, Tai et al. [10] further achieved a compact probe for microflows. A careful arrangement of multiple wires enables this technique to perform multidirectional velocity measurements. However, the invasive nature of the probe and poor spatial resolution inevitably lead to relatively large errors in measurements.

In the biological domain, demand for an appropriate measurement tool is always focused on high resolution, noninvasiveness, and dynamic capabilities as the size of target samples can easily range from submicron to hundreds of microns. However, conventional imaging approaches do not provide efficient quantitative analyses. In contrast, μPIV provides a two-dimensional (2-D) quantitative measurement, making it a valuable device for such biological investigations. It is noteworthy that μPIV features several unique differences from its conventional macroscopic counterpart (i.e., PIV), including (1) use of fluorescent particles, (2) micron resolution, and (3) volume illumination. Fluorescent imaging, although not a necessity, is typically used to overcome the poor signal and serious diffraction effects due to small particle size (200 nm–1 μm). Note that Brownian motion becomes quite significant for particles sizes below 1 μm. With the use of a microscope, the resolution can be drastically improved to less than micron. The light sheet illumination common in macroscopic PIV is replaced with volume illumination. Consequently, the theoretical contribution of an out-of-plane particle is estimated by the depth of correlation, which combines the effects of diffraction, geometric optics, and finite size of the particle.

A rapidly increasing literature has proven μPIV as an influential tool in various biological applications. Bitsch et al. [11] observed a so-called plug flow and found 3-μm wide cell-free boundary layer in the vicinity of channel walls. The measured capillary was 32.5 μm in height and 360 μm in width. Moreover, the capability of flow profile measurements has also been utilized to study the shear stress distribution over endothelial cells. Rossi et al. [12] hence discovered a connection between the cardiovascular conditions and cardiogenesis. His method can also be extended to atherosclerosis research. With a deliberate setup, μPIV can even engage with in vivo measurements. Vennemann et al. [13, 14] explored a live heart in a chicken embryo with a μPIV system. Comparing the fluorescent visualization of gene expression with a quantitative measurement of the instantaneous flow field, a relationship between the abnormal placental blood flow and cardiovascular malformations was established.

In the following text, we will first introduce the fundamental principles of a typical μPIV setup including hardware requirements and data processing. In Sections 12.5 and 12.6, a troubleshooting table and a summary of the μPIV technique are provided, respectively. The table serves as a practical guideline for some of the commonly encountered problems and their corresponding solutions. Finally, we provide a summary of the salient features of this methodology. Two bio-related applications based on μPIV will be reviewed for enhancing readers' understanding. It should be emphasized that the chosen applications are merely few examples from the vast literature that is associated with this topic.

12.2 Materials and methods

12.2.1 Experimental setup

Figure 12.1 depicts a typical μPIV setup consisting of a microscope, image acquisition facilities (e.g., CCD camera, computer, synchronizer) and microfluidic devices (e.g., microchannel, syringe pump). In principle, image acquisition in μPIV is not much different from taking pictures with a digital camera in daily life. Two laser pulses separated by a time interval, Δt, are used to illuminate the flow with tracer particles. The selection of fluorescent particles is contingent upon the illuminating light source. For example, if a green (λ=532 nm) laser is utilized, red fluorescent particles (λ=542/612 nm) should be selected. Eventually, the magnified images are digitized and recorded with a CCD camera. During the acquisition process, both the laser pulses and the corresponding images in the camera should be well synchronized, so that each image is recorded accurately without overlapping. By applying an appropriate computational algorithm upon the captured images, a final 2-D velocity vector map can be obtained. In the following subsections, the essential hardware of a μPIV system is classified into three categories.

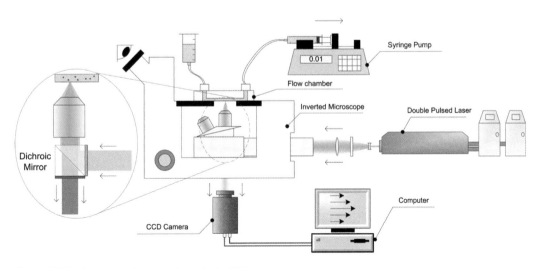

Figure 12.1 Schematic diagram of a typical μPIV setup.

Microfluidic apparatus

This portion is mainly used for transporting target samples or media in microchannels. A microchannel can be fabricated with a variety of techniques. One such technique, called "softlithography" [14] using poly(dimethylsiloxane) (PDMS), is a popular fabrication that has been widely adopted in many research laboratories. The natures of simplicity and flexibility of this technique make it very suitable for rapid prototyping prior to a mature product. Moreover, the optical transparency of PDMS makes it an ideal material for most optical diagnostics.

A syringe pump is often used to drive fluid flows in microchannels due to its programmability and broad output flow rates. However, for some flow rates below a certain threshold, a pulsatile flow is likely to occur due to the step motor. This limitation is sometimes adverse to some biological applications as their operational flow rates could be as slow as a few micro liters per minute [15]. In such circumstances some other alternative, such as gravity or capillary electrophoresis (CE), as the driver force should be considered.

Seeding working fluid with tracer particles is an essential means in PIV technique to visualize the fluid behavior. For particle selection, the following points can serve as guidelines for first-time users: (1) high image contrast (e.g., fluorescence), and (2) faithful flow tracking capability. Fluorescence is typically utilized to reduce the background noise and partially occlude serious diffraction effects. By incorporating an appropriate filter set, image contrast can be drastically improved. A dimensionless index, Stokes number (St), can be used as a reference to determine the feasibility of a particular tracer particle. Stokes number (St) is defined as

$$St = \frac{\tau_p}{\tau_f} \tag{12.1}$$

where $\tau_p = \rho_p d_p^2 / 18\mu_f$ is the particle response time, ρ_p is the density of the particle, d_p is the particle diameter, $\tau_f = L / u_f$ is the fluid response time, L is a characteristic length of the channel, and u_f is the fluid velocity. A simple rule of thumb is if St $\ll$ 1 then the particle will be able to follow the fluid streamlines. If we consider a typical polystyrene particle of 1-μm diameter running in a microchannel with a moderate fluid flow (several tens of mm/s) then the particle response time is around several microseconds and the fluid response time is around milliseconds, yielding a relatively small Stokes number ($<10^{-3}$).

Optical assembly

The purpose of this assembly is to provide sufficient illumination for capturing clear and instantaneous images. In general, the task may include transmitting laser beams to excite fluorescent particles at short wavelength and then an appropriate image acquisition device, such as a CCD camera, collects the emitted long wavelength. In order to capture instantaneous velocity fields, illumination usually comes from a double-pulsed laser system, which should be able to generate a wavelength that falls within the exciting bandwidth of the fluorescent particles (e.g., λ_{laser}=532 nm & λ_{exit}=542±20 nm). The time interval (Δt) between two successive pulses determines the maximum flow rate. However, for some biological measurements with slow flow rates, an alternative using a continuous wave (CW) laser modulated with a mechanical chopper or an internal circuit can also be explored. Before entering a microcope, the laser beam is usually expanded or reshaped using a multilens assembly. Additionally, a small-angle diffuser can be added to reduce the coherent nature of the laser, thus preventing unwanted interference patterns. The adjusted laser beams are then directed into a microscope system for volume illumination.

Volume illumination in μPIV is an important departure from the conventional light sheet illumination in a macroscopic PIV setup. The central idea of this unique illumination will be elaborated in Section 12.2.2. Here a commercial inverted epi-fluorescent microscope is employed to avoid misalignment. It plays a role in image magnification, transferring, and filtering off different wavelengths with a well-defined filter cube (e.g., exciter filter: 542 nm, barrier filter: 590 nm, dichroic mirror: 580 nm). The filter cube serves as a part of the optical unit and, as mentioned previously, helps boost the signal-to-noise ratio (SNR) and the contrast in the captured images.

Data acquisition system

Data acquisition system is utilized to capture particle images and convert them from analog to digital format for later computational processing, such as spatial cross-correlation. The overall system may include a CCD camera, a computer, and a synchronizer. Coordination between laser pulses and images captured by a CCD camera is of critical importance. The task is to trigger the laser pulses and camera shutter simultaneously, such that the double image frames can receive signals from different pulses. S1, S2, and S3 are control signals generated from a synchronizer. Laser pulses 1 and 2 are triggered by S1 and S2, respectively. To collect the emitted light from samples, the shutter should be opened long enough to contain the entire duration of the laser pulses 1 and 2 (~10 ns). S3 is another triggering signal defining the duration of first exposure of an image pair. A double-exposure camera is used to capture images corresponding to the two pulses. Similar to most digital cameras, double-exposure camera opens its shutter only once by the external triggering signal and will keep exposing till the end of its repetition rate (fps). Due to such a unique feature, a built-in "dead time" (200 ns) will arise after S3 to distinguish the second exposure from the first one. Consequently, μPIV measurements with a double-exposure camera are always undertaken in a darkroom in order to suppress the background noise in the second image frame.

Each complete exposure captures an image pair (two instantaneous images separated by Δt) for a μPIV analysis. Theoretically, one image pair is enough for a complete PIV computation based on spatial cross-correlation. However, more image pairs are always recommended for SNR enhancement. Once the images are acquired, the velocity vector field is obtained through the use of suitable computational algorithm. A variety of computational algorithms could be employed depending on requirement. Some basic algorithms will be thoroughly discussed in Section 12.2.3.

For readers' convenience, Table 12.1 lists the components of a μPIV system and their possible suppliers. Note that the table is only provided as an example. The associated models and suppliers are not necessarily limited to the listed information.

12.2.2   Volume illumination

As noted, a significant difference between a macro and a μPIV system is the type of illumination used. Volume illumination refers to the fact that all particles in the viewing

Table 12.1   Various μPIV Components and Their Possible Suppliers

Self-Established μPIV System

Item	Model	Function	Supplier
PDMS	Sylgard 184	Rapid chip prototyping	Dow Corning
Syringe Pump	552226	A driving force for fluids	Harvard Apparatus
Microfluidic Accessories	N/A	Tubing, fitting, adapters, etc.	Upchurch Scientific
Fluorescent Particle	Fluorescent Red	Tracking fluid flows	Duke Scientific
Laser System	Solo PIV 120XT	Double pulsed light source	New Wave Research
Inverted Epi-Fluorescent Microscope	TE 2000 or IX71	A nice housing and well-aligned optics for micro images and fluorescence handling	Nikon or Olympus
Optical Accessories	N/A	Reshaping and adjusting the incident laser beams	Thorlabs, Edmund Optics, etc.
CCD Camera	PCO 1600	Fast image acquisition	PCO
Synchronizer	BNC 575	Coordination of motions between devices (CCD and laser)	Berkeley Nucleonics Co.

Turnkey μPIV System

	Model	Supplier
1	Micro PIV system	TSI (U.S. and worldwide)*
2	Micro PIV system	Dantec Dynamics (worldwide except U.S.)
3	FlowMaster Micro PIV	Lavision (worldwide except U.S.)

*TSI holds the exclusive license to sell μPIV in the United States.

volume are illuminated. Depth of correlation (DOC) is defined as the axial distance from the object plane in which a particle becomes sufficiently out of focus, so that the intensity of the particle no longer contributes significantly to the signal peak in the correlation function. Unlike DOC, depth of field (DOF) is mainly concerned with the quality of the obtained image. Following the analysis of Bourdon et al. [16], the expression for DOC is derived as

$$z_{corr} = 2 \left\{ \frac{\left(1 - \sqrt{\varepsilon}\right)}{\varepsilon} \left[\frac{n_0^2 d_p^2}{4NA^2} + \frac{5.95(M+1)^2 \lambda^2 n_0^4}{16M^2 NA^4} \right] \right\}^{\frac{1}{2}} \tag{12.2}$$

where λ is the collected light wavelength, n_0 is the refractive index of medium the lens is immersed in, NA is the numerical aperture of the optical system, M is the magnification, and d_p is the diameter of the tracer particle and e represents the relative contribution of a particle displaced a distance z from the object plane, compared to a particle located in the object plane. The DOC, z_{corr}, is strongly dependent on numerical aperture, NA, and particle size, d_p, and is weakly dependent upon magnification, M.

For readers' information, some estimated values of DOC are listed in Table 12.2. Evidently, the DOC for a larger particle increases rapidly as the magnification decreases. In contrast with DOF, DOC also induces more uncertainties due to a relatively thick correlation depth. A simple solution to reduce the depth is an employment of a confocal microscope [17]. For more working principles and details of the confocal μPIV, readers may refer to [17, 18].

Table 12.2   Measurement Plane Thickness for Typical Experimental Parameters, z_{corr} (μm)

Particle Size, d_p	Microscope Objective Lens Characteristics				
	M = 60 NA = 1.4	M = 40 NA = 0.75	M = 40 NA = 0.6	M = 20 NA = 0.5	M = 10 NA = 0.25
0.01 μm	2.1	2.1	2.1	2.2	2.3
0.10 μm	2.1	2.2	2.2	2.3	2.9
0.20 μm	2.2	2.4	2.6	2.8	4.3
0.30 μm	2.3	2.8	3.1	3.5	5.9
0.50 μm	2.6	3.7	4.3	5.0	9.4
0.70 μm	3.1	4.7	5.7	6.7	13
1.00 μm	3.9	6.4	7.9	9.3	18
3.00 μm	10	18	23	27	55

12.2.3   Processing algorithms

12.2.3.1   Spatial cross-correlation

To obtain velocity vectors corresponding to different interrogation regions, a statistical algorithm called spatial cross-correlation is commonly applied to the captured image pairs. Typically, higher correlation can be achieved with a smaller time intervals. Let these two images be denoted by $I_1(X)$ and $I_2(X)$, the cross-correlation function can be obtained using the convolution integral [19]:

$$R(S) = \int I_1(X) I_2 (X + s) dX \tag{12.3}$$

$R(S)$ can be decomposed into three components as

$$R(S) = R_C(s) + R_F(s) + R_D(s) \tag{12.4}$$

where $R_C(s)$ is the convolution of the intensities of the two images and is a function of s with its diameter equal to the particle diameter, $R_F(s)$ is the fluctuating noise component, and $R_D(s)$ is the displacement component of the correlation function and gives the distance traveled by the particle during time Δt. Hence $R_D(s)$ is the component of the correlation function that contains the velocity information.

In practice, each image pair is divided into many small pieces called interrogation windows (typically 32×32 pixels and 50% overlap between two adjacent interrogation windows) for computation. The size of the interrogation window influences the accuracy of PIV results. If the window size is too large, the particles will not have a homogenous displacement. However, if it is too small, the number of particles will not be sufficient to give a distinct correlation peak. Spatially averaged cross-correlation is usually employed to predict the particle displacement in each small interrogation region. In practice, a discrete correlation function derived from (12.3) is applied for computations.

$$\Phi_k(m,n) = \sum_{j=1}^{Q} \sum_{i=1}^{P} f_k(i,j) \cdot g_k(i+m, j+n) \tag{12.5}$$

where $f_k(i,j)$ and $g_k(i,j)$ are the gray value distributions of the first and the second exposures, respectively, in the kth image pair. The size of the interrogation window is expressed as $P \times Q$ pixels. Considering computation speed and simplicity, most commercial PIV programs adopt fast Fourier transform based cross-correlation (FFT-CC) calculation. Computational time can be immensely reduced from N^4 to N^2 as compared with direct cross-correlation (D-CC) [20]. The central idea is the transformation of the images from spatial domain to frequency domain using 2-D FFT and a subsequent search for the frequency shift. After inversing the 2-D FFT, the correlation peak then provides the necessary information regarding the amount and direction of a displacement in an interrogation window.

In the microfluidic regime, strong background noise usually results in low SNR in the correlation domain. To enhance the signal, an ensemble-averaged correlation function [21] as shown in Figure 12.2 is preferred when the fluid flow is laminar and steady. The function is given as

$$\Phi_{ens}(m,n) = \frac{1}{N} \sum_{k=1}^{N} \Phi_{k}(m,n) \tag{12.6}$$

where N is the number of the image pairs. Figure 12.2 illustrates that the SNR is significantly enhanced when 101 image pairs are used instead of one. A Gaussian curve fit is subsequently applied to yield the correlation peak with sub-pixel accuracy. Eventually, the velocity is derived from dividing the displacement by the time interval, Δt. Figure 12.3 shows the overall velocity vectors are significantly improved after the ensemble-averaged processing is performed.

In order to account for the various possible experimental and flow conditions, more advanced algorithms like central difference interrogation (CDI), central difference image correction (CDIC), continuous window shift (CWS), and correlation averaging are used to obtain accurate velocity vectors. CDI and CDIC will be further elucidated upon in the subsequent section. More details regarding the other advanced image processing techniques can be found in [23–27].

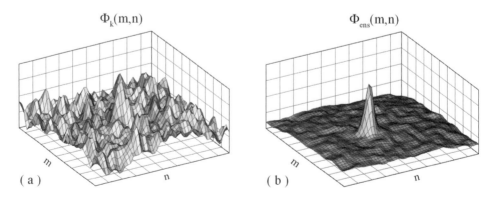

Figure 12.2 Effect of ensemble correlation: (a) results with conventional correlation for one of the PIV recording pairs, and (b) results with ensemble correlation for 101 PIV recording pairs [22]. (Reprinted with permission of the American Institute of Aeronautics and Astronautics, Inc.)

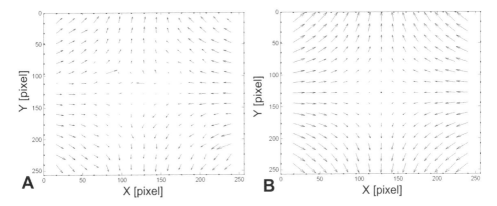

Figure 12.3 Comparison of the four-roll mill velocity vectors of (a) a single PIV recording pair and (b) with the ensemble average of 101 recording pairs. The red color indicates the spurious vectors.

12.2.3.2 Central difference image correction (CDIC)

For most PIV computational processing, a straightforward first-order scheme, forward difference interrogation (FDI), is commonly used due to its simplicity. The velocity derived from the FDI scheme is approximately written as

$$V_{FDI} = \frac{dX(t)}{dt} \approx \frac{X(t+\Delta t) - X(t)}{\Delta t} - \frac{(\Delta t)}{2}\frac{d^2 X(t)}{dt^2} \tag{12.7}$$

where Δt is the time interval between two instantaneous images. Despite the simple realization, FDI generally suffers from a vulnerability to errors when long exposure time measurements are involved. Alternatively, a second-order accurate processing using central difference interrogation (CDI) is thus proposed to improve this deficiency. CDI is an advanced numerical strategy used to minimize the large bias error in the previous FDI computation. By expanding the forward-time and the backward-time first-order derivatives of the location $X(t)$ in Taylor series, their summation constitutes a second-order accurate formula

$$V_{CDI} = \frac{dX(t)}{dt} \approx \frac{X(t+\Delta t) - X(t-\Delta t)}{2\Delta t} - \frac{(\Delta t)^2}{6}\frac{d^2 X(t)}{dt^2} \tag{12.8}$$

A special curvature flow, four-roll mill, is utilized to demonstrate the improvement [22]. The velocity field is axisymmetric about the x and y axes and are expressed as $u=Ax$ and $v=-Ay$, where A denotes the rotational constant. Therefore, the rotational velocity increases as the location is moving away from the origin. As shown in Figure 12.4, the bias error escalates in FDI as the radius is increasing. In contrast, the bias error in CDI is well suppressed and remains the same regardless of the radius. Precision error is a kind of random error, thus can be theoretically removed by averaging with numerous measurement data.

Even though CDI effectively reduces the bias error as compared with FDI, the precision error still limits the improvement of the total error. In practice most fluid flows have velocity gradient and curvature, thus distorting the particle images. With a

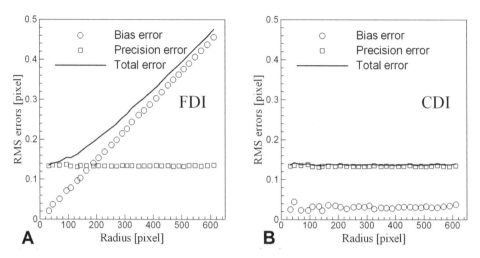

Figure 12.4 Dependences of evaluation errors on the location (radius) of the evaluation with (a) FDI and (b) CDI for the four-roll-mill test [22]. (Reprinted with permission of the American Institute of Aeronautics and Astronautics, Inc.)

deformed flow, the minimum precision error could remain at a high level despite no background noise and a perfect particle distribution. To overcome this problem, a practical image correction technique combined with CDI, named central difference image correction (CDIC), has been proposed. The principle of this algorithm is to transform the image pattern in each interrogation window based on calculations at four corners. Subsequently, a smoothly deformed image pattern in an interrogation window is obtained using bilinear interpolation. Compared with other image correction technique [28], CDIC features a rapid computation as realized by a FFT-based correlation. Details are also discussed in [29]. The precision errors derived by both FDI and CDI schemes are significantly reduced, implying less total errors.

12.2.3.3 Single-pixel evaluation (SPE)

Resolution and accuracy are always critical issues in microfluidic measurements. In some cases (e.g., boundary layer flow or atherosclerosis) the resolution may play an important role in determining the measurement derivatives, such as wall shear stress [30–32]. Currently, the best in-plane spatial resolution that a μPIV can achieve is around 1 μm. The major challenge is minimizing the size of an interrogation window because a spurious vector occurs when the spatial information is insufficient to sustain a meaningful SNR. A novel algorithm named single pixel evaluation (SPE) was proposed [33] to overcome the limitation by supplying the information along the time domain instead of the spatial domain. A two-layer shear flow is used to demonstrate the improvement as shown in Figure 12.5 (a). Obviously, SPE approximates the sharp step change better than conventional spatial cross-correlation algorithm since SPE reaches single-pixel resolution. In an extreme case where the projected particle image diameter (without diffraction) is smaller than the camera pixel size, the actual resolution can even reach nanoscale level (<100 nm).

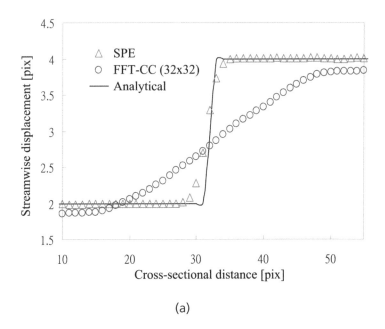

(a)

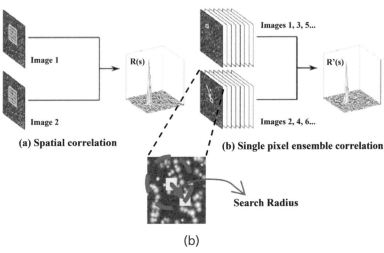

(b)

Figure 12.5   (a) A comparison of the velocity profiles derived from the CDI SPE and the FFT-CC for a two-layer shear flow. Δ denotes the CDI SPE; O denotes the FFT-CC with a 32×32 interrogation window; and the solid line is the exact solution [34]. (Reprinted with permission of the ASME.) (b) The illustration of the implementation for the central difference SPE. The red vector is the forward scheme, the blue vector is the backward scheme, and the green vector represents the resultant velocity [34]. (Reprinted with permission of the ASME.)

The principle of SPE is illustrated in Figure 12.5(b). The spatially-averaged cross-correlation function in (12.3) can be rewritten for an interrogation region comprising a single pixel averaged over many time steps as follow

$$\Phi_{spe}(m,n) = \sum_{k=1}^{N} f_k(x,y) \cdot g_k(x+m, y+n) \tag{12.9}$$

where N is the number of image pairs. As illustrated in Figure 12.5(b), the main concept is shrinking the interrogation window to its physical limitation given the pixelated format in which the images are recorded: one pixel and collecting the spatial information by searching the second pixel within a certain radius. Using the concept of ensemble correlation average [21], the SNR can be enhanced by increasing the number of image pairs. For instance, if a density of 10 particles over a 32×32 pixel window (9.77×10^{-3} particle/pixel) is used in conventional cross-correlation, at least 1,024 image pairs will be needed in the SPE computation in order to reach the same SNR in cross-correlation.

Obviously, (12.9) is based on FDI scheme, thus leading to only first-order accuracy. A second-order scheme based on CDI can improve the accuracy. For a forward-time displacement vector d_f, the correlation peak of SPE is derived from a particle search in the second image frame. Similarly, the correlation peak of the backward-time displacement vector d_b is derived from a particle search in the first image frame. After rotating the vector d_b by 180° and superposing these two correlation peaks, a new vector with second-order accuracy is obtained. The modified mathematical expression for the CDI SPE is thus rewritten as

$$\Phi^*{}_{spe}(m,n) = \sum_{k=1}^{N} \left[f_k(x,y)g_k(x+m,y+n) + g_k(x,y)f_k(x-m,y-n) \right] \tag{12.10}$$

Compared with the original SPE in (12.9), an additional term regarding the displacement vector d_b is added on the right-hand side. Besides the accuracy improvement, an additional advantage of this approach is the rapid elimination of the background noise benefited by the doubled image pairs (forward-time and backward-time displacements). In general, CDI scheme superposes both forward-time and backward-time correlations for an ensemble average, enabling a fast convergence and second-order accuracy.

12.3   Measurement procedures

Various μPIV systems are used today. Due to the diversity in cost and function, the various μPIV systems might be quite different in shape and operation. However, in order to provide readers a hands-on guideline, here we attempt to describe general operational principles based on the system configuration in Section 12.2.1. Hopefully, these may lead users to get their first μPIV system working step by step.

12.3.1   Step-by-step operations

Hardware operations

1. Turn on all equipments (microscope, CCD camera, computer, laser system, etc.), but do not activate the laser system and CCD camera.
2. Prepare particle solution ($5 \times 10^7 \sim 2 \times 10^8$ #/mL) in a syringe and set up a syringe pump.
3. Start driving fluid and wait till flow rate stabilizes.
4. Select an appropriate objective lens for a microchip (e.g., 10X objective, 300-μm microchannel, 1-μm particle).

5. Start the control software interface and set a time interval (Δt).

6. Execute the image acquisition by activating the laser system and CCD camera. Both equipments are synchronized by an internal timing device.

7. Count the maximum displacement of a single particle in an image pair. The goal is to keep the maximum displacement less than 15 pixels (i.e., for a 32×32 interrogation window). Repeat step 4 to step 5 until the requirement is met.

8. Adjust image contrast and balance two laser pulses if possible.

9. Start image acquisition after parameters are all set.

10. Stop the acquisition after desired number of image pairs is collected and save the data in a proper image format (e.g., B16, TIF, BMP).

11. Turn off unused equipment.

Basic image processing

1. Import the recorded image pairs to a μPIV software.[1]

2. Set time interval and conversion scale. The conversion scale transfers the scale from virtual image to real-world unit.

3. Open a setting window and select cross-correlation for evaluation.

4. Select ensemble average if more than one image pairs are involved.

5. Set the interrogation window size (e.g., 32×32 pixels, 64×64 pixels) and overlap percentage (e.g., typically 50%). A search radius might be required. It should not be larger than half of the length of the interrogation window.

6. A band-pass filter (also called a μPIV filter) is usually chosen to minimize noise. If possible, employ Gaussian curve fit to avoid peak locking effect.

7. Select a region of interest (ROI).

8. Run the computation.

9. Examine the raw velocity vector map and remove spurious vectors.

10. Replace blanks and removed holes in the vector map with interpolated vectors.

11. It is optional to smooth the entire vector map if it looks very discontinuous.

12. Save the processed vector data file for further analysis.

12.4 Discussion and commentary

12.4.1 Diffraction limit

The particle image diameter (d_e) is a combination of the diffraction size (d_s) and the actual geometric diameter (d_p). This optical limitation causes the particle image diameter always larger than the pixel size even when an extremely small particle is used. Olsen et al. [35] gave the following relation

---

[1] Readers may download the free processing software, EDPIV, from the following Web site: https://www.edpiv.com.

$$d_e = \left[M^2 d_p^2 + 1.49(M+1)^2 \lambda^2 \left[(n/NA)^2 - 1 \right] + \frac{M^2 D_a^2 z^2}{(s_o + z)^2} \right]^{\frac{1}{2}} \qquad (12.11)$$

where M is the magnification of the objective lens, D_a is the aperture diameter of the microscope objective, z is the distance of the particle away from the focal plane, s_0 is the object distance, and λ is the collected wavelength. Since the numerical aperture is inversely proportional to the particle image diameter (d_e), one can expect less diffraction effects with a high NA lens. However, optical diffraction limits the minimum resolution a *μ*PIV can achieve and the smallest particle diameter that a microscope can resolve. An issue regarding the ultimate spatial resolution in a *μ*PIV system will be discussed in Section 12.4.3.

12.4.2 Particle size effects

Although decreasing physical particle diameter (d_p) improves spatial resolution, growing side effects, such as Brownian motion, may inhibit the endeavor. As a result, it is important to select tracer particles that are small enough to attain the desired spatial resolution and follow the flow faithfully, yet are large enough not to be substantially affected by Brownian motion. As shown in previous literature [5], the first-order estimation of the relative error due to Brownian motion in the x-direction is written as

$$\varepsilon_B = \frac{\langle s^2 \rangle^{1/2}}{\Delta x} = \frac{1}{u} \sqrt{\frac{2D}{\Delta t}} \qquad (12.12)$$

where $D = kT/3\pi\mu d_p$ is the diffusion coefficient, k is Boltzmann's constant, μ is the dynamic viscosity of fluid, s is a random particle displacement associated with Brownian motion, Δx is the characteristic particle displacement, and u is the characteristic velocity of the fluid. In practice the error due to Brownian motion sets a lower limit on the particle diameter. For example to restrict the maximum error, $\varepsilon_B = 0.1$, the maximum displacement of 11.6 μm, and minimum displacement of 0.64 μm, the smallest particle diameter is estimated to be 50 nm.

Another important particle imaging issue is that as particle diameter decreases, the light scattered or fluoresced by the particle also decreases. For a particle larger than the wavelength of the scattering light, the intensity is proportional to the effective surface area of the particle. In contrast, for a particle much smaller than the wavelength of the illuminating light, the scattered intensity varies proportional to the sixth power of the diameter [36]. To keep a distinct particle image, a compromise between Brownian motion, light intensity, and the diffraction limit must be carefully considered.

12.4.3 Ultimate spatial resolution

Ultimate spatial resolution is discussed from two different perspectives: out-of-plane and in-plane. Out-of-plane resolution is determined in terms of the previous depth of correlation (DOC), hence the limitation is associated with particle diameter (d_p), numerical aperture (NA), magnification (M), and emitted wavelength (λ). Although the out-of-plane resolution can be enhanced by improving the hardware, optical diffraction limits improvement above a certain threshold. For example, the minimum out-of-plane resolution is estimated to be 2.1 μm for d_p=10 nm, NA=1.4, M=60X, and λ=560 nm.

The effective particle image diameter (d_e/M) places a bound on the in-plane spatial resolution in a μPIV system. For a pursuit of high spatial resolution, the particles should be imaged with high-resolution optics and with high magnification so that each particle can be resolved with at least 3 to 4 pixels in a CCD camera [37]. Table 12.3 gives a list of the effective particle image diameters recorded through a circular aperture and then projected back into the flow. Assuming that the particle images are sufficiently resolved by the CCD array, the location of the correlation peak can be sufficiently resolved to within 1/10th the particle image diameter [38]. Therefore, the uncertainty of the correlation peak location for a 0.2-mm diameter particle recorded with a NA = 1.4 lens is $d_e/10M = 35$ nm. Double of this uncertainty plus an effective particle image diameter will be the lower bound of an interrogation window. Doubling the displacement ensures Nyquist theorem is satisfied. In fact, however, the interrogation window can not be too small in order to include enough spatial information. The number of particles in an interrogation window should be at least 10 according to Keane and Adrian's study [19]. Accordingly, the minimum in-plane resolution in a typical μPIV system is about 1 μm.

Due to the limitation mentioned previously, measurements that go beyond 1 μm usually depend on interpolation assuming continuum of fluid mechanics. However, the assumption breaks when the characteristic length falls in nanoscale since the physics becomes more complex. Alternatively, SPE lifts this limitation and takes advantage of a CCD camera, thus achieving the ultimate in-plane resolution: single pixel. Here the spatial resolution is defined by a relationship of the magnified particle diameter ($\hat{d}_p$, without diffraction effect) and the pixel size (Δ_p) of a CCD camera. The magnified particle diameter is a nominal size derived from an actual particle diameter multiplied by the magnification factor without considering diffraction effect. A simple rule of thumb is expressed as follows

$$\Delta d_{\min} = \begin{cases} \hat{d}_p & if \quad \hat{d}_p > \Delta p \\ \Delta p & otherwise \end{cases} \tag{12.13}$$

For instance, a 1-μm particle with an M=20X objective lens yields a magnified diameter of 20 μm. Consequently, the magnified particle diameter is larger than the pixel size, implying the spatial resolution is determined by the particle itself. In contrast, consider-

Table 12.3 Effective Particle Image Diameters When Projected Back into the Flow, d_e/M (μm)

Particle size, d_p	Microscope Objective Lens Characteristics				
	M = 60 NA = 1.4	M = 40 NA = 0.75	M = 40 NA = 0.6	M = 20 NA = 0.5	M = 10 NA = 0.25
0.01 μm	0.29	0.62	0.93	1.24	2.91
0.10 μm	0.30	0.63	0.94	1.25	2.91
0.20 μm	0.35	0.65	0.95	1.26	2.92
0.30 μm	0.42	0.69	0.98	1.28	2.93
0.50 μm	0.58	0.79	1.06	1.34	2.95
0.70 μm	0.76	0.93	1.17	1.43	2.99
1.00 μm	1.04	1.18	1.37	1.59	3.08
3.00 μm	3.01	3.06	3.14	3.25	4.18

ing an M=100X objective lens, a CCD camera with a pixel size of 6.45 μm, and a particle diameter of 60 μm, the resolution can even reach 64.5 nm. Note that the tracer particles need to be carefully selected if a single pixel spatial resolution is to be achieved. For simplicity, we assume each particle as an infinitesimal point moving along the streamline and no Brownian motion, reactions, or elastic collisions between particles. Real flow is likely to deal with more uncertainties that may impact the image quality.

12.4.4 Velocity errors

Ideally, tracer particles in a μPIV system are expected to follow the streamlines of a fluid flow for a period of time (Δt), causing displacements in x and y directions as

$$\Delta x = u \Delta t \tag{12.14}$$

$$\Delta y = v \Delta t \tag{12.15}$$

where u and v are the x- and y- components of the time-averaged local fluid velocity, respectively. However, as the particle size diminishes, Brownian motion tends to push particles away from their deterministic pathlines. Assuming the flow field is steady over the time of measurement and the local velocity gradient is small, noise due to Brownian motion can be considered as a fluctuation about a streamline that passes through the particle's initial location. The relative errors, ε_x and ε_y, incurred as a result of imaging the random particle displacements in a two-dimensional measurement of the x- and y-components of particle velocity, are given as:

$$\varepsilon_x = \frac{\sigma_x}{\Delta_x} = \frac{1}{u}\sqrt{\frac{2D}{\Delta t}} \tag{12.16}$$

$$\varepsilon_x = \frac{\sigma_y}{\Delta_y} = \frac{1}{v}\sqrt{\frac{2D}{\Delta t}} \tag{12.17}$$

The errors estimated by (12.16) and (12.17) show that the relative error decreases as the measurement time increases. Experimentally, the interference from Brownian motion significantly rises as particle size is less than hundreds of nanometers (<500 nm) with flow velocities of less than about 1 mm/s. For a velocity of 0.5 mm/s and a 500-nm tracer particle, the lower limit for the time interval is approximately 100 μs for a 20% error due to Brownian motion. Since this error is random, it can be reduced either by averaging over several particles in a single interrogation window or by ensemble averaging over several image pairs. The error decreases as $1/\sqrt{N}$, where N is the total number of particles in the average [39].

The velocity errors in (12.16) and (12.17) show that the effect due to Brownian motion is relatively less important for a fast flow. It is common to set Δt small for such a high-velocity flow in order to acquire a highly correlated image pair. Although increasing Δt can minimize the relative error, the accuracy of the results may also decreases because conventional PIV computations are based on a first-order accurate approximation. Introducing a second-order accurate algorithm (CDIC) allows a longer Δt without increasing this error [22, 26].

12.4.5 Other flow visualization techniques based on μPIV

Conventional μPIV basically shows good performance in spatial resolution and planar flow quantification. Based on the fundamental configuration, more flow visualization techniques can be further developed to extend the capability of the current system in various regimes. The following briefly introduces some of the extensive works.

Confocal μPIV

Confocal microscope is a prevailing tool used to reduce the depth of field in a focal plane. The principle is placing a pinhole in the focal point of an image plane, thus eliminating the scattered light from out-of-plane regions. Park et al. [17] firstly applied the idea in μPIV, creating an optical slice with an estimated thickness of 2.8 μm. Comparisons between the particle images obtained by confocal laser scanning microscopy (CLSM) and conventional epi-fluorescent microscopy were exhibited in their paper. Obviously, the confocal images show distinct particle outlines due to a significant removal of the background noise. For a higher-speed acquisition, a pinhole array disk (Nipkow disk) is employed instead of a galvanometric mirror that is used in a CLSM, therefore the entire image can be captured at a time without scanning. Later Lima et al. [18] utilized the same device for characterizing a steady-state Poisueille flow in a square microchannel as well. Eventually, a three-dimensional two-component (3D2C) parabolic velocity profile was obtained.

Three-dimensional (3-D) μPIV

Biological entities usually have sophisticated geometries, resulting in a complicated flow pattern in space. However, the configuration of conventional μPIV limits its applications only in 2D2C measurements. To overcome this inherent limitation, numerous 3-D techniques based on μPIV have been developed. One such method utilizes a stereoscope. Macrostereoscopic PIV is a mature 2D3C measurement technique and has been adopted in many areas. The central concept is employing two cameras to view the same object in two different angles, thus providing 3-D information for computation. Integrating a stereoscope with a μPIV system deliberately, Lindken et al. [40] thus obtained a 3-D measurement in a T-mixer (Figure 12.6). The spatial resolution achieved was 44×44×15 μm³ and a full 3-D velocity vector map was thus generated by stacking multiple scanned planes.

By distinguishing a diffraction ring pattern of a fluorescent particle, it is also possible to extract the spatial location of a suspended particle without additions to the regular μPIV hardware. Recently, Park et al. [41] and Peterson et al. [42] came up with such an algorithm and got a proof-of-concept for this idea with a channel flow. A calibration curve regarding a relationship between the ring pattern and the out-of-plane distance is generated prior to experimental data acquisition. During each measurement, sparse particle images will be captured and a particle tracking algorithm will applied. A 3-D microvortex studied using this deconvolution microscopy is depicted in Figure 12.7.

Another promising approach in 3-D microfluidic measurement is microdigital-holographic particle-tracking velocimetry. Holography is a special photography used to capture the 3-D image of an object. Instead of recording intensity in a traditional picture, a hologram records the phase information of an object and hence is able to reconstruct

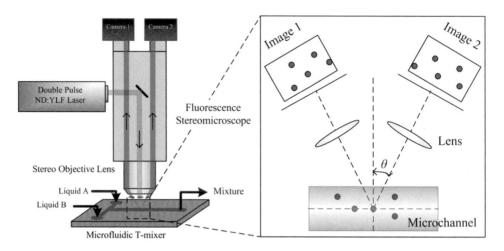

Figure 12.6 Schematic diagram of the experimental setup. The upper half of the setup shows the measurement system, the lower half shows the microfluidic chip. The close-up illustrates the principle of the stereoscopic images.

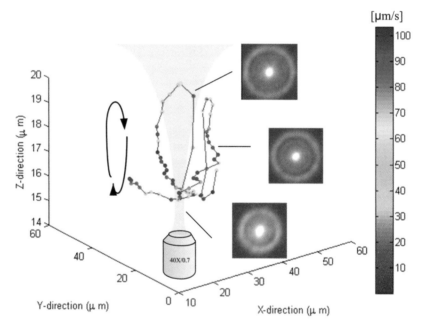

Figure 12.7 Pathline of a 3-D vortex measured by a deconvolution miscroscopy. The z location of a particle is indicated by the radius of the outermost ring. (Data courtesy of Dr. Sean Peterson [42].)

the image in three dimensions. Replacing a film with a digital CCD camera, an in situ and real-time measurement is achievable. Such a device setup was proposed by Satake et al. [43]. A flow rate of 1.1 mL/h in a rectangular micrchannel seeded with 1-*µ*m particles was measured. A fast Fourier transform coupled with a Fresnel transform function is employed to reconstruct the 3-D particle locations and can be expressed as

$$\Phi(I,J) = \int\limits_{-N_1/2}^{N_1/2} \int\limits_{-N_2/2}^{N_2/2} H(x,y)G(I,J)e^{-2\pi(Ix+Jy)}dxdy \tag{12.18}$$

where $\Phi(I,J)$, $H(x,y)$, $G(I,J)$ are reconstruction image in Fourier space, hologram image, and transform function using the Fresnel approximation, respectively. I, J, N_1, and N_2 are pixel point in x-direction, pixel point in y-direction, pixel number in x-direction, and pixel number in y-direction, respectively. Note that the last two approaches are fully 3D3C measurements in contrast with the first one.

Evanescent wave μPIV

Evanescent wave is a phenomenon that a near-field illumination occurs when a light beam undergoes total internal reflection (TIR) at a refractive index interface. The illumination from a small portion of the incident light beam penetrates through the interface and decays exponentially. The penetration depth is typically a few hundreds of nanometers away from the wall and can be expressed in a relation as follows [44, 45]

$$I = I_0 e^{-(z/z_p)}; \quad z_p = \frac{\lambda_0}{4\pi\sqrt{n_2^2 \sin^2\theta - n_1^2}} \tag{12.19}$$

where I_0 is the incident light intensity, z is the distance from the wall, λ_0 is the vacuum wavelength, θ is the incident angle, and n_1 and n_2 denote the refractive indices of the sparse and dense media, respectively. Zettner et al. [46] first integrated this phenomenon with a PIV system and successfully measured near-wall velocities within 380 nm. A schematic diagram showing the operation is illustrated in Figure 12.8. The fluid containing a suspension fluorescent particles (250 nm in radius) is sandwiched between two glass plates and driven by a rotating disk. An out-of-plane resolution of 380 nm was

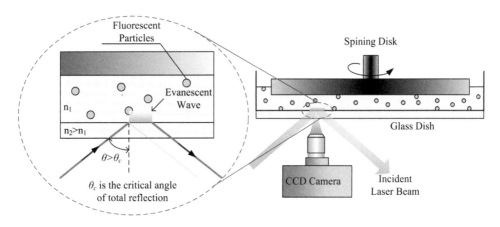

Figure 12.8   Experimental setup of the evanescent wave μPIV.

Troubleshooting Table

Problem	Significance	Possible Solutions
Nonmoving objects (chip defects, background noise, impurities, particle clumps) are present in the region of interest (ROI). Due to surface stiction, tracer particles are likely to aggregate on wall surface as measurement time increases.	Presence of nonmoving objects in the ROI decreases the SNR and increases the possibility of spurious cross-correlation peaks and hence the chances of incorrect vectors.	It is difficult to provide a generic solution. Possible solutions might be (a) use of background subtraction (b) Use of image masks (c) image correction.
Nonuniformity of illumination in ROI. This might be caused by inherent problems in the illumination system or by an improperly aligned lens.	Again this decreases the signal-to-noise ratio (SNR) and increases the possibility of spurious cross-correlation peaks and hence the chances of incorrect vectors.	If nonuniformity cannot be experimentally corrected, image correction algorithms might be applied.
Improper particle concentrations/interrogation window size.	The interrogation window affects spatial resolution and its minimum size depends in part on the particle concentration.	Increase particle concentration or use interrogation window size such that on an average 10 particles are present in a window. If higher spatial resolution is desirable, then use of interpolation might be advisable.
Large velocity gradients are present in ROI.	Velocity gradients introduce bias errors in spatial cross-correlation.	Reduce ROI size or time between images.
Significant Brownian motion is present.	Brownian motion is usually significant for sub-micron particles and the fluctuations increase with decrease in particle size.	Ensemble averaging can greatly enhance the SNR ratio in such cases. Also using a second-order accurate scheme will increase the accuracy.
Low image SNR or blurred streaks are present.	Image saturation and weak light intensity might be factors for low image SNR. Particle streaks can result due to a continuous light source or high exposure time.	Larger beads will have higher emission/scattering light intensity. To correct streaking stroboscopic illumination might be extremely helpful.
Significant out-of-plane velocity exists.	Correlation peak strength is decreased due to loss of particle pairs, increasing chances of spurious vectors.	Steps to increase the correlation depth may be necessary to minimize the loss of particle pairs. If the 3-D component is still very significant then alternate velocimetry techniques, capable of resolving the 3-D component should be considered.

achieved in this work. Due to the nature of the nanoscale resolution, this approach is also called nano-PIV by some authors [47].

12.6   Summary points

• PIV refers to a particular set of flow visualization tools used to quantitatively analyze fluid velocity fields. Typically, PIV utilizes tracer particles to follow fluid streamlines, a pulsed light source to illuminate the particles, an acquisition device to capture the particle images, and uses statistical algorithms, such as spatial cross-correlation, to compute velocities. This method when applied to microscale fluid flow phenomena by integrating with a microscope is known as μPIV.

• Several dissimilarities make μPIV differ from its macroscopic counterpart: (1) volume illumination, (2) use of fluorescent particles, and (3) integration with an epi-fluorescent microscope. The volume illumination replaces a conventional light sheet and only the particles within depth of correlation (DOC) are computed. Fluorescent particles are excited at short wavelength and emit at long wavelength, therefore effectively increasing the SNR and contrast in a microflow. A microscope provides a capability of visualizing microscale images.

- The major contributions from μPIV are an improvement of spatial resolution and quantification of microscale flow fields. However, limitations, such as poor out-of-plane resolution, cumbersome facilities, and potential contamination due to particle seeding, might need special care when dealing with some sophisticated analyses.

- Image processing is often applied before a computational algorithm is performed. Basic procedures comprise spurious vector removal, interpolation for vacancies, velocity vector smoothing, and so on. The computational algorithms are typically spatial cross-correlation with some image filters or corrections, such as CDIC.

- Typically, μPIV can be applied in any microscale phenomenon where optical access is available and the tracer particles are expected to follow particle streamlines faithfully. If necessary, μPIV possesses the flexibility of integration with other techniques, thus forming a multifunctional platform. However, sometimes exceptions may occur, for example, strong electric fields can cause movement of tracer particles independent of fluid flow. μPIV may not be appropriate in such situations.

- Some conditions should be met before μPIV is employed. (1) An optically transparent window for visualization, (2) a working fluid that allows seeded by tracer particles, (3) reasonable flow rates (i.e., less than subsonic speed), and (4) a dimension that is compatible with the microscope. For example, a focal plane cannot be reached if the channel is too deep and the working distance of an objective lens is too short.

- To compensate for the deficiencies of the current μPIV technology, researchers have extended its capability. For nanoscale resolution, there are evanescent wave mPIV and SPE PIV. For opaque objects, there are IR-μPIV and X-ray PIV. For increasing out-of-plane spatial resolution, there are confocal μPIV and evanescent wave μPIV. For 3-D meauusrements, there are stereoscopic μPIV, digital holographic μPIV, defocused μPIV, and deconvolution microscopy.

12.7   Application notes

12.7.1   Principles of diffusometry

Although Brownian motion is often seen as a source of unwanted noise, it can often be used to extract information regarding important physical quantities. Particle image diffusometry was proposed by Gorti et al. [48] to directly measure the diffusion coefficients of submicron particles for pathogen detection. The mean square displacement of a particle with diffusivity D due to diffusion over a time period Δt is given by

$$< s^2 > = 2nD\Delta t \tag{12.20}$$

where n is the number of translational degrees of freedom. Thus displacement statistics of microparticles can be utilized to compute quantities like diffusion coefficient. At a constant temperature, the diffusion coefficient is a function of the drag on the particle.

The basis for the use of μPIV in Brownian motion studies was laid by Olsen and Adrian [49], who in their work derived analytical expressions to define the shape and height of the correlation function in the presence of Brownian motion for light-sheet and volume-illumination in PIV system. From an analysis of cross-correlation algorithm, Olsen and Adrian [49] found that for light-sheet PIV, the width of the correlation peak, Δs_o, taken as the diameter of the Gaussian function measured at a height of $1/e$ times the peak value, can be expressed as

$$\Delta s_{o,a} = \sqrt{2d_e^2 / \beta^2} \tag{12.21}$$

in the absence of Brownian motion. When Brownian motion is present, the peak width $\Delta s_{o,c}$, can be expressed as

$$\Delta s_{o,c} = \sqrt{2\left(d_e^2 + 8M^2\beta^2 D\Delta t\right)/\beta^2} \tag{12.22}$$

Equation (12.22) reduces to (12.21) when Brownian motion ($D\Delta t$) is considered to be zero. The constant β is a parameter arising from approximating the Airy function as a Gaussian function and was found to be $\beta^2 = 3.67$ [50]. Figure 12.9 illustrates this broadening of the correlation peak due to Brownian motion.

For the case of μPIV type of illumination (i.e., volume illumination), (12.21) and (12.22) still hold, except that the term for particle image diameter, d_e must be replaced by

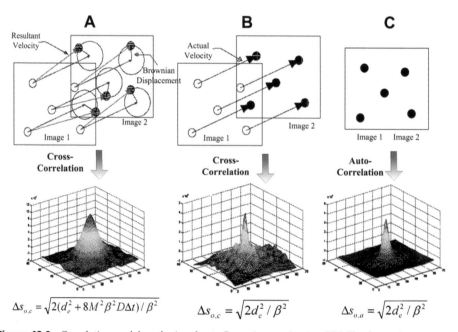

$$\Delta s_{o,c} = \sqrt{2(d_e^2 + 8M^2\beta^2 D\Delta t)/\beta^2} \qquad \Delta s_{o,c} = \sqrt{2d_e^2 / \beta^2} \qquad \Delta s_{o,a} = \sqrt{2d_e^2 / \beta^2}$$

Figure 12.9 Correlation peak broadening due to Brownian motion in μPIV. The figure shows a comparison of (a) cross-correlation function with Brownian motion, (b) cross-correlation function without any Brownian motion, and (c) autocorrelation.

the integral over the depth of the device. The cumbersome calculation of the integral term for d_e can be avoided by manipulation of (12.21) and (12.22), thus yielding the following relation:

$$\left\langle s^2 \right\rangle = \frac{\Delta s_{0,c}^2 - \Delta s_{0,a}^2}{8M^2} = \frac{2k\Delta t}{3\pi d_p} \cdot \frac{T}{\mu} \tag{12.23}$$

which can be directly used to calculate temperature. Note that $\Delta s_{0,a}$ can be determined by computing the autocorrelation of one of the PIV image pairs and $\Delta s_{0,c}$ can be determined by computing the cross-correlation of the PIV image pairs.

12.7.2 μPIV-based thermometry

Microscale temperature measurement techniques have being drawing much attention since microtechnology started booming. Some bioassays, such as PCR, especially rely on the accurate temperature control for optimal throughput. Among numerous temperature measurements, Brownian motion method is one technique capable of making nonintrusive and whole-field measurements in a fluid flow with high spatial resolution. Since the method can be considered a branch of μPIV, the same experimental setup and method can be applied to obtain simultaneous measurement of velocities and temperature without any significant added cost.

Microscopic particles suspended in a liquid exhibit random movement due to Brownian motion. The mean square of this displacement is proportional to the temperature of the liquid and inversely proportional to the liquid viscosity as well as the size of the particles. By measuring the displacements of suspended particles and holding all other parameters constant, the temperature of the liquid can be estimated. Based on this idea, Hohreiter et al. [51] measured the broadening of the μPIV correlation function due to the Brownian motion of suspended particles for temperature in a stationary fluid. An experimentally measured calibration constant was used to relate the broadening of the correlation function to temperature values.

12.7.3   Biosensing

The use of μPIV in biosensing was explored by Kumar et al. [52]. In their work, polymer beads were functionalized with antibodies against a specific analyte (M13 phage virus) and introduced into a sample containing the analyte. The binding of the analyte onto the beads led to an increase in the hydrodynamic drag with a subsequent reduction in the diffusion coefficients of the beads. A correlation was obtained between the concentration of viruses used to functionalize the beads and this formed the basis of their biological agent detection technique. Several methodologies exist for experimental measurement of diffusion coefficients of particles in microscale fluid flow [53], however, Kumar et al. [52] used multiple particle tracking and μPIV for achieving the same.

12.7.4   Wall shear stress (WSS) measurement

Wall shear stress is an important fluid mechanical property for the research of hemodynamics. The abnormal wall shear stress acting on the endothelial cells in blood vessels for instance has been proven to be a strong link to some cardiovascular diseases [54, 55]. Accurate WSS measurement is able to help model and predict the vascular lesion, such as atherogenesis, in early stage. Conventionally, the noninvasive detection of WSS in live arteries relies on the use of either ultrasound or magnetic resonance imaging (MRI). However, limited spatial resolution prevents these techniques from being used in smaller arteries and constraints their accuracy. Lately Rossi et al. [12] showed excellent experimental results as compared to their simulations using μPIV. With a modified μPIV system, the same research group [67, 68] came up with a capability of measuring WSS from the blood circulation in an embryonic chicken heart. The demonstration implied a potential development in biological applications. Based on a long-distance μPIV device, Kähler et al. [56] successfully performed a near-wall turbulence and a wall shear stress in a wind tunnel with a resolution up to a single pixel.

Obviously, an accurate WSS measurement is highly dependent on the spatial resolution since the shear rate is mainly derived from a near-wall velocity profile. In this section, WSS measurements using CDI SPE μPIV will be demonstrated, such that both the resolution and the accuracy can be improved at least one order of magnitude as compared to the conventional FFT-based spatial cross-correlation (FFT-CC). Accordingly, the measured velocity profile can further approach the exact solution due to the inherent improvement in the algorithm itself.

12.7.4.1 Analytical fit

Currently, there is still no universal approach to approximate the true velocity profile. For point measurements, such as LDV or ultrasound, the WSS is usually obtained from Poiseuille's law based on a fully developed parabolic flow profile [57, 58], which is a huge deviation as dealing with complicated geometries. Recent research works [59, 60] show a trend of introducing simulation for aiding the measurements. For this sort of technique, computer tomography (CT) scan is utilized to reconstruct a duplicated numerical model for a target object. A simulated environment is thus run in such a model in order to obtain reliable WSS distributions. However, a realistic measurement is still needed as a reference for the purpose of calibration. In order to fit a more representative curve, most techniques use a second-degree polynomial to predict the possible velocity profile.

$$u = C_1 x^2 + C_2 x + C_3 \tag{12.24}$$

where x is the distance across the channel starting from the wall and C_1, C_2, and C_3 are empirically-determined constants. Quantitatively, the shape of the second-order polynomial corresponds more closely to the shape measured data than does the shape of the analytical solution.

12.7.4.2 Experimental technique

To demonstrate the performance of the proposed CDI SPE in actual microfluid flows, the WSS distribution along a serpentine microchannel was measured. To ensure the inlet flow is a fully developed profile, a 7-mm leading straight channel is designed in front of the serpentine section. A fixed flow rate of 0.57 mL/h, 1-μm fluorescent particles, and a 10X objective lens were applied throughout the entire measurement. The particle

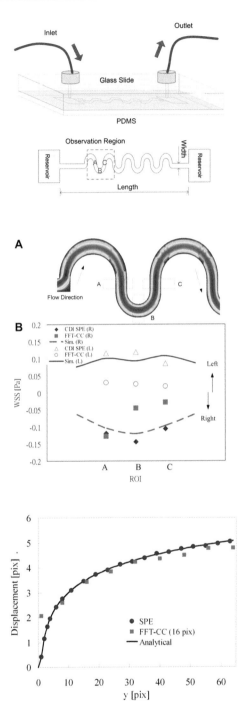

Figure 12.10 Top: The schematic diagram of the experimental microchannel. The device is basically a combination of a glass slide and a PDMS patterned with a channel. The lower graph shows the geometric structure of the serpentine channel and the five observation regions during measurements. Middle: (a) Velocity contour simulated by COMSOL Multiphysics.Regions A to C show three different regions of interest (ROI). (b) Comparisons of the WSS measurements derived from CDI SPE and FFT-CC algorithms at the three ROIs. The solid lines represent simulation results providing a comparable reference. Bottom: Flow profiles for both data obtained from SPE and FFT-CC algorithms. The solid line represents the analytical solution and the data points represent the computational results from the simulated images.

images were acquired from the midplane of the microchannel. As shown in Figure 12.10 (top), there were three regions of interest from A to C selected for WSS measurements. In each region, 1,024 image pairs were acquired for ensemble average in the CDI SPE computation while only one image pair was required in the FFT-CC for the equivalent SNR. The same images were also fed into computations with conventional FFT-CC algorithm. An interrogation window size yielding the same SNR was 32×32 pixels and the minimal resolution reached 16 pixels (10.32 μm) with 50% window overlap. During computation, a least-square polynomial curve fit was applied in order to minimize the disturbances resulted from some incorrect fluctuations.

Meanwhile, a 2-D fluid simulation was also conducted by COMSOL Multiphysics under the same conditions for later comparison. A module with Navier-Stokes equation and steady state flow were adopted in this study. An automatic mesh and a refined mesh were applied over the entire geometry. A total of 25,764 meshed triangles were constructed. A uniform velocity was set as an inflow and the outflow was specified equivalent to the ambient atmospheric pressure (1.05 bars). The resultant WSS data was obtained from the composite effect of both x and y shear stress components. Since the geometry is not a straight channel, the simulation is essential for aiding the understanding the possible flow profiles in advance. Later a trial-and-error approach may be applied to the numerical settings in order to match both the experimental data and simulations. Ideally, as a perfect match is done, the simulation can be used to diagnose the WSS variations over the entire channel without actual measurements. It should be noted that this final step is not accomplished in this measurement yet.

12.7.4.3 Discussion

Figure 12.10 (middle) shows the measured planar velocity in each region of interest. As focusing on the velocity profiles in each turn section, the maximal velocity tends to close the inner radius instead of the outer radius (regions C). Accordingly, this phenomenon induces higher shear rates on the inner side wall than the outer side wall. For the other regions B and D, the maximal velocity transits from one side to another due to the shift of the inner wall and the outer wall. Since the general slip length (~20 nm) for smooth hydrophobic surface is negligible [61] as compared to the current minimal resolution (1 μm), the boundary condition was assumed to be a no-slip flow condition. All of the shear rates were derived from the curve fitting addressed in Section 12.7.4.1. As compared with the results from FFT-CC, the results from CDI SPE show better accuracy and stability (see Figure 12.10 (bottom)). The results show an increase of WSS on the inner wall, which is consistent with the observation of velocity distribution in the three regions. Considering the simulation as a reference, the correlation coefficients, R^2, of the CDI SPE corresponding to right and left wall were 0.996 and 0.803, respectively, while the R^2 of the FFT-CC were only 0.003 for the right wall and 0.338 for the left wall.

The CDI SPE has been shown to produce more accurate velocity measurements up to second-order accuracy as well as single-pixel resolution. Overall, the observed WSS distributions in a serpentine microchannel indicated an apparent bias to the inner wall at each turn section. As compared to the simulation result, CDI SPE tended to show better consistency than FFT-CC. The R^2 of the CDI SPE and the simulation were 0.996 and 0.803 for the WSS on the right-hand side and left-hand side, respectively. Meanwhile, the R^2 of the FFT-CC and the simulation were only 0.003 and 0.338 for the

WSS on the right-hand side and left-hand side, respectively. Generally, this measurement exhibits significant improvements in accuracy, resolution, and stability when the CDI SPE is involved; however, there still exists opportunities for more work. One possibility could be an increase of the image data, thus yielding more reliable outputs in computation.

12.8 Future developments

A substantial amount of research effort has been invested in μPIV technique since it made its debut in 1998. Due to the prosperous development over the past 10 years, μPIV, as a flow visualization technique, has provided a broad diversity of applications and has abundantly benefited the microfluidic and biochemical societies. Reviewing the past research works, a clear trend indicates that the main streams are focused on high accuracy, high resolution, and 3-D capability. Envisioning the future, we believe the advancement of these targets will continue since they are the fundamental values of μPIV. In addition, advancement will definitely concentrate on in vivo, real-time, and tiny measurements in response to the rising of biotechnology and nanotechnology. Some of them, for example, are Infrared (IR) PIV [36] and magnetic resonance imaging (MRI) PIV [62]. IR-PIV employs an IR as a light source, so that the system is able to visualize an opaque flow chamber. MRI, which is a well-known medical instrument used for in vivo scanning, is able to create a cross-sectional image for PIV processing. To exploit more research areas, a μPIV sometimes could be enhanced by integrating with other powerful tools, therefore forming a multifunctional platform, such as LIF μPIV [63] and optical traps/μPIV [64, 65]. Overall, the tremendous flexibility of the μPIV makes it a promising and intriguing tool today. Since this technique is constantly growing and reaching out to more applications, we can anticipate inccreasing and flourishing development of μPIV taking place in the near future.

Acknowledgments

The authors are grateful to the technical discussions regarding diffusometry with Dr. Pramod Chamarchy at GE and the discussions regarding WWS measurements with Dr. Ralph Lindken at Delft University of Technology. We also thank Dr. Lichuan Gui at University of Mississippi for his EDPIV software and technical support associated with the numerical computation. Aloke Kumar acknowledges partial support from the Adelberg Fellowship, Purdue University. Finally, a special thanks to Birck Nanatechnology Center at Purdue Discovery Park for the support of facilities and instruments.

References

[1] Curtin, D., D. Newport, and M. Davies, "Utilizing Micro-PIV and Pressure Measurements to Determine the Viscosity of a DNA Solution in a Microchannel," *Experimental Thermal and Fluid Science*, Vol. 30, No. 8 2006, pp. 843-852.

[2] Hoffmann, M., M. Schluter, and N. Rabiger, "Examination of the Mixing Process Through the Use of Micro-LIF and Micro-PIV," *Chemie Ingenieur Technik*, Vol. 79, No. 7, 2007, pp. 1067-1075.

[3]   Wang, C., N. T. Nguyen, and T. N. Wong, "Optical Measurement of Flow Field and Concentration Fields Inside a Moving Nanoliter Droplet," *Sensors and Actuators A: Physical*, Vol. 133, 2007, pp. 317-322.

[4]   Meinhart, C. D., and H. Zhang, "The Flow Structure Inside a Microfabricated Inkjet Printhead," *Journal of Microelectromechanical Systems*, Vol. 9, No. 1, 2000, pp. 67-75.

[5]   Santiago, J. G., S. T. Wereley, C. D. Meinhart, D. J. Beebe, and R. J. Adrian, "A Particle Image Velocimetry System for Microfluidics," *Experiments in Fluids*, Vol. 25, 1998, pp. 316-319.

[6]   Mishina, H., T. Ushizaka, and T. Asakura, "A Laser Doppler Microscope: Its Optical and Signal-Analyzing Systems and Some Experimental Results of Flow Velocity," *Optics & Laser Technology*, Vol. 8, 1976, pp. 121-127.

[7]   Chuang, H. S., and Y. L. Lo, "Microfluidic Velocity Measurement Using a Scanning Laser Doppler Microscope," *Optical Engineering*, Vol. 46, No. 2 2007, p. 024301.

[8]   Park, S., J. Kim, and B. Lee, "Hot-Wire Measurements of Near Wakes Behind an Oscillating Airfoil.," *American Institute of Aeronautics and Astronautics*, Vol. 28, No. 1, 1990, pp. 22-28.

[9]   Wittmer, K., W. Devenport, and J. Zsoldos, "A Four Sensor Hot-Wire Probe System for Three-Component Velocity Measurement," *Experiments in Fluids*, Vol. 27, No. 4, 1999, pp. U1-U1.

[10] Tai, Y., and R. Muller, "Lightly Doped Polysilicon Bridge as an Anemometer," *Sensors and Actuators A: Physical*, Vol. 15, No. 1, 1988, pp. 63-75.

[11] Bitsch, L., L. H. Olesen, C. H. Westergaard, H. Bruus, H. Klank, and J. P. Kutter, "Micro PIV on Blood Flow in a Microchannel," in *MicroTAS 2003*, 2003.

[12] Rossi, M., Lindken, R., Hierck, B. P., and J. and Westerweel, J., "Tapered Microfluidic Chip for the Study of Biochemical and Mechanical Response of Endothelial Cells to Shear Flow at Subcellular Level," *Lab Chip*, Vol. 9, 2009, pp.1403-1411.

[13] Poelma, C., Vennemann, P., Lindken, R., and J. Westerweel, "In Vivo Blood Flow and Wall Shear Stress Measurements in the Vitelline Network," *Exp Fluids*, Vol. 45, 2008, pp. 703-713.

[14] Vennemann, P., Lindken, R., and J. Westerweel, "In Vivo Whole-Field Blood Velocity Measurement Techniques," *Exp Fluids*, Vol. 42, 2007, pp. 495-511.

[15] Easley, C. J., J. M. Karlinsey, J. M. Bienvenue, L. A. Legendre, M. G. Roper, S. H. Feldman, M. A. Hughes, E. L. Hewlett, T. J. Merkel, J. P. Ferrance, and J. P. Landers, "A Fully Integrated Microfluidic Genetic Analysis System with Sample-In–Answer-Out Capability," *Proceedings of the National Academy of Sciences*, Vol. 103, No. 51, 2006, pp. 19272-19277.

[16] Bourdon, C. J., M. G. Olsen, and A. D. Gorby, "The Depth of Correlation in Micro-PIV for High Numerical Aperture and Immersion Objectives," *Journal of Fluids Engineering*, Vol. 128, No. 4, 2006, pp. 883-886.

[17] Park, J. S., C. K. Choi, and K. D. Kihm, "Optically Sliced Micro-PIV Using Confocal Laser Scanning Microscopy (CLSM)," *Experiments in Fluids*, Vol. 37, 2004, pp. 105-119.

[18] Lima, R., S. Wada, K. Tsubota, and T. Yamaguchi, "Confocal Micro-PIV Measurements of Three-Dimensional Profiles of Cell Suspension Flow in a Square Microchannel," *Measurement Science and Technology*, Vol. 17, 2006, pp. 797-808.

[19] Keane, R. D., and R. J. Adrian, "Theory of Cross-Correlation Analysis of PIV Images," *Applied Scientific Research*, Vol. 49, No. 3 1992, pp. 191-215.

[20] Willert, C. E., and M. Gharib, "Digital Particle Image Velocimetry," Experiments in Fluids, Vol. 10, 1991, pp. 181-193.

[21] Meinhart, C. D., S. T. Wereley, and J. G. Santiago, "A PIV Algorithm For Estimating Time-Averaged Velocity Fields," *Journal of Fluids Engineering*, Vol. 122, 2000, pp. 285-289.

[22] Wereley, S. T., L. C. Gui, and C. D. Meinhart, "Flow Measurement Techniques for the Microfrontier," in *American Institute of Aeronautics and Astronautics Annual Meeting*, Reno, NV, 2001.

[23] Cowen, E. A., and S. G. Monismith, "A Hybrid Digital Particle Tracking Velocimetry Technique," *Experiments in Fluids*, Vol. 22, 1997, pp. 199-211.

[24] Gomez, R., R. Bashir, A. Sarikaya, M. Ladish, J. Sturgis, J. Robison, T. Geng, A. Bhunia, H. Apple, and S. T. Wereley, "Microfluidic Biochip for Impedance Spectroscopy of Biological Species," *Biomedical Microdevices*, Vol. 3, No. 3, 2001, pp. 201-209.

[25] Huang, H., D. Dabiri, and M. Gharib, "On Errors of Digital Particle Image Velocimetry," *Measurement Science and Technology*, Vol. 8, 1997, pp. 1427-1440.

[26] Gui, L., R. Lindken, and W. Merzkirch, "Phase-Separated PIV Measurements of The Flow Around Systems of Bubbles Rising in Water," in *ASME-FEDSM97-3103*, ASME, New York, 1997.

[27] Wereley, S. T., and C. D. Meinhart, "Adaptive Second-Order Accurate Particle Image Velocimetry," *Experiments in Fluids*, Vol. 31, 2001, pp. 258-268.

[28] Huang, H. T., H. E. Fiedler, and J. J. Wang, "Limitation and Improvement of PIV. Part II: Particle Image Distortion, a Novel Technique," *Experiments in Fluids*, Vol. 15, 1993, pp. 263-273.

[29] Wereley, S. T., and L. Gui, "A Correlation-Based Central Difference Image Correction (CDIC) Method and Application in a Four-Roll Mill Flow PIV Measurement," *Experiments in Fluids*, Vol. 34, 2003, pp. 42-51.

[30] Gidden, G., C. Zarins, and S. Glagov, "Response of Arteries to Near-Wall Fluid Dynamic Behavior," *Applied Mechanics Reviews*, Vol. 43, 1990, pp. S98-S102.

[31] Long, D. S., M. L. Smith, A. R. Pries, K. Ley, and E. R. Damiano, "Microviscometry Reveals Reduced Blood Viscosity and Altered Shear Rate and Shear Stress Profiles in Microvessels after Hemodilution," in *Proceedings of the National Academy of Sciences USA*, 2004, pp. 10060-10065.

[32] Jin, S., J. Oshinski, A. Tannenbaum, J. Gruden, and D. Giddens, "Flow patterns and wall shear stress distributions at atherosclerotic-prone sites in a human left coronary artery—An exploration using combined methods of CT and computational fluid dynamics," in *Proceedings of the 26th Annual International Conference of the IEEE EMBS*, San Francisco, CA, 2004.

[33] Westerweel, J., P. F. Geelhoed, and R. Lindken, "Single-pixel resolution ensemble correlation for micro-PIV applications," *Exp. Fluids*, Vol. 37, 2004, pp. 375-384.

[34] Chuang, H. S., and S. T. Wereley, "In-Vitro Wall Shear Stress Measurements for Microfluid Flows by Using Second-Order SPE Micro-PIV," in Proceedings of IMECE2007, *2007 ASME International Mechanical Engineering Congress and Exposition*, , Seattle, WA, 2007.

[35] Olsen, M. G., and R. J. Adrian, "Out-of-Focus Effects on Particle Image Visibility and Correlation in Microscopic Particle Image Velocimetry," *Experiments in Fluids*, Vol. 29, 2000, pp. S166-S174.

[36] Liu, D., S. V. Garimella, and S. T. Wereley, "Infrared Micro-Particle Image Velocimetry in Silicon-Based Microdevices," *Experiments in Fluids*, Vol. 38, 2005, pp. 385-392.

[37] Lee, S. Y., J. Jang, and S. T. Wereley, "Optical Diagnostics to Investigate the Entrance Length in Microchannels." in *The MEMS Handbook: MEMS Design and Fabrication*, M. Gad-el-Hak (eds.), CRC Press, 2005.

[38] Prasad, A. K., and R. J. Adrian, "Stereoscopic Particle Image Velocimetry Applied to Liquid Flows," *Experiments in Fluids*, Vol. 15, 1993, pp. 49-60.

[39] Bendat, J. and A. Piersol, Random Data-Analysis and Measurement Procedures, New York: Wiley Interscience., 1986.

[40] Lindken, R., J. Westerweel and B. Wieneke, "Stereoscopic Micro Particle Image Velocimetry," *Experiments in Fluids*, Vol. 41, 2006, pp. 161-171.

[41] Park, J. S., and K. D. Kihm, "Three-Dimensional Micro-PTV Using Deconvolution Microscopy," *Experiments in Fluids*, Vol. 40, 2006, pp. 491-499.

[42] Peterson, S., H. S. Chuang, and S. T. Wereley, "A Simple Method for Extracting Three Velocity Components Using a Standard Micro-Particle Image Velocimeter," *Measurement Science and Technology*, Vol. 19, 2008, pp. 115406.

[43] Satake, S., T. Kunugi, K. Sato, T. Ito and J. Taniguchi, "Three-Dimensional Flow Tracking in a Micro Channel with High Time Resolution Using Micro Digital-Holographic Particle-Tracking Velocimetry," *Optical Review*, Vol. 12, No. 6 2005, pp. 442-444.

[44] Axelrod, D., T. P. Burghardt and N. L. Thompson, "Total Internal Reflection Fluorescence (in biophysics)," *Annual Review of Biophysics and Bioengineering*, Vol. 13, 1984, pp. 247-268.

[45] Prieve, D. C. and N. A. Frej, "Total Internal Reflection Microscopy: A Quantitative Tool for the Measurements of Colloidal Forces," *Langmuir*, Vol. 6, 1990, pp. 396-403.

[46] Zettner, C., and M. Yoda, "Particle Velocity Field Measurements in a Near-Wall Flow Using Evanescent Wave Illumination," *Experiments in Fluids*, Vol. 34, No. 1, 2003, pp. 115-121.

[47] Hohenegger, C., and P. J. Mucha, "Reconstruction of Velocity Profiles for Nano-PIV," in *57th Annual Meeting of the Division of Fluid Dynamics*, Seattle, WA, American Physical Society, 2004.

[48] Gorti, V. M., H. Shang, S. T. Wereley, and G. U. Lee, "Immunoassays in Nanoliter Volume Reactors Using Fluorescent Particle Diffusometry," *Langmuir*, Vol. 24, No. 6, 2008, pp. 2947-2952.

[49] Olsen, M. G., and R. J. Adrian, "Brownian Motion and Correlation in Particle Image Velocimetry," *Optics & Laser Technology*, Vol. 32, No. 7-8 2000, pp. 621-627.

[50] Adrian, R. J., and C. S. Yao, "Pulsed Laser Technique Application to Liquid and Gaseous Flows and the Scattering Power of Seed Materials," *Applied Optics*, Vol. 24, No. 1, 1985, pp. 44-52.

[51] Hohreiter, V., S. T. Wereley, M. G. Olsen, and J. N. Chung, "Cross-Correlation Analysis for Temperature Measurement," *Measurement Science and Technology*, Vol. 13, No. 7, 2002, pp. 1072-1078.

[52] Kumar, A., V. M. Gorti, H. Shang, G. U. Lee, N. K. Yip, and S. T. Wereley, "Optical Diffusometry Techniques and Applications in Biological Agent Detection," *Journal of Fluids Engineering*, Vol. 130, No. 11, 2008, pp. 111401.

[53] Chamarthy, P., A. Kumar, J. Cao, and S. T. Wereley, "Fundamentals of Diffusion in Microfuidic Systems." in *Encyclopedia of Microfluidics and Nanofluidics*, D. Li (eds.), New York: Springer, 2008.

[54] Chiu, J. J., C. N. Chen, P. L. Lee, C. T. Yang, H. S. Chuang, S. Chien, and S. Usami, "Analysis of The Effect of Disturbed Flow on Monocytic Adhesion to Endothelial Cells," *Journal of Biomechanics*, Vol. 36, 2003, pp. 1883-1895.

[55] Reneman, R. S., T. Arts, and A. P. G. Hoeks, "Wall Shear Stress—An Important Department of Endothelial Cell Function and Structure in the Arterial System In Vivo," *Journal of Vascular Research*, Vol. 43, 2006, pp. 251-269.

[56]   Kähler, C. J., U. Scholz, and J. Ortmanns, "Wall-Shear-Stress and Near-Wall Turbulence Measurements up to Single Pixel Resolution by Means of Long-Distance Micro-PIV," *Experiments in Fluids*, Vol. 41, 2006, pp. 327-341.

[57]   Malek, A. M., S. L. Alper, and S. Izumo, "Hemodynamic shear Stress and Its Role in Atherosclerosis," *Journal of the American Medical Association*, Vol. 282, No. 21, 1999, pp. 2035-2042.

[58]   Shaaban, A. M., and A. J. Duerunckx, "Wall Shear Stress and Early Atherosclerosis: A Review," *American Journal of Roentgenology*, Vol. 174, 2000, pp. 1657-1666.

[59]   Bonert, M., R. L. Leask, J. Butany, C. R. Ethier, J. G. Myers, K. W. Johnston, and M. Ojha, "The Relationship Between Wall Shear Stress Distributions and Intimal Thickening in the Human Abdominal Aorta," *Biomedical Engineering Online*, Vol. 2, 2003, pp. 1-14.

[60]   Jin, S., J. Oshinski, A. Tannenbaum, J. Gruden, and D. Giddens, "Flow Patterns and Wall Shear Stress Distributions at Atherosclerotic-Prone Sites in a Human Left Coronary Artery—An Exploration Using Combined Methods of CT and Computational Fluid Dynamics," in *Proceedings of the 26th Annual International Conference of the IEEE EMBS*, San Francisco, CA, 2004.

[61]   Eijkel, J., "Liquid Slip in Micro- and Nanofluidics: Recent Research and Its Possible Implications," *Lab on a Chip*, Vol. 7, 2007, pp. 299-301.

[62]   Heese, F., P. Robson, and L. Hall, "Magnetic Resonance Imaging Velocimetry of Fluid Flow in a Clinical Blood Filter," *American Institute of Chemical Engineers*, Vol. 51, No. 9, 2005, pp. 2306-2401.

[63]   Chamarthy, P., S. V. Garimella, and S. T. Wereley, "Non-Intrusive Temperature Measurement Using Microscale Visualization Techniques," *Experiments in Fluids*, Vol. 47, No. 1, July, 2009.

[64]   Leonardo, R. D., J. Leach, H. Mushfique, J. M. Cooper, G. Ruocco, and M. J. Padgett, "Multipoint Holographic Optical Velocimetry In Microfluidic Systems," *Physical Review Letters*, Vol. 96, 2006, pp. 134502.

[65]   Nève, N., J. K. Lingwood, J. Zimmerman, S. S. Kohles, and D. C. Tretheway, "The μPIVOT: An Integrated Particle Image Velocimeter and Optical Tweezers Instrument for Microenvironment Investigations," *Measurement Science and Technology*, Vol. 19, 2008, pp. 095403.

[66]   Yeh, Y., and H. Cummings, "Localized Fluid Flow Measurements with a He-Ne Laser Spectometer," *Applied Physical Letters*, Vol. 4, 1964, pp. 176–178.

[67]   Vennemann, P., Kiger, K.T., Lindken, R., Groenendijk, B. C. W., Stekelenburg-de Vos, S., ten Hagen, T. L. M., Ursem, N. T. C., Poelmann, R. E., Westerweel, J., and B. P. Hierck, "In Vivo Micro Particle Image Velocimetry Measurements of Blood-Plasma in the Embryonic Avian Heart," *J. Biomech.*, Vol. 39, 2006, pp. 1191-1200.

[68]   Groenendijk, B. C. W., Stekelenburg-de Vos, S., Vennemann, P., Wladimiroff, J. W., Nieuwstadt, F. T. M., Lindken, R., Westerweel, J., Hierck, B. P., Ursem, N. T. C., and R. E. Poelmann, " The Endothelin-1 Pathway and the Development of Cardiovascular Defects in the Hemodynamically Challenged Chicken Embryo," *J. Vascular Research*, Vol. 45, 2008, pp. 54-68.

Microtubule Motors in Microfluidics

Maruti Uppalapati,[1] Ying-Ming Huang,[2] Shankar Shastry,[1] Thomas N. Jackson,[2] and William O. Hancock[1]*

[1] Department of Bioengineering and [2] Department of Electrical Engineering
Penn State University, University Park, PA 16802

* Corresponding author: William O. Hancock, 205 Hallowell Bldg., University Park, PA 16802, phone: 814-863-0492, fax: 814-863-0490, e-mail: wohbio@engr.psu.edu

Abstract

Kinesin motor proteins carry out a range of transport functions in eukaryotic cells including long-distance transport in neurons and the movement of chromosomes during cell division. Because emerging microfluidic devices utilize channel geometries similar to cellular scales and because these devices generally require transport through these channels, there is significant interest in incorporating biomotor-driven transport into microfluidic devices. Kinesin-driven transport has the advantage of being able to work against concentration gradients and bulk fluid flow, and microtubules can be functionalized to carry a range of cargo. This chapter describes the foundational methods for generating kinesin-driven transport in engineered microchannels. Protocols are described for expressing and purifying recombinant kinesin, purifying and fluorescently labeling bovine brain tubulin, and carrying out the microtubule gliding assay. General design considerations are presented for integrating kinesins and microtubules into microchannels, followed by a detailed description of a microfluidic device prototype that successfully integrates microtubule-based transport.

Key terms   kinesin
microtubule
transport
motor protein
bionanotechnology

13.1   Introduction

In emerging lab-on-a-chip technologies, the microfluidic devices designed to shrink analytical tests and biosensors down to microscale dimensions generally require some manner of mechanical actuation. While these mechanical forces can be achieved by pressure-driven fluid pumps, electrically or magnetically driven movement, electroosmotic flow, or other approaches, there is a continuing need in this area for reliable mechanical actuation and directed transport. In eukaryotic cells, kinesin motor proteins transport intracellular cargo and provide the mechanical forces underlying chromosome movements and other mechanical processes. These motor proteins utilize the chemical energy of ATP hydrolysis to walk on protein filaments, called microtubules. This biologically driven movement can be reconstituted under a microscope by immobilizing kinesin motors on a substrate (such as a glass coverslip) and observing the movement of microtubules over the surface. Building on the robust microscale transport in this "microtubule gliding assay," there has been considerable interest in utilizing kinesin-driven microtubule motion for nano/microscale transport applications in lab-on-a-chip devices [1–3]. While the development of lab-on-a-chip devices is primarily based on microfluidics technology, integration of kinesin-driven motion in microfluidic devices has several advantages. First, motors can transport cargo against concentration gradients. Second, no bulk fluid flow or pumping mechanisms are required, easing the fabrication and power requirements of such devices. Third, microtubules and kinesins can be modified to bind specific cargo, enabling their use in biosensing and separation processes.

A schematic of a potential device of this type is illustrated in Figure 13.1. This device is designed to detect viral RNA in biosensing applications, or to detect mRNA from a cell lysate, which would enable gene expression profiling, in principle down to single cell levels [1]. In the design, microtubules functionalized with single-stranded DNA constructs (molecular beacons) are transported through microchannels functionalized with kinesin motors [4]. These mobile probes capture and transport specific RNA and report this sequence-specific binding by unquenching their fluorescence. Integrated optical sensors and electrodes detect and guide the microtubules such that the cargo is separated from the heterogeneous sample and is concentrated in the collection chamber. If successful, this device could have a range of uses, and it is only one of many biomotor-driven devices that have been envisioned. However, successful integration of microtubules in such lab-on-a-chip devices requires (1) the ability to achieve directional motion of microtubules in enclosed microfluidic channels, (2) the ability to bind and transport cargo, and (3) the ability to detect and sort microtubules based on the presence of cargo. This is an active area of research and key blocks of technology have been developed by our group and others. In this chapter we detail the experimental methods involved in purifying kinesin motors and microtubules, techniques for modifying these proteins for use in microfluidic channels, and approaches for optimizing the design of microfluidic channels to achieve robust kinesin-driven transport.

13.2   Materials

Relevant materials, reagents, and equipment are listed below. Both the "Materials" and the "Methods" sections are divided into four subsections: (1) Kinesin expression and

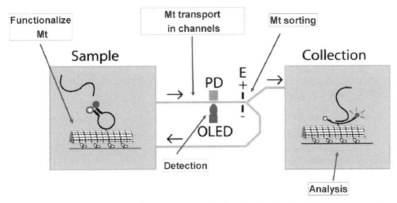

Figure 13.1 Schematic for biosensor/bioseparation device-based kinesin-driven transport. Microtubules are functionalized with molecular beacons that bind RNA and fluoresce upon binding. The beacon-functionalized microtubules are transported through microfluidic channels to which kinesin motor proteins are adsorbed. An integrated photodiode (PD) and organic light emitting diode (OLED) detect the presence of RNA cargo on the microtubule and electrodes (E) direct the microtubules at a bifurcation, directing cargo-loaded microtubules into the sample chamber for further analysis such as PCR amplification or sequencing. (Image taken, with kind permission from Springer Science+Business Media, from Jia et al. [1].)

purification, (2) tubulin purification and labeling, (3) microtubule gliding assay, and (4) integrating motility into microchannels. Unless specified, the reagents mentioned below can be obtained from alternative vendors, and for reagents where no source is specified, Sigma, Fischer, or any equivalent source are sufficient.

13.2.1 Kinesin expression and purification materials

13.2.1.1 Kinesin reagents

- Kinesin expression plasmid in BL21(DE3) bacterial cells
- LB media (Qbiogene, Cat # 3002-092)
- Ampicillin sodium salt (Sigma-Aldrich, Cat # A9518)
- Isopropyl β-D-1-thiogalactopyranoside (IPTG) (Sigma-Aldrich, Cat # I5502)
- Phenylmethanesulfonyl fluoride (PMSF) (Sigma-Aldrich, Cat # P7626)
- Adenosine-5'-triphosphate (ATP) (Plenum Scientific Research, Inc., Cat # 0163-04)
- ß-Mercaptoethanol (ßME) (Sigma-Aldrich, Cat # M3148)
- Dithiothreitol (DTT) (Bio-Rad, Cat # 161-0611)
- Sodium phosphate monobasic (Sigma-Aldrich, Cat # S3139)
- Sodium phosphate dibasic (Sigma-Aldrich, Cat # S3264)
- Sodium chloride (Sigma-Aldrich, Cat # S3014)
- Magnesium chloride (Sigma-Aldrich, Cat # 68475)
- Sodium hydroxide (NaOH)
- Sucrose (Sigma-Aldrich, Cat # S0389)
- Imidazole (Sigma-Aldrich, Cat # I5513)

13.2.1.2   Kinesin buffers

- Lysis buffer: 50-mM sodium phosphate, 300-mM NaCl, 40-mM imidazole, 5-mM βME, pH 8.0.
- Wash buffer: 50-mM sodium phosphate, 300-mM NaCl, 60-mM imidazole, 100-μM MgATP, 5-mM βME, pH 7.0.
- Elution buffer: 50-mM sodium phosphate, 300-mM NaCl, 500-mM imidazole, 100-μM MgATP, 5-mM βME, 10% (w/v) sucrose, pH 7.0.

Note: For making these phosphate buffers, either combine monobasic and dibasic forms to approximate the final pH and then adjust accordingly, or use monobasic form and pH with concentrated NaOH. pH buffers before adding βME and sucrose. βME has a lifetime of days in these buffers, so prepare accordingly: refrigerate sucrose containing solutions to prevent bacterial growth. One convenient approach is to make up buffers without ATP, βME, or sucrose, and add these just before using.

13.2.1.3   Kinesin equipment

- 2L Erlenmeyer flask (preferably baffled)
- Temperature-controlled incubator with orbital shaker
- Sonicator for lysing bacterial cells (Branson Ultrasonics)
- AKTA-FPLC chromatography system (or equivalent)
- Ni-NTA agarose chromatography column (5 ml)
- SDS-PAGE gel electrophoresis equipment

13.2.2   Tubulin purification materials

13.2.2.1   Tubulin reagents

- Cow brains from freshly slaughtered cows (obtain from local slaughterhouse)
- Leupeptin hydrochloride (Sigma-Aldrich, Cat # L9783)
- Trypsin inhibitor (Sigma-Aldrich, Cat # T9253)
- Pepstatin A (Sigma-Aldrich, Cat # P5318)
- Aprotinin (Sigma-Aldrich, Cat # A6103)
- N_α-p-Tosyl-L-arginine methyl ester hydrochloride (TAME) (Sigma-Aldrich, Cat # T4626)
- Tosyl-L-phenylalanyl-chloromethane (TCPK) (Sigma-Aldrich, Cat # T4376)
- Ethylene glycol-bis(2-aminoethylether)-N,N,N',N'-tetraacetic acid (EGTA) (Sigma-Aldrich, Cat # E3889)
- Ethylenediaminetetraacetic acid (EDTA) (Sigma-Aldrich, Cat # E9884)
- L-Glutamic acid monosodium salt monohydrate (sodium glutamate) (Sigma-Aldrich, Cat # G2834)
- L-Glutamic acid potassium salt monohydrate (potassium glutamate) (Sigma-Aldrich, Cat # G1501)
- Piperazine-N,N'-bis(2-ethanesulfonic acid) (PIPES) (Sigma-Aldrich, Cat # P6757)

- N-(2-Hydroxyethyl)piperazine-N'-(2-ethanesulfonic acid) (HEPES) (Sigma-Aldrich, Cat # H3375)
- Glycerol (Sigma-Aldrich, Cat # G5516)
- DMSO anhydrous (Sigma Aldrich, Cat # 276855)
- 2'-Guanosine 5'-triphosphate, sodium salt (GTP) (Jena Bioscience, Cat # NU-1012)
- Paclitaxel (Taxol) (Sigma Aldrich, Cat # T7402) (*Note:* Required for stabilizing microtubules. Dissolve at 1 mM in DMSO for use and store frozen aliquots. Taxol inhibits cell division and DMSO aids absorption into skin, so avoid contact with skin.)
- 5-(and-6)-carboxytetramethylrhodamine, succinimidyl ester (Rhodamine-NHS) (Invitrogen, Cat # C1171)

13.2.2.2   Tubulin buffers and solutions

- PBS buffer—20-mM sodium phosphate, 150-mM NaCl, pH 7.4
- Glutamate buffer—1.0M sodium glutamate, 2-mM EGTA, 0.1-mM EDTA, 2-mM $MgCl_2$, pH 6.6
- BRB80 buffer—80-mM K-PIPES, 1-mM EGTA, 1-mM $MgCl_2$, pH 6.85. (Can make with mixture of PIPES acid and salt forms or use all acid form and pH with KOH. PIPES is only minimally soluble below pH 6 and will dissolve as base is added.)
- 5x BRB80 buffer—400-mM K-PIPES, 5-mM EGTA, 5-mM $MgCl_2$, pH 6.85
- HEPES/40% glycerol (labeling buffer)—0.1 M HEPES, 1-mM $MgCl_2$, 1-mM EGTA, 40% glycerol (v/v), pH 8.6
- Quench solution—0.5 M potassium glutamate, pH 8.6
- Depolymerization solution—50-mM potassium glutatmate, 0.5-mM $MgCl_2$, pH 7.0

13.2.2.3   Tubulin equipment

- Waring blender
- Temperature-controlled 37°C water bath
- Beckman Ultracentrifuge, Type 19 and 50.2 Ti rotors, and centrifuge tubes/bottles
- Tissue tearor
- Beckman Airfuge
- UV-Vis Spectrophotometer

13.2.3   Microtubule gliding assay materials

- Fisher's Finest plain glass microscope slides (Fisher Scientific, Cat # 12-544-1)
- Corning 18 x18-mm coverslips (Fisher Scientific, Cat # 12-519A) (*Note:* Corning coverglass works best for gliding assay compared to other coverglasses. Different sizes can be used)
- Double-stick tape
- Casein from bovine milk (Sigma-Aldrich, Cat # C3400) (*Note*: This is an important reagent that is used to optimize surfaces for functional kinesin adsorption. Casein should be dissolved in BRB80 at ~20 mg/ml and stored at –20°C. Dissolve 2.5g

casein in 25-ml BRB80 buffer in 50-ml Falcon tube and rock for 4 hours to over-night to dissolve (keep at 4°C if > 4 hours). Next, spin in centrifuge (10 minutes at >100,000 x g is good, but less is acceptable) to pellet insoluble material. Finally, fil-ter using 0.2-μm syringe filters. You will need to go through a number of filters as they tend to clog. Assess concentration by UV absorbance (1 mg/ml = 1 A_{280}), if > 20 mg/ml, then dilute, aliquot, and store at –20°C.)

- D-glucose (Sigma-Aldrich, Cat # G7528)
- Glucose oxidase (Sigma-Aldrich, Cat # G2133)
- Catalase (Sigma-Aldrich, Cat # C515)

Note: Glucose, glucose oxidase, and catalase are parts of an antifade oxygen scavenging system. It is convenient to make up stock solutions in BRB80 at 2M, 2 mg/ml, and 0.8 mg/ml, respectively, store 10-μl aliquots at –20°C, and thaw and dilute 100-fold into motility solution for motility experiments.

- Fluorescence microscope with 60x or 100x objective.

Note: Generally a CCD camera, monitor, and recording device such as a VCR or com-puter is necessary for optimal visualization and analysis. While expensive enhanced CCD cameras such as the Roper Cascade 512 work well, we have also had success with a Genwac 902H CCD camera, which can be found online for less than $400.)

13.2.4   Microfabrication materials

13.2.4.1   Microfabrication reagents and supplies

- Borosilicate glass substrates (1-mm thick)
- ProSciTech 50 × 50-mm coverslips
- 996-kDa poly(methyl methacrylate) (PMMA)
- Chlorobenzene
- Acetone
- Isopropyl alcohol (IPA)
- Sulfuric acid (H_2SO_4)
- Hydrogen peroxide (H_2O_2)
- Piranha solution (H_2SO_4:H_2O_2 = 4:1)
- 1811 photoresist
- Tetramethylammonium hydroxide (TMAH) developer
- SU-8 photoresist
- SU-8 developer
- Chrome and gold for deposition
- Gold for electroplating
- Chrome etch solution
- Gold etch solution
- Hydrogen fluoride (HF)
- Ammonium fluoride (NH_4F)
- Buffered oxide etch (BOE) (NH_4F/HF 10:1)

Note: Piranha solution sulfuric acid, hydrogen fluoride, ammonium fluoride, Cr and Au etch solutions and BOE are extremely corrosive and precautions should be taken to prevent any contact with skin or clothing. Always use proper safety procedures when using these chemicals.

13.2.4.2 Microfabrication equipment

- Karl Suss MA-55 aligner
- Photoresist spinner
- Hotplate
- Chemical wet bench
- DC power supply (for Au electroplating)
- Resistive filament thermal evaporator
- RIE equipment (for oxygen plasma)
- Warner 100 hydraulic press laminator

13.3 Methods

13.3.1 Kinesin expression and purification

A number of kinesins have been investigated from many different organisms, but most of the biophysical investigations to date and virtually all of the microfluidics research using kinesins have employed conventional kinesin (also called Kinesin-1 [5]). Our investigations have employed full-length *Drosophila* conventional kinesin that contains a hexaHis-tag on its C-terminus for purification [6, 7]. The full-length gene can be truncated to remove the tail domain and the tail replaced with a biotinylation sequence, green fluorescent protein, or other sequences. In our work we also employ a headless kinesin construct that retains the rod and tail domains and hexaHis tag, but lacks its motor domain [6, 8]. Below, we provide a detailed protocol for expressing this kinesin in bacteria, purification by Ni column chromatography, and freezing for long term storage.

This protocol is for full-length *Drosophila* conventional kinesin in a pET plasmid expressed in BL21(DE3) *E. coli* cells, though it will work for other hexaHis-tagged kinesins. It generally follows protocols described in Hancock et al. [6] and in Coy et al. [7]. Related protocols can be found in a book of kinesin protocols [9] and on the Kinesin home page (http://www.cellbio.duke.edu/kinesin//).

The amino acid sequence of the protein and the nucleotide sequence of the entire plasmid (pPK113) are online (entry AF053733 at http://www.ncbi.nlm.nih.gov pubmed/). Furthermore, this plasmid is available to academic researchers from the plasmid depository Addgene (http://www.addgene.org). The method for preparing and purifying these proteins is described below:

1. Prepare LB media and autoclave to sterilize. After the media cools to room temperature, add 50 mg of ampicillin sodium salt (final concentration 100 mg/L).

2. Pipette 5 ml of LB+Amp into a 15-ml Falcon tube and inoculate it with a colony of plasmid-containing BL21(DE3) cells. Incubate overnight (16 hours) at 37°C while shaking at 250 rpm.

3. Transfer the 5-ml overnight culture to 500-ml LB medium in a 2L Erlenmeyer flask and incubate at 37 °C while shaking at 250 rpm until the culture grows to an optical density at 600 nm at of 0.5 to 1.0. This typically requires 3 to 6 hours of growth.

4. Cool the culture to room temperature on ice and add 45 mg of IPTG (final conc. 0.5 mM) to induce protein expression. Incubate at 20°C for 3 hours while shaking at 250 rpm to express protein. (*Note:* Expression at this temperature improves solubility of the expressed protein.)

5. Transfer culture to centrifuge bottles and harvest cells by centrifuging for 10 minutes at 6,000 x g. Resuspend pellet in 25 ml of lysis buffer. If desired, cells can be flash-frozen in liquid nitrogen and stored at -80°C at this point for later purification.

6. Thaw cells in 37°C water bath if necessary. Add 0.1-mM ATP and 1-mM PMSF to the solution and lyse the cells by sonicating the cells on ice. Use to 4 to 6 pulses of 20 seconds each on duty cycle 5 to completely lyse the cells. Cell lysis can also be carried out by French press [6]. (*Note:* one indicator of lysis is that the solution goes from cloudy to clear due to fragmenting the light scattering bacterial cells. Another test is to run a gel of the pellet and supernatant from step 7—incomplete lysis results in a large portion of the expressed protein in the pellet.)

7. To remove cell debris and insoluble components, centrifuge the cell lysate at 100,000 × g for 30 minutes at 4°C. The supernatant (clarified lysate) contains the soluble recombinant protein along with bacterial proteins.

8. While the cell lysate is spinning, equilibrate the Ni-NTA column by flowing through 10x column volumes (50 ml) of lysis buffer at a flow rate of 5 ml/min.

9. Apply clarified lysate to the column at a flow rate of 2 ml/min to bind hexaHis-tagged protein to the column. After loading, wash column with 10x column volume (50 ml) wash buffer at a flow rate of 5 ml/min to remove unwanted bacterial proteins. The UV absorbance readout on the chromatography system (280 nm) should fall from a peak during loading to a flat baseline at the end of the wash. (*Note:* if greater purity of final eluted protein is desired, imidazole in wash buffer can be raised to 80 or 100 mM.)

10. Elute the kinesin by flowing elution buffer over the column at a flow rate of 1 ml/min. Collect the elution in 0.5-ml fractions. Monitor elution by UV absorbance —because imidazole has an absorbance at 280 nm, there will be an absorbance jump due to the high imidazole concentration in the elution buffer, but a peak or at least a shoulder from the eluted protein should be observable before the absorbance plateaus. The fractions corresponding to the peak absorbance will contain the purified protein. The peak fractions can also be determined and the protein concentration quantified by a UV-Vis spectrophotometer or by running an SDS-PAGE gel of the elution fractions. Typical yields are 1 to 2 ml at 100 µg/ml concentration for full-length kinesin, though they can be significantly higher for truncated constructs. (*Note:* microtubule gliding assays using 5x-diluted elution fractions can also be run to identify the peak fractions.)

11. The purified protein can be flash-frozen in liquid nitrogen and stored at –80°C. (*Note:* the 10 % sucrose in the elution buffer acts as a cryoprotectant.)

Notes: Kinesin frozen at –80°C is stable for years, but samples stored at 4°C only last for days and samples at room temperatures lose activity over hours. Low-concentration samples can be stabilized somewhat by adding 0.5-mg/ml casein or BSA—part of the activity loss in samples with low kinesin concentrations is due protein adsorption to the sides of the tubes, which is minimized by adding these inert proteins.

13.3.2   Tubulin purification and labeling

Microtubules, which are 25 nm in diameter and can be tens of microns in length, are made up of α-β heterodimers of the protein tubulin [10]. There are no established techniques for bacterial expression of tubulin, so the standard approach is to purify tubulin from cow or pig brains. Microtubules are then polymerized from this purified tubulin and stabilized against depolymerization using the drug Taxol. There are published protocols for large scale tubulin purification, including those by Williams and Lee [11] and Castoldi and Popov [12], as well as online protocols from the Salmon Lab (http://www.bio.unc.edu/Faculty/Salmon/lab/protocolsporcinetubulin.html) and Mitchison Lab (http://mitchison.med.harvard.edu/protocols/tubprep.html) Web sites. Here, we describe an adapted protocol for isolating bovine brain tubulin that has worked reproducibly in our laboratory. The procedure involves polymerizing microtubules, pelleting them by centrifugation to remove unwanted soluble proteins, and then depolymerizing and centrifuging to remove unwanted insoluble proteins. The key hurdle in isolating pure tubulin is removing unwanted microtubule associated proteins (MAPs). Earlier protocols used a final phosphocellulose cation exchange column step to remove these positively charged MAPs from the negatively charged tubulin, while this protocol uses high concentrations of glutamate in the buffers to reduce the affinity of the MAPs for the microtubules.

Note that tubulin is commercially available from Cytoskeleton, Inc. (www.cytoskeleton.com). For small-scale studies, this is a reasonable source and is cost-effective. However, for full control over the materials, to enable the ability to carry out diverse modifications of the protein, and for applications where the large material requirements make purchasing it prohibitively expensive, it is desirable to purify the tubulin from the source.

13.3.2.1   Protocol for tubulin purification

The procedure starts with harvesting cow brains, homogenizing them to break open the cells, and clarifying the homogenate by centrifugation to obtain a solution of soluble intracellular proteins. This is followed by two cycles of polymerization/centrifugation/depolymerization/centrifugation to purify tubulin away from unwanted cellular proteins and MAPs. When carried out successfully, the protocol yields roughly 500 mg of pure tubulin at >99% purity, though yields will vary.

1.   Prepare 2L each of PBS buffer and glutamate buffer and chill them to 4°C.
2.   Obtain two cow brains from freshly slaughtered cows. Transport the brains submerged in ice-cold PBS buffer in an ice-filled cooler. (*Note:* It is critical to process the brains within ~1 hour of slaughter. Delays in processing reduce yield due to degradation of tubulin.)

3. Dissect the brains on ice, discarding brain stem, cerebellum, and corpus collosum. Strip meninges (tough filaments on outside of brain) and remove blood clots, then rinse dissected cerebrums with ice-cold PBS. Work quickly and try to balance speed with thoroughness of cleaning—90% is a good rule here.

4. Prepare protease inhibitor cocktail by dissolving the protease inhibitors listed in Table 13.1 in 1 ml of their respective solvents.

5. Weigh cerebrums (in preweighed beaker) and transfer to a chilled Waring blender. Add 50% v/w of chilled glutamate buffer (for example: add 350 ml for 700g of brains). Add protease cocktails and 1-mM ATP, 0.25-mM GTP, 4-μM DTT, 0.1% βME.

6. Homogenize the brains (blending in 4×10-second high-speed pulses, with 15-second pauses between pulses to minimize heating), and distribute the homogenate into 250 ml centrifuge bottles.

7. Spin the homogenate at 50,000 x g (i.e., Beckman type 19 rotor, 19,000 rpm [53,900 × g, k = 951]) for 60 minutes at 4°C to clarify the solution. Pour the supernatants gently into a 1L graduated cylinder without disturbing the soft pellets. Discard pellets.

8. To polymerize tubulin add prewarmed (37°C) glycerol to the supernatant to a final concentration of 33% (v/v) glycerol, and add 0.5-mM GTP, 1.5-mM ATP each dissolved in ~1 ml of buffer. Mix well by inverting (cover with Parafilm and hold in place with your hand) and transfer the contents to a 500-ml stainless steel beaker. Incubate Parafilm-covered beaker in 37°C water bath for 1 hour to polymerize microtubules. (*Note:* Solution should become more viscous and take on a subtle opalescence if polymerization is proceeding.)

9. In the meantime, warm the centrifuge and rotor to 37°C. After polymerization, transfer microtubule solution to prewarmed centrifuge tubes and centrifuge at 37°C at 300,000 × g for 30 minutes to pellet microtubules. We used a Beckman 50.2 Ti rotor at 50,000 rpm [302,000 × g, k = 133]).

10. Discard supernatant containing unwanted proteins, weigh the pellets, and resuspend pellets in ice-cold glutamate buffer. Add 1 ml of cold buffer to each centrifuge tube and physically dislodge the pellet from the tube using a spatula. Pool the pellets in a 250-mL glass beaker on ice and add ice-cold buffer. Choose volume for resuspension to be roughly 1/3 of polymerization volume to maintain high tubulin concentrations. To depolymerize microtubules, homogenize pellets with tissue tearor and incubate 30 minutes on ice with additional tissue tearor homogenization during the incubation.

Table 13.1   Protease Inhibitors

Component	Final	Solvent
Leupeptin	5 mg/mL	DI water
Trypsin inhibitor	10 mg/mL	DI water
Pepstatin	1 mg/mL	DMSO
Aprotinin	1 mg/mL	DMSO
TAME	1 mg/mL	DMSO
TPCK	1 mg/mL	DMSO
PMSF	100 mM	Ethanol

11. In the meantime, chill the rotor and centrifuge to 4°C. Following incubation, clarify the depolymerized tubulin by centrifuging at 300,000 × g for 30 minutes at 4°C. Save supernatant and discard pellets.

12. Repeat tubulin cycling steps 8 to 11, with the exception that the pellets are resuspended in cold BRB80 buffer instead of glutamate buffer in step 10. This "twice-cycled tubulin," cycled in high (glutamate) buffer to remove microtubule associated proteins, is equivalent to "PC tubulin" obtained by passing the solution over a phosphocellulose column.

13. The tubulin concentration in the final supernatant can be quantified by UV absorbance at 280 nm (1 A_{280} = 1 mg/ml tubulin). Aliquot tubulin into cryotubes, flash-freeze on liquid nitrogen, and store at –80°C. A helpful rule to remember is that 1 mg/ml of the 100-kD tubulin dimer equals 10-μM tubulin.

13.3.2.2   Protocol for labeling tubulin with rhodamine

The following protocol describes labeling of pure tubulin (obtained in the previous step) by rhodamine succinimidyl ester (rhodamine-NHS), a reactive form of the rhodamine fluorophore. There are a range of fluorophores that can be purchased in this reactive form as well as biotin and other compounds. These can be purchased from Molecular Probes/Invitrogen or other sources, and this protocol is suitable for microtubule labeling using these compounds as well. This protocol is adapted from a widely used protocol in the literature [13].

1. Warm the rotor (i.e., Beckman 50.2 Ti) and centrifuge to 37°C. Thaw a 100-mg aliquot of purified tubulin at 37°C (can be scaled up or down). Add prewarmed glycerol to 33% final v/v, along with 1-mM GTP and 5-mM $MgCl_2$ and polymerize for 30 minutes in a 37°C water bath.

2. Spin the polymerized solution at 300,000 x g for 30 minutes to pellet microtubules. Discard the supernatant and resuspend the pellet in 3.5-ml prewarmed HEPES/40% glycerol (with 1-mM GTP and 5-mM $MgCl_2$), and maintain the solution at 37°C. (*Note:* Make sure the microtubules are well suspended by pipetting up and down repeatedly.) Quickly measure the tubulin concentration using UV absorbance.

3. Just prior to labeling, dissolve the rhodamine-NHS dye (5-(and-6)- carbo-xytetramethylrhodamine, succinimidyl ester, Invitrogen) in DMSO to a final concentration of 50 mM. Based on the microtubule concentration add the dye in the ratio of 20 dye molecules per tubulin heterodimer and incubate at 37°C for 15 minutes. (*Notes*: (1) It is advisable to briefly centrifuge the dissolved dye solution (i.e., 1 minute in any bench top centrifuge) to remove any undissolved dye crystals that will later pellet with the microtubules. (2) HEPES buffer at pH 8.6 is used because NHS reactive groups label ionized (NH_3^+) amino groups and this is closer to the pKa for the lysine side chains that are labeled. We have, however, had success with labeling in BRB80 at pH 6.8, so this is an option for optimizing labeling.)

4. Stop reaction by adding K-glutamate to a final concentration of 50 mM. Centrifuge the solution at 300,000 x g for 30 minutes at 37°C to pellet microtubules away from unincorporated dye.

5. Discard supernatant and resuspend the pellet in 2 ml of cold depolymerization buffer. Incubate on ice for 30 minutes, with occasional gentle vortexing to depolymerize.

6. Centrifuge the solution in a Beckman Airfuge at 30 psi for 10 minutes to remove insoluble material and protein aggregates. Remove the supernatant and add 20% volume of 5x BRB80 buffer.

7. Polymerize microtubules by adding 1-mM GTP, 5-mM $MgCl_2$ and 5% DMSO and incubating at 37°C for 30 minutes. Centrifuge the solution 5 minutes in the Airfuge, remove the supernatant, resuspend the microtubules in 1.5 ml of cold BRB80, and depolymerize on ice for 30 minutes.

8. Centrifuge the solution one last time in the Airfuge and harvest supernatant containing the labeled tubulin. To quantify the yield and the dye:tubulin ratio, measure the absorbance in a UV/Vis spectrophotometer at 280 and 555 nm. The dye has absorbance at both 280 and 555 nm (absorption coefficient $\varepsilon_{555} = 49{,}291$ $M^{-1}cm^{-1}$ and $\varepsilon_{280}/\varepsilon_{555} = 0.2476$) and tubulin has an absorbance at 280 nm ($\varepsilon_{280} = 101{,}900$ $M^{-1}cm^{-1}$). The dye concentration is obtained at 555 nm, and the tubulin concentration is obtained at 280 nm after subtracting out the dye absorbance at this wavelength. A 1:1 dye:tubulin ratio is a good target, though even minimally labeled tubulin is usually useful. Aliquot the labeled tubulin, flash-freeze in liquid nitrogen, and store at −80 °C.

13.3.2.3 Cycling tubulin and mixing labeled unlabeled tubulin for optimal functionality

For visualizing microtubules by epifluorescence microscopy, we have found that the optimum dye:tubulin ratio is 1:5. This ratio gives microtubules that are bright enough to visualize, and the dye has no apparent effect on microtubule or kinesin function. The approach is to combine labeled and unlabeled tubulin at specified ratios and to aliquot, freeze and store this tubulin for use in motility experiments. We have observed that microtubules containing 1:1 dye:tubulin ratios have different polymerization character-istics, so caution should be used when working with highly labeled microtubules, but the optimum dye:tubulin ratio may vary for different applications, and can be modi-fied. Furthermore, multifunctional microtubules that contain tubulin labeled with combinations of fluorophores or with biotin can be assembled (i.e., biotinylated, rhodamine-labeled microtubules). When making up tubulin solutions, unlabeled tubulin is typically cycled, quantified, and then combined with labeled tubulin. The following cycling protocol ensures that all of the tubulin is active and removes any dead protein.

1. Thaw a 50-mg aliquot of tubulin in a 37°C water bath. Add 1-mM GTP, 5-mM $MgCl_2$ and 5% DMSO and incubate at 37°C for 30 minutes to polymerize.

2. Centrifuge the polymerized microtubules at 300,000 × g for 30 minutes (i.e., 50,000 rpm in a 50.2Ti rotor) at 37°C. Discard supernatant, resuspend the pellet in 1.5 ml of cold BRB80 buffer, and depolymerize on ice for 30 minutes.

3. Centrifuge the solution in an Airfuge for 10 minutes and harvest the supernatant. Measure the concentration of cycled tubulin using UV absorbance.

For normal rhodamine-labeled microtubules, the cycled tubulin and rhoda-mine-labeled tubulin are mixed to yield a dye:tubulin ratio of 1:5, and the solution is diluted using BRB80 buffer to a final concentration of 4 mg/ml. The solution is aliquotted into 10-μl aliquots, flash-frozen in liquid nitrogen, and stored at −80°C for later use.

Table 13.2 Microtubule Polymerization Recipe

Vol.	Reagent (Stack Conc.)	Final Conc.
10 μl	Rho-tubulin (4 mg/ml, 1:5 rho:tubulin)	32 μM
0.5 μl	MgCl$_2$ (100 mM)	4 mM
0.5 μl	MgGTP (25 mM)	1 mM
0.6 μl	DMSO	5%
0.9 μl	BRB80	
Total	12.5 μl	

13.3.2.4 Polymerizing and stabilizing microtubules for microtubule gliding assays

To polymerize microtubules for kinesin motility experiments, thaw a 10-μl aliquot of rhodamine tubulin and add the components listed in Table 13.2.

Mix by briefly flicking tube or pipetting, and incubate the solution at 37°C for 20 minutes to polymerize the microtubules. Meanwhile, make a 1-ml solution of BRB80 + 10-μM Taxol (BRB80T) by adding 10 μl of 1-mM Taxol in DMSO to 990-μl BRB80 and rapidly vortexing to disperse the Taxol. Following polymerization, add the entire microtubule solution to the BRB80T solution and disperse by inverting the tube repeatedly. This results in the Taxol binding to the microtubules and stabilizing in their polymer form. These microtubules (called MT100 for 100-fold diluted microtubules) are stable for a week on the bench at room temperature. However, placing the tube on ice for a few minutes will depolymerize the microtubules. This microtubule solution can be flash-frozen on liquid nitrogen and stored at –80°C; thaw rapidly (warm hands work well) to prevent depolymerization.

13.3.3 Standard protocol for the microtubule gliding assay

The microtubule gliding assay involves immobilizing kinesin motors on a glass coverslip, introducing fluorescent microtubules in the presence of ATP, and observing their kinesin-driven transport. Versions of this assay are published in [14, 15] and online (http://www.cellbio.duke.edu/kinesin//), and there are many potential modifications that can be made. This assay is included here because it provides the foundational experiments for achieving kinesin-driven transport in microfluidic devices. It is also routinely used as a control experiment to assess the function of motors, microtubules, and other reagents. The protocol for the microtubule gliding assay are as follows:

1. Prepare a flow cell using a glass microscope slide and 1½ coverslip, using double-stick tape as a spacer. The depth of the cell will be ~100 μm, and the width can be set by where the tape is placed.

2. Prepare 1 ml each of three solutions. BRB80CS is a casein-blocking solution consisting of 0.5-mg/ml filtered casein in BRB80. BRB80CA, made of 0.2-mg/ml casein and 1-mM ATP in BRB80 is used for motor dilutions. BRB80CT, made of 0.2-mg/ml casein and 10-μM paclitaxel in BRB80 is used for microtubule solutions.

2. Block the surface by flowing 50 μl of BRB80CS into the flow cell and incubating for 5 minutes. By carefully pipetting the solution into the edge of the experimental chamber, capillary action will draw the solution into the flow cell, filling it. The casein adsorbs to the glass surface and prevents denaturation of the motors on the surface.

3. Dilute the purified kinesin to 1 μg/ml in BRB80CA, and introduce 50 μl of this motor solution into the flow cell. Solution exchange is achieved by introducing the solution into one end of the flow cell using a pipette while simultaneously wicking the solution out of the other end of the flow cell using a Kim Wipe, filter paper, or other absorbent material. Incubate for 5 minutes to allow motor binding to the glass surface. (*Note:* Useful starting motor concentrations are from 0.05 to 5 μg/ml, and can be optimized empirically.)

4. For making motility solution, use stock solutions described in the "Materials" section with the recipe listed in Table 13.3.

   This motility solution contains microtubules, MgATP, Taxol to stabilize the microtubules, casein to maintain an adsorbed layer on the surface, and an oxygen scavenging system consisting of glucose, glucose oxidase, catalase, and β-mercaptoethanol to prevent photobleaching. Flush the flow cell with 50 μl of this motility solution.

5. Observe microtubule movement using epifluorescence microscopy. It is possible to observe microtubule movements by eye, but optimal performance is achieved by using a CCD camera attached to the microscope.

13.3.4 Design considerations for integrating motor proteins into microfluidic devices

To successfully integrate motor proteins in microfluidic devices, the following issues need to be considered in the design of the device.

1. Compatibility of materials. The chosen materials should be transparent and support kinesin-driven motion. Some traditional photoresists such as 1811 are autofluorescent and as such they hinder the visualization of fluorescent microtubules. We have assessed the activity of kinesin motors on a range of surfaces and found that motors retain their function when adsorbed to a number of different hydrophilic surfaces including glass, gold and chrome electrodes, plasma-treated silicon, and plasma-treated SU-8. However, motors lose their function and/or denature on hydrophobic surfaces such as hydrophobic silanes and untreated SU-8 [16, 17]. In addition, some materials such PDMS cannot be used with fluorescent microtubules, as microtubules are photobleached and depolymerize under the microscope even in the presence of an antifade system. This behavior has been attributed to the high oxygen permeability of PDMS, which overwhelms the oxygen scavenging activity of the antifade system [18, 19].

Table 13.3 Mobility Solution Recipe

Vol.	Reagent (Stock Conc.)	Final Conc.
85 μl	BRB80CT	
1 μl	MgATP (100 mM)	1 mM
1 μl	D-glucose (2 M)	20 mM
1 μl	glucose oxidase (2 mg/ml)	20 mg/ml
1 μl	catalase (0.8 mg/ml)	8 mg/ml
0.5 μl	β-mercaptoethanol	0.5%
10 μl	MT100 (0.32 μM)	~32 nM
100 μl		

2. Controlling microtubule movement. In the standard gliding assay, microtubules move in all directions across the surface, and the directionality is primarily determined by the orientation of the microtubules when they land on the surface. A number of groups have shown that microfabricated surface features such as walls and channels can be used to control the trajectory of kinesin-driven microtubules [17–22] (see Figure 13.2). One design feature that comes out of these studies is that controlling the direction of microtubule movement requires channels with widths less than 10 μm (the average length of microtubules polymerized in vitro). Due to the high flexural rigidity of microtubules, confinement in this size range forces the microtubules to follow the microchannel trajectory. Smaller channels provide better confinement, but control of fluid flow and solution exchange is increasingly challenging as the channel width is decreased.

3. Fluid flow and solution exchange. When enclosed microfluidic channels are used, fluidic connections from a syringe pump to the microchannels are required to enable the sample introduction into the device. The fluidic connections should be designed for fast solution exchange (the typical gliding assay requires two solution exchanges) and the fluidic connections should allow the sample to be mounted on a microscope stage.

4. Depletion of fuel and buffer. The surface of the typical gliding assay as described above is open to bulk solution so the ~100-μm layer of solution contains sufficient ATP to power the motors for many hours. However, in enclosed microchannels the high surface-to-volume ratio can lead to depletion of ATP fuel and loss in the buffering capacity of the solution over time [8]. Any design should therefore consider diffusion of these species into the channels, for instance by placing reservoirs along the length of the channel to replenish the fuel. Alternately, slow bulk flow in the channels can replenish the fuel without affecting microtubule movement.

 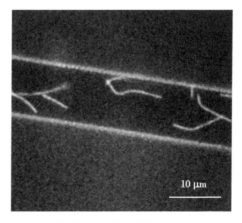

Figure 13.2 Confinement of microtubule motion using microfabricated walls. SU-8 is deposited on a glass substrate to create micron-scale walls (left). Under the fluorescence microscope with simultaneous low-level bright field illumination, moving microtubules and the microchannel walls can be observed (right). Microtubules that collide with the walls buckle and are redirected, resulting in transport along the channel axis.

13.3.5 Fabricating enclosed glass channels for microtubule transport

We have developed an approach for fabricating microchannels that address all of the issues discussed in the previous section. The design consists of a three-tier hierarchical structure (Figure 13.3) that links functional microchannels to macroscopic fluid connections. Shallow microchannels (~5-μm wide and ~ 1-μm deep) for microtubule confinement connect to intermediate channels (100-μm wide) that serve as reservoirs and also connect to 250-μm deep macrochannels that hold fine gauge tubing for simple external fluid connections. The micro-, intermediate-, and macroscale channels are etched in a 1-mm thick borosilicate glass substrate and bonded to a cover glass using poly(methyl methacrylate) (PMMA) as an adhesive [8]. Figure 13.4 shows the processing steps involved in fabrication of the hierarchical channels.

The materials used (glass and PMMA) support microtubule motility and do not interfere with visualization of fluorescent microtubules. PMMA is used as an adhesive to bond a glass coverslip to the glass substrate, a bonding approach that is more tolerant of particulates than anodic bonding or fusion bonding, and can be carried out at lower temperatures.

The macrochannel is designed in a U-shape with two tubing inserts connected to syringe needles to allow rapid solution exchange (See Figure 13.3). If a macrochannel with a single tubing is employed then each solution exchange requires a significant time to flush the solution present in the tubing and macrochannel through the narrow cross-section (1 × 5 μm) microchannel. Instead, in the U-shaped macrochannel one needle can be used to introduce solution and the other needle can be blocked to direct solution into the microchannel. For solution exchange, the second needle is unblocked to flush the U-shaped macrochannel with a new solution, and then is reblocked to direct the new solution into the microchannel.

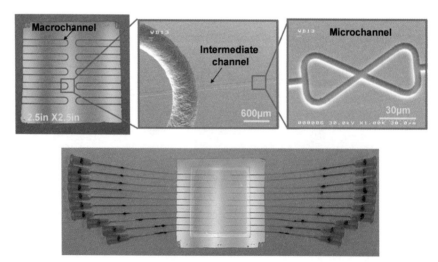

Figure 13.3 Hierarchical microchannel design. Macrochannels (250-μm deep) enable sample introduction through attached syringe needles. Intermediate channels (100-μm wide and 1-μm deep) connect to microchannels (5-μm wide and 1-μm deep) where microtubule motility is observed. The bottom panel shows a completed sample including the coverglass bonded using PMMA adhesive, and tubing for sample injection. (Image adapted, with kind permission from Springer Science+Business Media, from Huang et al. [8].)

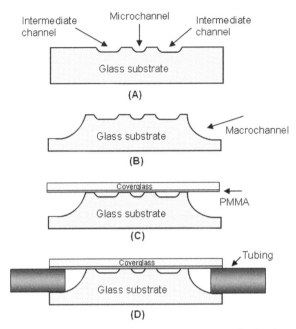

Figure 13.4 Schematic of fabrication process for enclosed glass microfluidic channels. Micro- and intermediate channels are first etched in the glass substrate (a), followed by etching deep macrochannels (b). A glass coverslip is then bonded using PMMA as an adhesive (c) and tubing for fluid connections are epoxied into the macrochannels (d). See text for fabrication details. (Image taken, with kind permission from Springer Science+Business Media, from Huang et al. [8].)

13.3.5.1 Microchannel fabrication protocol

The fabrication process is described below. For detailed protocols on making masks, photolithography steps, and photoresist development, consult the microfabrication protocols detailed in earlier chapters.

1. Clean glass substrates using acetone and isopropyl alcohol in an ultrasonic bath for 10 minutes each, followed by cleaning in Piranha solution (H_2SO_4:H_2O_2 = 4:1) for 20 minutes. The Piranha treatment both cleans the glass and results in a high-surface energy OH-saturated surface that improves the adhesion of deposited layers.

2. Deposit a 10-nm Cr layer and a 100-nm Au layer on the cleaned glass substrate using a resistive filament thermal evaporator. Micro- and intermediate channel designs are then patterned on the metal layers by photolithography. First, spin on Shipley 1811 photoresist, then expose the sample through a mask (we use a Karl Suss MA-55 aligner), and finally develop the photoresist. (*Note:* The Cr/Au layer eliminates photoresist lifting that can be a problem for direct etching of glass with only a photoresist mask.)

3. Etch shallow micro- and intermediate channels using a buffered oxide etch solution (NH_4F/HF 10:1) at room temperature with moderate agitation. A 20-minute etch results in 1-μm deep channels, which work well for this application. After etching, strip the photoresist, Au, and Cr layers by immersing in acetone, Au etchant, and Cr etchant, respectively.

4. Etch U-shaped macroscale channels into the glass to connect the macroscopic fluid connections to the microchannels. The glass etching uses a robust Cr/Au/SU-8

protection layer and concentrated hydrofluoric acid (HF). First, deposit a thermally evaporated layer of Cr (10 nm) and Au (100 nm) to achieve good adhesion to the glass surface and provide a seed layer for gold electroplating. Second, deposit 1811 photoresist on this metal layer, pattern the U-shaped macrochannels in it by photolithography using a dark field mask, and then use Cr and Au etching to expose the glass in these regions. Third, electroplate a 1.5-μm layer of gold (Techni gold 25ES, Technic Inc.) on the remaining Cr/Au. (*Note:* Active filtering of the electroplating solution during the electroplating process helps to remove particulates that can lead to pinhole defects in later steps.) Finally, spin on a 40-μm layer of SU-8 photoresist, align to the U-shaped channels using a light field mask, and pattern the macrochannels in this SU-8 to expose the glass. The thick electroplated Au provides good protection against concentrated HF with reduced defect density and cost compared to evaporation or sputtering, and the SU-8 photoresist provides additional protection against pinholes.

5. Etch glass for 45 minutes in 49% HF at room temperature, which results in ~300-μm deep macrochannels. After etching, the SU-8 layer can be removed by piranha cleaning and the electroplated Au and evaporated Cr/Au layers removed by wet etching. Alternatively, the sample can be simply soaked in Cr etchant to lift off the whole Cr/Au/Au/SU-8 layer.

6. Spin-coat a 500-nm thick PMMA layer from chlorobenzene on 48 × 50-mm coverglass (Gold seal #1, thickness: 0.13–0.17 mm) and bake on a 180°C hot plate for 20 minutes to remove the solvent and allow the PMMA layer to flow. Then, expose both the PMMA and glass microchannel surfaces to oxygen plasma for 1 minute to increase their adhesivity. Finally, laminate the PMMA-coated coverslip to the glass substrate using a Warner 100 hydraulic press laminator at 50 psi and 120°C for 10 minutes, and cool to room temperature while maintaining pressure. (*Note:* 996 kDa PMMA is chosen because it is sufficiently adhesive at convenient bonding temperatures to bond the two layers and is sufficiently viscous to prevent filling of the microchannels.)

7. Insert stainless-steel tubing into the macrochannels and bond using epoxy. Connect the other end of this tubing to syringe needles for sample introduction.

13.3.5.2 Integrating motors and microtubules into microfluidic channels

The microtubule gliding assay protocol can be integrated into microfluidic channels by sequentially flowing casein, kinesin, and microtubule solutions into the channels. However, to account for the high surface-to-volume ratios in microchannels, the reagent concentrations should be increased. The high surface-to-volume ratio also leads to gradients in motor adsorption in the channels, with higher motor densities upstream and lower motor densities downstream due to depletion of motors. While this gradient can be eliminated by flowing in large amounts of kinesin solution to saturate the surface with motors, the resulting high motor densities in the intermediate channels lead to dense microtubule meshes forming at the macrochannel/intermediate channel junction that eventually clog the intermediate channels [8]. This problem can be solved using a headless kinesin construct [6] that competes with the full-length kinesin for surface adsorption to the channel walls but does not interact with microtubules. By varying the ratio of full-length kinesin to headless kinesin in the motor solution, an optimal surface density of full-length kinesin can be obtained such that robust transport is achieved without dense networks of microtubules clogging the channels. The protocol for achieving functional microtubule transport in microfluidic channels is described below:

1. To immobilize motors inside the microchannels, introduce a solution containing 4-mg/ml casein, 7.5-μg/ml conventional kinesin and 32-μg/ml headless kinesin (KRT) in BRB80 buffer. (*Note:* Here we eliminated one solution exchange step by combining casein and kinesin solutions, using a large excess of casein such that casein adsorption will be much faster than kinesin adsorption. Alternately, casein solution can be introduced first, followed by the kinesin solution.)

2. Flush in a microtubule solution containing 0.64-μM rhodamine-labeled microtubules, 10mM ATP, 50 μM paclitaxel and antifade reagents (0.1 M D-glucose, 0.1 mg/ml glucose oxidase, 0.04mg/ml catalase, and 0 c.35M β-mercaptoethanol) in BRB80 buffer. (*Note:* The concentrations of ATP and antifade omponents are 10 to 20 times higher than standard assay to account for depletion in the microchannels.)

3. Visualize microtubule movements in the microchannels by epifluorescence microscopy. To simultaneously image the microchannels and the fluorescent microtubules, use simultaneous bright field microscopy with low illumination levels, which results in an overlaid image of the channels and the moving microtubules (as in Figure 13.2).

13.4   Results

Successful integration of microtubule-based motility into microfluidic channels requires both maintaining functional activity of the kinesin and microtubule proteins, as well as designing and fabricating biocompatible microchannels in geometries that generate useful transport. Here, we describe results from a standard microtubule gliding assay and then demonstrate two microchannel applications; one that generates unidirectional microtubule transport in enclosed channels and a second that accumulates dense bundles of uniformly aligned microtubules. These examples are from our published work [8], and for the interested reader there are a number of other published strategies from our group and others for controlling the kinesin-driven transport of microtubules in both open top and enclosed microchannels [1, 16, 17, 20–28].

The microtubule gliding assay is both the experimental foundation for integrating microtubule transport into microfluidic channels, as well as a helpful research tool for assessing the activity of kinesin and microtubule proteins and for testing the

Figure 13.5   Microtubules moving over a kinesin-coated surface in the microtubule gliding assay. Image acquired by fluorescence video microscopy with frames 5 seconds apart.

biocompatibility of different surface chemistries and materials. Figure 13.5 shows frames from a standard microtubule gliding assay, and a video is available on the Artech House Web site (http://www.methodsinbioengineering.com). Immediately following the introduction of microtubules, filaments land and move across the kinesin-covered surface in the range of 0.5 to 1 μm/s at room temperature. Occasionally, microtubules will dissociate from the surface and diffuse away, but when high motor densities (1-μg/ml kinesin and above) and long microtubules (> 5 μm) are used this dissociation is rare. When very low kinesin concentrations are used in the assay, transient events can be observed in which microtubules land, are moved by individual kinesin motors, and then diffuse away when the motor reaches the minus-end of the microtubule [14]. Investigators who are exploring novel microchannel geometries, exploring new materials for such microchannels, or otherwise exploring this research area are advised to use the microtubule gliding assay to fully explore experimental variables before integrating the motility into channels.

When these same motors are integrated into enclosed microfluidic channels (fabricated according to the protocol in Section 3.5), the channel walls guide the microtubules and specific channel geometries can be used to achieve unidirectional microtubule transport. As a microtubule is being transported across a kinesin-functionalized surface, the leading end of the microtubule is continually encountering new motors. In the absence of any external forces the path the microtubule takes will tend to be fairly straight, since on these micron dimensions microtubules are quite stiff [29]. However, if the tip is biased to encounter motors in one specific direction, such as when a microtubule encounters a microfabricated barrier at an angle, the entire microtubule will tend to follow that path. One useful feature for microtubule-based microfluidic devices is a directional rectifier that causes microtubules to all move in the same direction. Figure 13.6 shows the design and performance of one such directional rectifier. Microtubules that enter from the left are guided either to the top or the bottom channel, and they travel around the structure and exit from the left. In contrast, microtubules that enter from the right collide with the channel wall, are guided either up or down, and then they continue through the channels and exit to the left. Thus, this structure generates uniform transport directions for a population of filaments. We found that

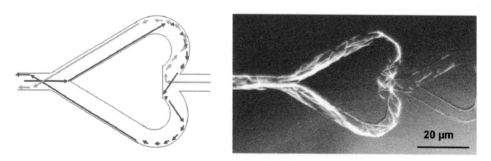

Figure 13.6 A microtubule directional rectifier fabricated in enclosed glass microchannels. Schematic at left shows expected microtubule paths. Microtubules entering from the left (red trace) travel around the structure and exit to the left, while microtubules that enter from the right (green trace) pass through the structure. Image at right shows microtubules moving through the rectifier. Microtubule originate from a reservoir at left side of image; the small number of microtubules to the right of the structure demonstrate the successful redirection. (Image adapted, with kind permission from Springer Science+Business Media, from Huang et al. [8].)

roughly 95% of the filaments are redirected in the correct orientation, and if higher tolerances are required, multiple rectifiers can be placed in series.

Another useful structure is one that both reorients microtubules to achieve uniform transport and acts to corral these microtubules to generate a collection of uniformly polarized filaments. We have accomplished this by fabricating enclosed glass microchannels to create a ring or "roundabout" structure. The fabrication approach detailed in Section 3.5 was used to create macrochannels for fluid introduction, intermediate channels that act as microtubule reservoirs (see Figure 13.3), and a circular microchannel was created that contains input and exit ports oriented tangentially (Figure 13.7). Microtubules that enter from the reservoir at the right travel around the ring, guided by the channel walls, while any filaments that change direction exit through one of the ports. Over time, thousands of uniformly oriented filaments can be collected that all move together through the channel. Among other uses, this system could serve as a starting point for cargo attachment, and if a gate were created, the filaments could be simultaneously directed at the chosen time into new channel.

The two microchannel structures described above are important building blocks towards realizing the device envisioned in Figure 13.1. There is clearly room for further developments in this area, and the successful strategies presented here can be considered a starting point for future studies in this area. In other work, we and others have developed approaches for attaching molecular beacons [4], magnetic nanoparticles [30, 31], quantum dots [32], and antibodies [33] to microtubules for carrying cargo. Most of these strategies rely on using biotinylated tubulin and using the biotin binding protein streptavidin as a bridge to attach the cargo. Tubulin can be biotinylated using the same protocol as described above for rhodamine-labeling.

13.5 Discussion of pitfalls

While there is considerable potential for incorporating microtubule motility into microfluidic channels, because this interdisciplinary area relies on both maintaining

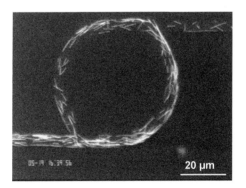

Figure 13.7 Microtubule concentration and unidirectional transport in an enclosed glass microchannel ring structure. Microtubules enter from reservoirs at both the left and right and either travel around the ring or are guided out of the ring. At left, microtubule movements shortly after introduction into the channel. Microtubules move in both directions at this point. At right, an image of concentration ring after 90 minutes of accumulation. Microtubules are moving counterclockwise around the ring; fluorescence intensity measurements estimate that thousands of microtubules are accumulated. (Image adapted, with kind permission from Springer Science+Business Media, from Huang et al. [8].)

functional proteins and fabricating useful microdevices, there are a number of points in the process that can fail. A number of potential pitfalls and alternate strategies are discussed here, following the same sequence used in the "Materials" and "Methods" sections.

13.5.1 Kinesin purification

The bacterial expression and column purification of kinesin has a number of potential pitfalls, and the protocol generally requires a few runs to achieve success. In the bacterial work, standard sterile technique is important to prevent bacterial contaminations. In assessing protein expression, running SDS-PAGE gels of bacteria before and after inducing with IPTG is very helpful. As the protein is hexaHis tagged, it is possible to use commercial kits (Novagen is one source) for detecting His-tagged proteins. If a centrifuge is not available, the bacterial lysate can be clarified by passing through a 0.45-micron syringe filter (though multiple filters are generally needed due to clogging). For the chromatography steps, other chromatography systems besides the FPLC will work fine and it is even possible to carry out batch (centrifuge-based) purifications using the Ni-NTA resin. As with expression, using SDS-PAGE gels to assay the motor concentrations in the clarified lysate, column flow through, washes, and elution is very helpful for troubleshooting. Gels of the elution fractions will also show the relative purity and the final concentration of the motor sample.

Following purification, it is generally a good idea to assess motor function using the microtubule gliding assay. Because this assay is relatively quick and is sensitive, it can be used for very low concentration samples to determine if any motor is present below the threshold of gel detection. One mode of failure is that the purified motors are nonfunctional. It is important to remember that retaining kinesin functionality requires keeping some nucleotide in the buffers (including all column buffers) at all time ($10\text{-}\mu$M MgATP minimum is a good rule of thumb). Besides the microtubule gliding assay, another method for assessing motor function is to carry out a microtubule binding step in the presence of AMP-PNP, a nucleotide analog that locks the motors onto microtubules. Motor function can be assessed by incubating motors and microtubules in 1 mM AMP-PNP, pelleting the sample in the Airfuge, and analyzing the pellet and supernatant by SDS-PAGE. Nonfunctional (nonbinding) motors will remain in the supernatant, while functional motors will pellet with the microtubules. If column-purified motors are nonfunctional, it is advised to repeat expression and test the motility of the clarified lysate. Although there are a number of bacterial proteins, bacteria have no microtubule motors and if motors are functional in the bacteria, this lysate should work in gliding assays.

13.5.2 Tubulin

The tubulin purification protocol involves a number of steps that can go awry and success requires determination and persistence. On the plus side, it is reasonably easy to troubleshoot and determine which steps fail. As discussed, using fresh brains transported rapidly from the slaughterhouse is important for achieving high yields. As an aside, we have noticed that yields are reduced in the summer, and a colleague Richard Cyr has also reported this seasonal variation. As cows are homeotherms, the reason for this is not at all clear, but if preps can be carried out in other seasons, that is ideal. It should be noted that there have been unpublished reports of tubulin preps starting with frozen brains, but we have no experience with this.

The first hurdle of the tubulin prep is obtaining the first microtubule pellet (following polymerization of the clarified brain homogenate). These pellets should be clearly observable by eye and have a somewhat opalescent quality to them. Very small pellets in this step almost always spell doom for the rest of the purification. A second indicator of success is an increase in viscosity as the tubulin solutions polymerize. While these indicators are diagnostics and not fixes, they are very helpful for assessing if the prep is working. One step toward avoiding failure in the tubulin prep is to ensure that the GTP is good. Purchase from a reliable supplier and store properly.

When labeling tubulin, losses are unavoidable—a 10% yield from the thawed tubulin sample is routine. One way to avoid further losses is to maintain the tubulin concentration above its critical concentration of ~1 mg/ml. In fact, maintaining the tubulin concentration above 5 mg/ml is recommended. Because the reactive dyes are fairly unstable, care should be taken to use dry (ideally newly purchased) DMSO for dissolving the reactive dyes. Also, excessive labeling can reduce yields of active tubulin. The degree of labeling can be optimized by changing either the dye concentration or the incubation time, and different fluorophores may have different optimal labeling times and concentrations.

For both the tubulin prep and tubulin labeling, three important points to keep in mind are (1) start with sufficient amounts of tubulin, (2) reduce volumes through the procedures to maintain sufficiently high tubulin concentrations, and (3) work to minimize the time that tubulin is sitting on the bench or in an ice bucket as the protein activity falls over time. While losses in tubulin activity are unavoidable, the good news is that active tubulin can easily be recovered by cycling such as in the tubulin labeling protocol (Section 3.2.3, steps 1–3). By polymerizing the active tubulin and pelleting it, any nonfunctional tubulin can be removed from the sample and, because the resuspension volume can be chosen, this is also a way to concentrate tubulin samples.

13.5.3 Motility assays

When working properly, the microtubule gliding assay serves as an important tool for testing and optimizing motor and microtubule function for microchannel experiments. However, the microtubule gliding assay itself can fail for a number of reasons. First, it is important to start with a sufficient concentration of stabilized microtubules. The presence of microtubules in the MT100 sample can be assessed by dropping a few microliters of the solution onto a microscope slide, covering with a glass coverslip, and visualizing the microtubules in this "squash" by epifluorescence microscopy. The next crucial reagent in the experiment is casein. Casein must be prepared as described in Materials and stored frozen. Beware that casein solutions left on the bench for days serve as excellent incubators for bacteria (think spoiled milk). If there are questions with the casein, it is best to make it up new from powder. The third important component of the gliding assay is the antifade system. The enzymes scavenge dissolved oxygen in the solution and the reducing agent further prevents free radical formation resulting from illumination of the fluorophores by the microscope illumination. Failure of the antifade system is fairly easy to assess because motility quickly ceases and the microtubules photobleach and then fragment due to free radical attack on the proteins. This can occur at the edges of the flow cell or near any bubbles in the flow cell. If this is air contact and is ruled out, the motility solution should be made from scratch using newly thawed aliquots of all antifade components. Note also that β-mercaptoethanol can oxidize over time and should be replaced roughly every 6 months.

Motility assays can fail in two ways—either no microtubules land on the surface or microtubules land on the surface but they don't move. With proper casein treatment, microtubules do not bind to the glass coverslip surface in the absence of motors, and so any bound microtubules can be attributed to kinesin activity. If, after 5 minutes no microtubules are seen on the surface, this generally means either that the motor concentration is low or the motor activity is low or absent. The first option can be tested by repeating with a higher motor concentration. One way to enhance the signal from a small population of active motors is to repeat the motility experiment with the modification that the ATP in the motility solution is replaced with AMP-PNP. This nonhydrolyzable ATP analog causes motors to bind to microtubules irreversibly, so any filaments that do land are stuck and they accumulate over time. If no microtubule binding is seen after 5 to 10 minutes in AMP-PNP, this means there is no motor activity and suggests that the motors are completely inactive.

Another mode of failure for kinesins is irreversible binding to microtubules. If these so-called "deadheads" are present in a motor sample, the gliding speed will generally be reduced, spiraling microtubules will be observed at high motor concentrations (due to the front end getting stuck), and over time an increasing amount of bound but immobile microtubules will be observed. In very poor motor samples, it is possible to obtain only microtubule binding and no microtubule movement; there is generally no hope of rescuing these samples. If kinesin samples have a small number of deadheads, the activity can be improved by simply using lower motor concentrations on the surface. An alternative step is to "clean up" the motors by incubating motors with 1-μM microtubules in 1-mM ATP for 10 minutes, pelleting the microtubules in the Airfuge to remove any motors that irreversibly bind, and collecting the supernatant that should contain active motors. In conclusion, because achieving kinesin-driven motility in channels rests upon reliable kinesin and microtubule samples, it is worth spending time working with gliding assays to fully characterize the functional activities of the motors and microtubules.

13.5.4 Motility in microchannels

Achieving robust microtubule transport in microfluidic channels requires proper introduction of solutions into the channels, proper adsorption of motors to the channel walls, and good functional activity of the kinesin motors and microtubules. Before introducing motor or microtubule samples, it is often worthwhile to introduce a solution of fluorescent dye or labeled protein such as rhodamine-BSA into the channels to confirm that the channels are not occluded, that they don't leak, and that solution can be introduced into all regions of the device. Pumping a solution that is very easy to visualize also confirms that the bonding process can withstand the hydrostatic pressures created by the pump or syringe used to introduce the sample.

As discussed above, motility in microchannels differs from the standard gliding assay in the high surface to volume ratios involved. For this reason, it is often important to increase not only the motor concentrations introduced into the channels but also the

Troubleshooting Table

Problem	Explanation	Potential Solutions
Kinesin purification failed.	No kinesin expressed.	Do miniprep and plasmid digest to confirm plasmid is correct. Make sure BL21(DE3) cells are used for expression. Try motility assay with clarified lysate to look for any motor activity.
	No kinesin purified.	Confirm that Ni column is properly charged. Perform motility assay with elution peak using AMP-PNP to look for any microtubule binding. Confirm kinesin expression (above).
	Kinesin is inactive.	Confirm that there is at least 1-μM MgATP in all column buffers and that expression was carried out at 20°C.
Tubulin prep failed.	Low yield or activity.	Minimize time from cow to blender and maintain on ice. Check GTP by using it to polymerize some control tubulin. Run gels of all steps to determine where protein was lost.
Tubulin labeling failed.	Tubulin is present but not labeled.	Buy new rhodamine-NHS dye. Start with new bottle of dry DMSO.
	Low or zero tubulin yield.	Decrease labeling time and/or label concentration to control at ~1 label/tubulin. Spec solutions to determine where protein lost.
Microtubule gliding assay fails.	No microtubules on surface.	Confirm that microtubules are present by visualizing squash (see text) or focusing into solution above surface. Confirm kinesin concentration and use > 1 μg/ml. Replace ATP with AMP-PNP to check microtubule binding.
	Microtubules bind to surface but don't move.	Dilute motors to minimize dead head influence. Confirm that ATP is present. Confirm that antifade is working and that βME is < 6 months old (will likely see fast bleaching). May need to do new kinesin prep to obtain functional motors.
No motility in microchannels.	Microtubules aren't seen.	Flow through dye to confirm channels are open. Visualize upstream to confirm microtubules aren't trapped. Use headless kinesin at entry to minimize microtubule sticking.
	Microtubules don't move.	Use higher kinesin concentrations to compensate for large surface area/volume. Increase concentration of antifade components. Use standard gliding assay to confirm that all reagents are working properly.

casein, ATP, and antifade concentrations (increases up to 10× generally don't have any detrimental effects). As noted in Section 3.5.2, when high concentrations of kinesin adsorb to long channels, this can prevent microtubules from entering the channels because the microtubules bind and become tangled at the channel entry. Combining the active kinesin with headless kinesin reduces the maximal kinesin surface activity and helps to alleviate this problem. The optimum active:headless kinesin ratio is best determined empirically.

Generally, troubleshooting problems with kinesin-driven motility in channels is best achieved using the standard microtubule gliding assay. For instance, if novel substrates or photoresists are being used to create the microfluidic channels, flat surfaces can be created from these materials (such as by spinning the material onto coverslips), flow cells can be assembled on them, and the gliding assay can be carried out on the modified surface. We have found that for a range of surfaces, plasma treatment renders them hydrophilic and compatible with kinesin motors.

13.5.5 Final comments

The integration of biologically-driven transport into microengineered devices is a relatively new and emerging area. To date, applications have focused on microscale transport in channels, but there are a number of potential directions for this research. Here, we have provided the foundation protocols for harvesting and modifying kinesin motors and microtubules as well as successful approaches for integrating these cellular proteins into engineered microdevices. These principles should provide a solid foundation for future investigations and will hopefully lead to functional devices based on these technologies.

Acknowledgments

The authors thank members of the Hancock and Jackson labs, particularly Lili Jia, Samira Moorjani, Yangrong Zhang, Gayatri Muthukrishnan, and Zach Donhauser. This project was funded by the Penn State Center for Nanoscale Science (NSF MRSEC DMR0213623) and by an NSF Biophotonics Grant (0323024) to W.O.H. and T.N.J. funded jointly by the NSF and NIH/NIBIB.

References

[1]   Jia, L., et al., "Microscale transport and sorting by kinesin molecular motors," *Biomedical Microdevices*. Vol. 6, No. 1, 2004, pp. 67–74.

[2]   Hancock, W.O., "Protein-based nanotechnology: Kinesin-microtubule driven systems for bioanalytical applications," in *Nanodevices for Life Sciences*, pp. 241–271, C. Kumar, (ed.), Wiley-VCH: Weinheim, Germany, 2006.

[3]   Hess, H., "Materials science. Toward devices powered by biomolecular motors," *Science*. Vol. 312, No. 5775, 2006, pp. 860–861.

[4]   Raab, M., and W.O. Hancock, "Transport and detection of unlabeled nucleotide targets by microtubules functionalized with molecular beacons," *Biotechnol Bioeng*. Vol. 99, No. 4, 2008, pp. 764–773.

[5]   Lawrence, C.J., et al., "A standardized kinesin nomenclature," *J Cell Biol*. Vol. 167, No. 1, 2004, pp. 19–22.

[6]   Hancock, W.O., and J. Howard, "Processivity of the motor protein kinesin requires two heads," *J Cell Biol*. Vol. 140, No. 6, 1998, pp. 1395–1405.

[7]   Coy, D.L., M. Wagenbach, and J. Howard, "Kinesin takes one 8-nm step for each ATP that it hydrolyzes," *J Biol Chem*. Vol. 274, No. 6, 1999, pp. 3667–3671.

[8]   Huang, Y.-M., et al., "Microtubule transport, concentration and alignment in enclosed microfluidic channels," *Biomed Microdevices*. Vol. 9, 2007, pp. 175–184.

[9]   Vernos, I., *Kinesin Protocols. Methods in Molecular Biology*, Vol. 164. Totowa, NJ: Humana Press, 2000.

[10]   Desai, A., and T.J. Mitchison, "Microtubule polymerization dynamics," *Annu Rev Cell Dev Biol*. Vol. 13, 1997, pp. 83–117.

[11]   Williams, R.C., Jr., and J.C. Lee, "Preparation of tubulin from brain," *Methods Enzymol*. Vol. 85 Pt B, 1982, pp. 376–385.

[12]   Castoldi, M., and A.V. Popov, "Purification of brain tubulin through two cycles of polymerization-depolymerization in a high-molarity buffer," *Protein Expr Purif*. Vol. 32, No. 1, 2003, pp. 83–88.

[13] Hyman, A., et al., "Preparation of modified tubulins," *Methods Enzymol.* Vol. 196, 1991, pp. 478–485.

[14] Howard, J., A.J. Hudspeth, and R.D. Vale, "Movement of microtubules by single kinesin molecules," *Nature.* Vol. 342, No. 6246, 1989, pp. 154–158.

[15] Howard, J., A.J. Hunt, and S. Baek, "Assay of microtubule movement driven by single kinesin molecules," *Methods Cell Biol.* Vol. 39, 1993, pp. 137–147.

[16] Huang, Y., et al., "Microfabricated capped channels for biomolecular motor-based transport," *IEEE Transactions on Advanced Packaging.* Vol. 28, No. 4, 2005, pp. 564–570.

[17] Moorjani, S., et al., "Lithographically patterned channels spatially segregate kinesin motor activity and effectively guide microtubule movements," *Nano Lett.* Vol. 3, No. 5, 2003, pp. 633–637.

[18] Brunner, C., et al., "Lifetime of biomolecules in polymer-based hybrid nanodevices," *Nanotechnology* Vol. 15, 2004, pp. S540–S548.

[19] Kim, T.S., et al. "Biomolecular motors as novel prime movers for microTAS: microfabrication and materials issues," in *7th Int. Conf. on Micro Total Analysis Systems.* 2003. Squaw Valley, CA: Transducers Research Foundation.

[20] Hiratsuka, Y., et al., "Controlling the direction of kinesin-driven microtubule movements along microlithographic tracks," *Biophys J.* Vol. 81, No. 3, 2001, pp. 1555–1561.

[21] Hess, H., et al., "Ratchet patterns sort molecular shuttles," *Applied Physics A-Materials Science & Processing.* Vol. 75, No. 2, 2002, pp. 309–313.

[22] Clemmens, J., et al., "Mechanisms of microtubule guiding on microfabricated kinesin-coated surfaces: Chemical and topographic surface patterns," *Langmuir.* Vol. 19, No. 26, 2003, pp. 10967–10974.

[23] Hess, H., et al., "Molecular shuttles operating undercover: A new photolithographic approach for the fabrication of structured surfaces supporting directed motility," *Nano Lett.* Vol. 3, No. 12, 2003, pp. 1651–1655.

[24] Clemmens, J., et al., "Motor-protein "roundabouts": microtubules moving on kinesin–coated tracks through engineered networks," *Lab Chip.* Vol. 4, No. 2, 2004, pp. 83–86.

[25] Cheng, L., et al., "Highly Efficient Guiding of Microtubule Transport with Imprinted CYTOP Nanotracks," *Small.* Vol. 1, No. 4, 2005, pp. 409–414.

[26] van den Heuvel, M.G., et al., "High rectifying efficiencies of microtubule motility on Kinesin-coated gold nanostructures," *Nano Lett.* Vol. 5, No. 6, 2005, pp. 1117–1122.

[27] van den Heuvel, M.G., M.P. de Graaff, and C. Dekker, "Molecular sorting by electrical steering of microtubules in kinesin-coated channels," *Science.* Vol. 312, No. 5775, 2006, pp. 910–914.

[28] Lin, C.T., et al., "Efficient designs for powering microscale devices with nanoscale biomolecular motors," *Small.* Vol. 2, No. 2, 2006, pp. 281–287.

[29] Gittes, F., et al., "Flexural rigidity of microtubules and actin filaments measured from thermal fluctuations in shape," *J Cell Biol.* Vol. 120, No. 4, 1993, pp. 923–934.

[30] Platt, M., et al., "Millimeter scale alignment of magnetic nanoparticle functionalized microtubules in magnetic fields," *J Am Chem Soc.* Vol. 127, No. 45, 2005, pp. 15686–15687.

[31] Hutchins, B.M., et al., "Directing transport of CoFe2O4-functionalized microtubules with magnetic fields," *Small.* Vol. 3, No. 1, 2007, pp. 126–131.

[32] Bachand, G., et al., "Assembly and transport of nanocrystal CdSe quantum dot nanocomposites using microtubules and kinesin motor proteins," *Nano Lett.* Vol. 4, No. 5, 2004, pp. 817–821.

[33] Bachand, G.D., et al., "Active capture and transport of virus particles using a biomolecular motor-driven, nanoscale antibody sandwich assay," *Small.* Vol. 2, No. 3, 2006, pp. 381–385.

About the Editor

Jeffrey D. Zahn is an assistant professor of biomedical engineering at Rutgers University in Piscataway, New Jersey. He received his Ph.D. in bioengineering from the joint graduate group in bioengineering at the University of California at San Francisco and Berkeley in 2001. Dr. Zahn's research interests lie in microfluidics, microdevice design, and fabrication. Dr. Zahn also has expertise in microflow dynamics, two-phase and particulate flow, and optical flow diagnostics, as well as experience in molecular biology and biological preparation techniques.

List of Contributors

Nitin Agrawal
BioMEMS Resource Center
Center for Engineering in Medicine and Surgical Services
Massachusetts General Hospital
Boston, MA 02114

Han-Sheng Chuang
Birck Nanotechnology Center and School of Mechanical
Engineering
Purdue University
West Lafayette, IN 47907

Alex Fok
Department of Biomedical Enigneering
Rutgers, The State University of New Jersey
Piscataway, NJ 08854

Jennifer O. Foley
Department of Bioengineering
University of Washington
Seattle, WA 98195

Elain Fu
Department of Bioengineering
University of Washington
Seattle, WA 98195

Jason P. Gleghorn
Sibley School of Mechanical & Aerospace Engineering
Cornell University
Ithaca, NY 14853

William O. Hancock
Associate Professor
Department of Bioengineering
The Pennsylvania State University
205 Hallowell Bldg.
University Park, PA 16802
e-mail: wohbio@engr.psu.edu

Benjamin G. Hawkins
Department of Biomedical Engineering
Cornell University
Ithaca, NY 14853

Amy E. Herr
Assistant Professor
Department of Bioengineering
University of California, Berkeley
308B Stanley Hall, MC 1762
Berkeley, CA 94720-1762
e-mail: aeh@berkeley.edu

Ying-Ming Huang
Department of Electrical Engineering
The Pennsylvania State University
University Park, PA 16802

Zohora Iqal
Department of Bioengineering
University of California, Berkeley
Berkeley, CA 94720-1762

Daniel Irimia
Research Associate
BioMEMS Resource Center
Center for Engineering in Medicine and Surgical Services
Massachusetts General Hospital
Shriners Hospital for Children
Harvard Medical School
114 16th Street, Room 1239
Charlestown, MA 02129
e-mail: dirimia@hms.harvard.edu

Thomas N. Jackson
Department of Electrical Engineering
The Pennsylvania State University
University Park, PA 16802

Yang Jun Kang
School of Information and Mechatronics
Gwangju Institute of Science and Technology
Buk-gu Gwangju, 500-712 Republic of Korea

Myoung Gon Kim
School of Information and Mechatronics
Gwangju Institute of Science and Technology
Buk-gu Gwangju, 500-712 Republic of Korea

Brian J. Kirby
Assistant Professor
Sibley School of Mechanical & Aerospace Engineering
Cornell University
238 Upson Hall
Ithaca, NY, 14853
e-mail: bk88@cornell.edu

Aloke Kumar
Birck Nanotechnology Center and School of Mechanical
Engineering
Purdue University
West Lafayette, IN 47907

Abraham P. Lee
Professor
Department of Biomedical Engineering
Department of Mechanical & Aerospace Engineering
University of California at Irvine
3120 Natural Sciences II Engineering
Tower 716F
Irvine, CA 92697
e-mail: aplee@uci.edu

Luke P. Lee
Lloyd Distinguished Professor
Department of Bioengineering
University of California, Berkeley
408C Stanley Hall
Berkeley, CA 94720-1762
e-mail: lplee@berkeley.edu

Robert Lin
Department of Biomedical Engineering
University of California at Irvine
Irvine, CA 92697

Frank B. Myers
Department of Bioengineering
University of California, Berkeley
Berkeley, CA 94720-1762

Kjell E. Nelson
Department of Bioengineering
University of Washington
Seattle, WA 98195

Nam-Trung Nguyen
Associate Professor
School of Mechanical and Aerospace Engineering
Nanyang Technological University
50 Nanyang Avenue
Singapore 639798
e-mail: mntnguyen@ntu.edu.sg

Vamsee Pamula
Advanced Liquid Logic Inc.
Morrisville, NC 27560

Marcelo B. Pisani
Department of Electrical Engineering
The Pennsylvania State University
University Park, PA 16802

Michael Pollack
Advanced Liquid Logic Inc.
Morrisville, NC 27560

Shankar Shastry
Department of Bioengineering
The Pennsylvania State University
University Park, PA 16802

Ramakrishna Sista
Advanced Liquid Logic Inc
Morrisville, NC 27560

Vijay Srinivasan
Advanced Liquid Logic, Inc.
615 Davis Dr., Ste 800
Morrisville, NC, 27560
e-mail: VSrinivasan@Liquid-Logic.com

Dean Y. Stevens
Department of Bioengineering
University of Washington
Seattle, WA, 98195

Srinivas A. Tadigadapa
Associate Professor of Electrical Engineering
The Pennsylvania State University
121 Electrical Engineering West
University Park, PA 16802
e-mail: srinivas@engr.psu.edu

Shia-Yen Teh
Department of Biomedical Engineering
University of California at Irvine
Irvine, CA 92697

Mehmet Toner
BioMEMS Resource Center
Center for Engineering in Medicine and Surgical Services
Massachusetts General Hospital
Boston, MA 02114

Maruti Uppalapati
Department of Bioengineering
The Pennsylvania State University
University Park, PA 16802

Steven T. Wereley
Associate Professor
Birck Nanotechnology Center and School of Mechanical Engineering
Purdue University
1205 W. State Street
West Lafayette, IN 47907
e-mail: wereley@purdue.edu

Lavanya Wusirika
Department of Bioengineering
University of California, Berkeley
Berkeley, CA 94720-1762

Paul Yager
Professor and Acting Chair
Department of Bioengineering
University of Washington
Box 355061, Foege N530J
Seattle, WA 98195
e-mail: yagerp@u.washington.edu

Sung Yang
Assistant Professor
School of Information and Mechatronics
Graduate Program of Medical System Engineering
Department of Nanobio Materials and Electronics
Gwangju Institute of Science and Technology
1 Oryong-dong
Buk-gu Gwangju, 500-712 Republic of Korea
e-mail: syang@gist.ac.kr

Sang Youl Yoon
Graduate Program of Medical System Engineering
Gwangju Institute of Science and Technology
Buk-gu Gwangju, 500-712 Republic of Korea

Jeffrey D. Zahn
Assistant Professor
Department of Biomedical Engineering
Rutgers, The State University of New Jersey
599 Taylor Road Room 311
Piscataway, NJ 08854
e-mail: jdzahn@rci.rutgers.edu

Index

A

Absorbance detection, 185–88
AC fields, 145–46
Acoustic micromixers, 72–75
    defined, 72
    example, 74
    schematic, 74
    See also Micromixers
Active micromixers, 70–75
    acoustic, 72–75
    electrohydrodynamic mixing, 71–72
    temporal pulsing/Taylor dispersion, 70–71
    See also Micromixers
Actuators, 33–42
    bimetallic, 38
    capillary-force, 42
    chemical, 41–42
    electromagnetic, 39–40
    electromechanical, 40–41
    electrostatic, 39
    illustrated, 36
    operation parameters, 34
    piezoelectric, 38–39
    pneumatic, 35–36
    response time, 33
    shape-memory, 38
    solid-expansion, 37–38
    stack-type, 33
    thermopneumatic, 36–37
Additive processes, 6–10
    defined, 3
    deposition techniques, 7–10
    growth of SIO_2, 6–7
Alginate hydrogel microbeads, 253
Angled constriction, 164–65
Angled electrodes, 160–62
Antibody assays, 154
Automated glucose assays on-chip,
    123, 126–27

B

Backscattering interferometry, 193–94

Ball bonding, 17–18
Bead washing, 117
Benzocyclobutene, 21
Biconvex microlenses, 186
Bimetallic actuators, 38
Bioassays, 248–49
Biosensing, 303
Bonding processes, 3, 13–16
    lamination, 13–14
    polymer, 16
    wafer methods, 14–16
Brownian motion, 296, 301, 302

C

Capillary electrophoresis (CE),
    88, 189, 190–91
Capillary-force actuators, 42
Capillary pumps, 50–51
Castellated electrode traps, 159–60
CD-based microfluidics, 187
Central difference image correlation (CDIC),
    288, 289–90
Central difference interrogation (CDI),
    288, 289
    defined, 289
    precision errors, 290
    SPE, 304, 306
Channel surface modification, 255
Chaotic advection, 66–70
Chaotic micromixers, 61, 66–70
    geometric patterning, 69
    herringbone, 69
    types of, 68
Charge-carrying particles, 154
Check valve pumps, 43–45
    compression ratio, 44, 45
    defined, 43
    performance, 45
    structure, 44
    See also Micropumps
Chemical actuators, 41–42
Chemical-catalyzed polymerization, 250

Chemical gradients, 210–23
    chemotaxis assay, 216–20
    data acquisition, 220
    device design, 210–13
    device fabrication, 213–15
    generator networks, 210–11
    introduction to, 210
    materials, 213
    methods, 213–20
    surface treatment, 215
    troubleshooting tips, 220
Chemical-mechanical polishing (CMP), 13
Chemical resistance, 258
Chemical vapor deposition (CVD), 8–9
Chemiluminescence
    defined, 191
    detection, 191–92
    for immunoassays, 122
    measurements, 122
Chemokine gradients, 210
Chemotaxis assay, 216–20
    blood sample loading, 217–19
    device operation, 219–20
    device priming, 216–17
    media preparation, 216
    neutrophil isolation, 217–19
Cholestech GDX System, 187
Circle-dot geometry, 150
Circle-dot traps, 158–59
Clausius-Mossotti factor, 138, 147
Clinical diagnostics, 115
Colorimetry, 121–22
Computer tomography (CT), 304
Concentration gradient immunoassay
        (CGIA), 227, 229–30
    analyte concentration, 240–41
    assay shift, 237–39
    complex sample analysis, 241–42
    concentration determinations, 230
    data acquisition and results, 237–39
    defined, 229
    discussion, 240–41
    fabrication and flowcells, 232–33
    line profiles, 238
    mass flux, 230
    materials, 231
    operation, 229
    response magnitude, 240
    sample and reagent delivery, 233–35
    schematic, 229
    SPR difference image, 237, 238
    troubleshooting table, 241
    See also Immunoassays
Confocal μPIV, 297
Continuous-flow dielectrophoresis, 172

Continuous window shift (CWS), 288
Correlation averaging, 288
Coulomb force, 135
Cross-linked gels, 83
Curved constriction, 167–68
Curved electrodes, 160–62

D

DC fields, 146–47
Deadheads, 334
Deep UV (DUV), 6
Deposition techniques, 7–10
    chemical vapor deposition (CVD), 8–9
    electroless plating, 9–10
    electroplating, 9
    spin-on deposition, 7
    sputtering, 8
    vacuum evaporation, 7–8
Depth of correlation (DOC), 286, 294
Depth of field (DOF), 286
Dicing, 16–17
Dielectric deposition, 120
Dielectrophoresis (DEP), 112, 133–76
    application notes, 169–76
    continuous, 172
    as continuous-flow separation technique,
        175
    defined, 134
    device considerations, 152–56
    dipole approximation, 140–41
    electric field frequency, 145–47
    electric field phase, 147–49
    electrode-based, 152, 156–63
    electrode configurations, 149–51
    electrodeless, 151
    electrolysis, 152
    experimental parameters, 168
    fabrication uncertainty, 155–56
    for field flow fractionation (FFF-DEP), 173
    force, 174
    fractionation, 175
    frequency response characteristic, 146
    geometry, 149–52
    infinite domain assumption, 144
    insulative, 151–52, 163–68
    introduction to, 134–45
    isolated particle assumption, 144–45
    limiting assumptions, 138–45
    materials, 145–52
    Maxwellian equivalent body, 141–42
    Maxwell's stress tensor and, 135, 136–37
    media conductivity, 153
    methods, 152–68
    multiple-frequency (MFDEP), 172
    negative (nDEP), 149–50, 170

origination, 134
particle adhesion, 154
for particle and cell manipulations, 133–76
physical origins of, 134–35
positive (pDEP), 149, 170
for screening, 140
spherical approximation, 139–40, 142–44
theory, 135–45
thermal effects, 153–54
for transport, 140, 155
for trapping, 140
traveling-wave (twDEP), 138, 147–49
troubleshooting, 169
upstream electrode, 171
Dielectrophoretic mobility, 143
Diffraction limit, 293–94
Diffusional mixing, 60–61
Diffusometry, 301–3
Dipole approximation, 140
Direct cross-correlation (D-CC), 288
Drag forces, 138
Droplet arrays, 259–61
advantages, 259
droplets captured by, 260
large, 259
with reservoir, 259–60
Droplet-based immunoassay, 125
Droplet based microfluidics, 245–63
advantages, 246–48
aqueous reagents, 254
bioassays, 248–49
biomedical applications, 248–53
channel surface modification, 255
data acquisition, 259–62
device fabrication, 253
discussion and commentary, 262–63
fluid manipulation, 254
gas, 254
generation of droplets, 258–59
hydrophilic surface treatment, 255–57
hydrophobic surface treatment, 257
introduction to, 246–48
large numbers, 247
materials, 253–54
methods, 255–59
monodispersity, 247–48
oils, 254
particle formation, 249–52
rapid mixing, 246
schematic, 247
solution preparation, 257–58
solvents, 254
surfactants, 254
therapeutic delivery, 252–53

troubleshooting table, 263
Droplet pathways, 116–17
bead washing, 117
incubation and detection, 116–17
mixing, 116
Droplets
captured by array, 260
cell viability in, 252
generation of, 258–59
generation rates, 261
high density of, 260
manipulation, 113–14
mixing, 114
Dry etching, 11–13
Dynamic cytometer, 176

E

EK effects, 146
Elastomers, 20
Electoosmotic micropumps, 48–49
Electrical interconnection, 17
Electric double layer (EDL), 48
Electric field frequency, 145–47
AC fields, 145–46
DC fields, 146–47
Electric field phase, 147–49
electrorotation experimentation, 148
equipment and experimental setups, 148–49
spatially uniform, 147
Electroaffinity column, 171
Electrocapillarity, 42
Electrochemical assays, 184
Electrode array implementation, 175
Electrode-based DEP, 152, 156–63
electrode configurations, 153
fabrication uncertainty, 155
filtering/binary sorting, 156–57
media conductivity, 153
ROT spectra, 162–63
sorting, 160–62
thermal effects, 153
transport, 155
trapping, 157–60
See also Dielectrophoresis (DEP)
Electrodeless dielectrophoresis, 151
Electrohydrodynamic micropumps, 49–50
conduction pump, 49–50
defined, 49
injection pump, 49, 50
See also Micropumps
Electrohydrodynamic mixing, 71–72
Electroless plating, 9–10
Electrolysis, 152
Electromagnetic actuators, 39–40

configurations, 40
defined, 40
See also Actuators
Electromechanical actuators, 40–41
Electroosmotic flow (EOF), 272
Electroosmotic pumps, 48–49
defined, 48
interpretation, 102–3
optimization parameters, 49
See also Micropumps
Electrophoresis
anticipated results, 102–3
biomedical applications, 88–89
capillary (CE), 88
clinically relevant assays, 88–89
data acquisition, 102–3
design and performance attributes, 84
facilities/equipment, 91
introduction to, 84–89
materials, 89–91
methods, 91–105
microfluidic, 84–88
multidimensional analyses, 87–88
multiplexed analyses, 86–87
polyacrylamide gel (PAGE), 92–96
polyacrylamide gel (PAGE), based on
isolectric focusing, 96–102
proteomics, 88
separations benefit from microfluidics,
85–86
summary, 105
troubleshooting guide, 106–7
urea gradient, 103–4
Electroplating, 9
Electrorotation, 162–63
methods, 163
use of, 162–63
Electrostatic actuators, 39
Electrowetting, 111–31
applications, 112
automated glucose assays on-chip,
123, 126–27
chemicals, 117–18
chip loading, 125
detection setup, 121–22
device fabrication, 119–21
device testing, 125–26
dielectric deposition, 120
digital microfluidic LOC, 115–17
dilution factor, 129–30
droplet dispensing, 125–26
droplet manipulation with, 113–14
droplet mixing, 126
droplet pathways, 126
fabrication materials, 118

hydrophobic coating, 121
introduction to, 112–15
low interfacial tension liquids, 130
magnetic bead manipulation on-chip,
123–24
materials, 117–18
method challenges, 129–30
methods, 123–25
oil as filler fluid, 130
single layer chips, fabrication, 119
summary, 131
system assembly, 122–23
system integration, 130
theory, 112–13
top plate fabrication, 121
troubleshooting table, 130
two layer chips, fabrication, 119–20
Electrowetting-on-dielectric (EWOD),
246, 248
ELISA, 174, 226
bench-top devices, 201
diagnostics, 190
incubation times, 226
tests, 154
Enzyme-linked immunosorbant assay.
See ELISA
Epoxy, 20
Etching, 10–13
dry, 11–13
gas phase, 13
plasma, 11–13
reactive ion (RIE), 11–12
wet, 10–11
Evanescent wave μPIV, 299
Extent of mixing (EOM), 78

F

FFT-based spatial cross-correlation (FFT-CC),
304, 306
Field flow fractionation DEP (FFF-DEP), 173
Filtering/binary sorting, 156–57
Flow-through membrane immunoassay
(FMIA), 227, 228–29
advantages, 229
capture/labeling reagents, 240
card schematic/image, 232
complex sample analysis, 241–42
data acquisition and results, 236–37
defined, 228
discussion, 239–40
fabrication of flowcells, 231–32
materials, 230–31
parameters, 239
response, 237
sample and reagent delivery, 233

schematic, 228
signal obtained from, 236
troubleshooting table, 241
See also Immunoassays
Fluidic input port, 115–16
Fluidic interconnection, 18–19
Fluorescence
    detection, 188–91
    labeling, 270
    use of, 288
Fluorescence-activated cell sorting (FACS),
    173
    cell preparation, 268
    defined, 268
    See also MicroFACS system
Forward-scattered (FSC) light, 270, 271
Fractionation, 172–75

G

Gas, 254
Gas phase etching, 13
Glass, 19
Glucose assay, 117
Grid-electrode geometry, 150

H

High-speed cameras, 261–62
Human serum albumin (HSA), 242
Hydrodynamic focusing, 64–66
Hydrogels, 41
    alginate, microbeads, 253
    for chemical actuators, 41
Hydrophilic surface treatment, 255–57
    plasma surface oxidation, 255–56
    poly(vinyl alcohol) coating, 256–57
Hydrophobic coating, 118, 121
Hydrophobic surface treatment, 257

I

Immunoassay reagents, 117–18
Immunoassays, 115, 225–42
    chemiluminescence, 122
    complex sample analysis, 241–42
    data acquisition and results, 236–39
    design/operation considerations, 227
    discussion, 239–42
    droplet-based, 125
    example formats, 227–30
    fabrication of flowcells, 231–33
    flow rate, 227
    on human insulin and interleukin-6, 129
    introduction to, 226–30
    materials, 230–31
    microfluidic device, 230–31

parameters, 227
pumps and interconnections, 230
sample and reagent delivery, 233–35
troubleshooting table, 241
See also Concentration gradient
immunoassay (CGIA); Flow-through
    membrane immunoassay (FMIA)
Inductively coupled plasma (ICP), 12–13
Infinite domain assumption, 144
Infrared (IR) PIV, 307
Insulative dielectrophoresis (iDEP), 151–52
    angled constriction, 164–65
    compressions/expansions, 163
    curved constriction, 167–68
    defined, 151
    electrode configurations, 153
    equipment/experimental setup, 151–52
    fabrication uncertainty, 155–56
    perpendicular constriction, 164
    post-array, 165–67
    substrate choice, 152
    thermal effects, 153–54
    transport, 155
    See also Dielectrophoresis
Integrated optical systems, 185–95
    absorbance detection, 185–88
    chemiluminescence detection, 191–92
    fluorescence detection, 188–91
    interferometric detection, 192–94
    surface plasmon resonance (SPR)
        detection, 194–95
    See also Optical microfluidics
Interferometric detection, 192–94
Interferometry
    backscattering, 193–94
    reflectometric, 193
Isoelectric focusing (IEF), 88, 90–91
    acrylamide/bis-acrylamide solution
        preparation, 99
    analysis of isoforms, 104–5
    cross-linking acrylamide, 100
    fabrication, 96–98
    polyacrylamide gel electrophoresis,
        96–102
    polymerization preparation, 99–100
    silanization, 98–99
    storing device, 100–102
Isotachophoresis (ITP), 88

K

Kinesin, 311
    buffers, 314
    equipment, 314
    expression, 313–14, 317–19
    purification, 317–19, 332

reagents, 313
Kinesin-driven transport, 311

L

Lab-on-a-chip (LOC) devices, 32
  for clinical diagnostics, 115
  design, 115–17
  microvalves/micropumps in, 32–33
  pneumatic actuators for, 36
  in portable platforms, 33
Lamination, 13–14
Laser Doppler microscope (LDM), 282
Laser Doppler velocimetry (LDV), 282
Lateral-flow assays (LFAs), 184
LIF μPIV, 307
Limit of detection (LOD), 186, 227
Liquid-core waveguide arrangements, 188
Liquid reservoirs, 116
Localized surface plasmon resonance (LSPR),
    198–200
  biosensors, 199
  measurements, 198
  wavelength, 200
Low-pressure chemical vapor deposition
    (LPCVD), 9

M

Macrochannels, 326
Magnetic beads
  attraction, 123, 127–28
  manipulation on-chip, 123–24
  resuspension, 124, 129
  retention, 124, 128
  wash efficiency, 123–24
  washing optimization, 127–29
Magnetic resonance imaging (MRI) PIV, 307
Magnetohydrodynamic (MHD) pumps, 50
Materials
  chemical gradients, 213
  dielectrophoresis (DEP), 145–52
  droplet based microfluidics, 253–54
  electrophoresis, 89–91
  electrowetting, 117–18
  immunoassay, 230–31
  microfabrication, 19–22
  microFACS system, 269–70
  microfluidic interconnect, 62–63
  micromixer, 62
  microparticle image velocimetry (μPIV),
      283–92
MATLAB code, 212, 221–23
Maxwellian equivalent body, 141–42
  defined, 141–42
  uses, 142

Maxwell's stress tensor, 135, 136–37
Mechanical microvalves, 52–54
Mechanical pumps, 43–47
Media conductivity, 153
Metal/oxide/semiconductor field effect
    transistors (MOSFETs), 7
Metamorph imaging software, 220
Microcapsules, 251
Microchannels
  fabrication protocol, 327–28
  motility in, 334–36
Microelectromechanical systems (MEMS),
    2, 269
Microfabrication materials, 316–17
Microfabrication techniques, 1–25
  additive processes, 6–10
  bonding processes, 13–16
  materials, 19–22
  packaging processes, 16–19
  sacrificial layer, 16
  subtractive, 10–13
  summary, 22
  transfer processes, 4–6
  troubleshooting table, 22, 23–25
MicroFACS system, 267–79
  acousto-optic modulator (AOM), 274
  actuation model, 271
  application notes, 277
  cell manipulation, 272
  contents, 270
  defined, 267
  electroosmotic flow (EOF), 272
  experimental results, 275
  goal, 278
  introduction to, 268–69
  materials, 269–70
  measurement factors, 270–71
  methods, 270–79
  operating methods, 273
  performance, 274
  results, 274–75
  schematic, 271
  sorting, 267, 273
  statistical analysis, 276–77
  summary, 278–79
  tasks, 267
  troubleshooting table, 279
  two-dimensional data representation, 277
Microfluidic cards, 232
Microfluidic devices
  introduction to, 2–3
  microfabrication, 1–25
  micromixing within, 59–80
Microfluidic immunoassays.
      See Immunoassays

Microfluidic interconnects, 62–63
Microfluidic systems
  defined, 2
  fabrication techniques, 3–4
  fluidic interconnection, 18–19
  introduction to, 2–3
Micromixers, 59–80
  acoustic, 72–75
  active, 70–75
  chaotic, 61, 66–70
  computer acquisition, 77
  discussion and commentary, 78–79
  extent of mixing, 78
  introduction to, 60–62
  materials, 62–63
  multiphase, 75–77
  optical assembly, 63
  passive, 64–70
  performance metrics, 78
  reaction monitoring, 78
  reagents, 63
  summary, 79–80
  troubleshooting, 79
Microparticle image velocimetry (μPIV),
    281–307
  application, 301
  application notes, 301–7
  biosensing, 303
  in Brownian motion, 302
  CDI SPE, 304, 306
  central difference image correlation
    (CDIC), 289–90
  characteristics, 281
  confocal, 297
  data acquisition system, 285
  defined, 282
  diffraction limit, 293–94
  diffusometry and, 301–3
  discussion and commentary, 293–300
  evanescent wave, 299
  experimental setup, 283–85
  future developments, 307
  introduction to, 282–83
  LIF, 307
  macroscopic dissimilarities, 300–301
  materials and methods, 283–92
  measurement procedures, 292–93
  microfluidic apparatus, 284
  optical assembly, 284–85
  optical traps, 307
  particle size effects, 294
  processing algorithms, 287–92
  single-pixel evaluation (SPE), 290–92
  spatial cross-correlation, 287–88
  summary, 300–301

thermometry, 303
three-dimensional (3-D), 297–99
troubleshooting table, 300
ultimate spatial resolution, 294–96
velocity errors, 296
volume illumination, 285–87
wall shear stress (WSS) measurement,
    303–7
Micropumps, 42–51
  actuators for, 33–42
  capillary, 50–51
  check valve, 43–45
  electoosmotic, 48–49
  electrohydrodynamic, 49–50
  in LOC applications, 32–33
  magnetohydrodynamic (MHD), 50
  mechanical, 43–47
  nonmechanical, 48–51
  outlook, 55
  performance, 43
  peristaltic, 45–46
  summary, 56
  troubleshooting, 55–56
  valveless rectification, 46–47
Microtubule associated proteins (MAPs), 319
Microtubule motors, 311–36
  concentration/unidirectional transport,
    331
  confinement, 325
  glass channel fabrication, 326–29
  gliding assay materials, 315–16
  integrating into microfluidic channels,
    328–29
  introduction to, 312
  kinesin expression, 313–14, 317–19
  kinesin purification, 317–19
  materials, 312–17
  methods, 317–29
  microfabrication materials, 316–17
  motor proteins integration, 324–25
  pitfalls, 331–36
  purification materials, 313–14
  results, 329–31
  standard protocol for gliding assay,
    323–24
  troubleshooting table, 335
  tubulin purification and labeling, 319–23
  tubulin purification materials, 314–15
Microtubules
  directional rectifier, 330
  moving over kinesin-coated surface, 329
  MT100, 323
  polymerization, 319, 323
  rhodamine-labeled, 322
  size, 319

visualizing, 322
Microvalves, 51–55
    active, 51–52
    actuators for, 33–42
    functions, 51
    in LOC applications, 32–33
    mechanical, 52–54
    nonmechanical, 54–55
    outlook, 55
    performance, 52
    summary, 56
Miniaturization, 32
Mixing variance coefficient (MVC), 78
Mobility
    dielectrophoretic, 143
    in microchannels, 334–36
    PAGE protein, acquisition, 102
Molding, 6
Monodispersity, 247–48
Motility assays, 333–34
Multidimensional analyses, 87–88
Multilevel glass devices, 152
Multiphase mixers, 75–77
Multiple-frequency DEP (MFDEP), 172
Multiplexed analyses, 86–87

N

Nanoengineered optical probes, 195–202
    localized surface plasmon resonance,
        198–200
    quantum dots, 196–97
    silver-enhanced nanoparticle labeling,
        197–98
    up-converting phosphors, 197
    See also Optical microfluidics
Nanosphere lithography (NSL), 198–99
Negative dielectrophoresis (nDEP),
        149–50, 170
    cage, 171
    force, 161
    perpendicular force, 161
    use of, 170
Neutrophil chemotaxis assay, 216–20
    blood sample loading, 217–19
    device operation, 219–20
    device priming, 216–17
    media preparation, 216
    neutrophil isolation, 217–19
Neutrophils, 209
    capturing, 216
    isolation, 211, 217–19
    migration, 219–20
Nonmechanical microvalves, 54–55
Nonmechanical pumps, 48–51
    Norland Optical Adhesive (NOA), 249

Nyquist theorem, 295

O

Oils, 254
Optical microfluidics, 183–204
    conclusions, 203
    integrated systems, 185–95
    nanoengineered probes, 195–202
    summary, 203–4
Optical traps/μPIV, 307

P

Packaging processes, 16–19
    dicing, 16–17
    electrical interconnection and wire
        bonding, 17–18
    fluidic interconnection, 18–19
Parallel lamination, 64–66
Particle formation, 249–52
    chemical-catalyzed polymerization, 250
    photo-initiated polymerization, 249
    solvent extraction/evaporation
        polymerization, 250–52
Passive capillarity, 42
Passive micromixers, 64–70
    chaotic advection, 66–70
    creeping flow regime, 67
    hydrodynamic focusing, 64–66
    parallel lamination, 64–66
    serial lamination, 66
    See also Micromixers
Peristaltic pumps, 45–46
Perpendicular constriction, 164
Photo-initiated polymerization, 249
Photolithography, 4–6
    deep UV (DUV), 6
    defined, 4
    illustrated, 5
    radiation wavelength limitation, 5–6
Photomultiplier tube (PMT) luminescent
        reader, 192
Photonic multiplier tube (PMT), 268
Photoresists, 4
Piezoelectric actuators, 38–39
    defined, 38
    thin-film, 39
    See also Actuators
Plasma-enhanced chemical vapor deposition
        (PECVD), 9
Plasma etching, 11–13
Plasma surface oxidation, 255–56
Pneumatic actuators, 35–36
Point-lid geometry, 150
Point-of-care (POC) diagnostics, 184, 203

Polyacrylamide gel electrophoresis (PAGE),
    92–96
    chip preparation, 92–94
    fabrication, 92–95
    gradient generation, 94–95
    imaging measurements, 95–96
    measurements, 96
    microchannel loading, 94
    monomer solution preparation, 94
    overview illustration, 93
    protein mobility acquisition, 102–3
    system setup, 95
    troubleshooting guide, 106–7
    for urea gradient electrophoresis, 103–4
Polyacrylamide gels, 91, 92–96
Polycarbonate, 22
Polydimethylsiloxane, 20
Polyimides, 21
Polymerase chain reaction (PCR), 276
Polymer bonding, 16
Polymeric devices, 152
Polymerization
    chemical-catalyzed, 250
    microtubules, 319
    photo-initiated, 249
    solvent extraction/evaporation, 250–52
Polytetrafluoroethylene, 22
Poly(vinyl alcohol) coating, 256–57
Positive dielectrophoresis (pDEP), 149, 170
Post-array, 165–67
    mechanism of action, 165
    methods, 165–67
    safety, 167
Prevalidated calibration model, 235
Protease inhibitors, 320
Proteomics, 88
P-selectin, 215
Pyrex, 19

Q

Quadrupole traps, 157
Quantum dots (QDs), 196–97
    barcodes, 196
    defined, 196
    problems, 195
    See also Nanoengineered optical probes
Quartz, 19

R

Raman scattering, 200
Reactive ion etching (RIE), 11–12
Recirculating flow, 246
Reflectometric interferometry, 193
Rhodamine, 321–22

Rotating electric fields, 148

S

Sacrificial layer techniques, 16
Self-assembled monolayers (SAMs), 198
Serial lamination, 66
Shape-memory actuators, 38
Side-scattered (SSC) light, 270–71
Silicate glasses, 19
Silicon, 19–20
Silver-enhanced nanoparticle labeling,
    197–98
Single-particle trapping, 175–76
Single-pixel evaluation (SPE), 290–92
    defined, 290
    displacement vector term, 292
    principle, 291
Single tandem repeat (STR) typing, 191
SIO$_2$
    dielectric constant, 6
    growth, 6–7
Solid-expansion actuators, 37–38
Solvent extraction/evaporation
        polymerization, 250–52
Solvents, 254
Sorting, 160–62, 172–75
    angled/curved electrodes, 160–62
    filtering/binary, 156–57
    fluorescence-activated cell (FACS), 173
    microFACS system, 267, 273
    traveling-wave dielectrophoresis, 162
Spatial cross-correlation, 287–88
Spatial resolution, 294–96
Spherical approximation, 139–40, 142–44
Spin-on deposition, 7
Sputtering, 8
Staggered herringbone mixer (SHM), 69
SU-8 thick resists, 20–21
Subtractive processes, 10–13
    chemical-mechanical polishing (CMP), 13
    defined, 3
    etching, 10–13
Surface acoustic waves, 112
Surface enhanced Raman scattering (SERS),
    183
    arrays, 202
    biosensors, 200
    defined, 200
    substrate, 201
    variation, 201
Surface-enhanced Raman spectroscopy,
    200–202
Surface-enhanced resonance Raman
        scattering (SERRS), 201
Surface plasmon resonance (SPR)

localized (LSPR), 198–200
with nanohole gratings, 200
Surface plasmon resonance (SPR) detection,
    194–95
defined, 194
gold-coated, 195
laboratory-scale systems, 195
sensitivity, 195
Surfactants, 254

T

Taylor dispersion, 71, 72
Terminal particle velocity, 138
Thermal effects, 153–54
Thermocapillarity, 42, 112
Thermometry, μPIV-based, 303
Thermopneumatic actuators, 36–37
Thick positive resists, 21
Three-dimensional (3-D) μPIV, 297–98
Transfer processes, 4–6
defined, 3
molding, 6
photolithography, 4–6
Trapping, 169–72
application to biological problems, 172
with post-array type devices, 165
single-particle, 175–76
Traps, 157–60
castellated electrode, 159–60
circle/dot, 158–59
quadrupole, 157
Traveling-wave dielectrophoresis (twDEP),
    138
applications, 149
canonical example, 147
defined, 147
design and fabrication, 148
electrorotation and, 149
sorting, 162
See also Dielectrophoresis
Troubleshooting
chemical gradients, 220
dielectrophoresis, 169
droplet based microfluidics, 263
electrophoresis, 106–7
immunoassays, 241

microfabrication techniques, 22, 23–25
microFACS system, 279
micromixers, 79
microparticle image velocimetry (μPIV),
    300
micropumps, 55–56
microtubule motors, 335
polyacrylamide gel electrophoresis
    (PAGE), 106–7
Tubulin
buffers and solutions, 315
cycling, 322
equipment, 315
labeled unlabeled, missing, 322–23
labeling, 321–22, 333
pitfalls, 332–33
purification materials, 314–15
purification protocol, 319–21
reagents, 314–15
Two-layer crossing channels micromixers
    (TLCCM), 68

U

Upconverting phosphor technology (UPT),
    197
UV/visible absorption spectroscopy, 185

V

Vacuum evaporation, 7–8
Valveless rectification pump, 46–47
defined, 46
diffuser/nozzle structure, 47
pressure loss, 46
valvular conduit, 47
See also Micropumps
Velocity errors, 296

W

Wafer bonding methods, 14–16
Wall shear stress (WSS) measurement, 303–7
analytical fit, 304
discussion, 306–7
experimental technique, 304–6
Wet etching, 10–11
Wire bonding, 17–18